Biometeorology

Volume 7

This Series is now being indexed by SCOPUS.

Biometeorology has its linguistic origins in Greek; bio refers to life while meteoros makes reference to the study of phenomena near and above the earth's surface. These are the essential elements of biometeorology, such that it is concerned with the interaction between living organisms and variations in atmospheric processes at a range of time and space scales.

Since its emergence as a discipline in the mid 1950s, biometeorology has grown to have great societal relevance. Much of its subject matter focuses on how biophysical, human, animal and plant systems are influenced by change and variability in the atmospheric environment. Understanding such relationships provides insights into how the provisioning, supporting and regulating services provided by the biosphere may be affected by atmospheric variations and ultimately the societal consequences of atmosphere related changes in biophysical and ecosystem processes.

The purpose of this series in biometeorology is to:

- complement the material published in the International Society of Biometeorology's International Journal of Biometeorology,
- to communicate the interdisciplinary philosophy and science of biometeorology to as wide an audience as possible,
- introduce scientists and policy makers to the societal relevance of and recent developments in its subfields.
- demonstrate how a biometeorological approach can provide entry points to the understanding and possible solution of cross-cutting environmental issues such as climate change, desertification, deforestation, biodiversity loss, land degradation, emerging diseases and the availability of clean water and air.

The series in biometeorology comprises theme-based volumes organised around subject matter within the subfields of biometeorology, which broadly include human, animal, plant, ecosystems, tourism, artificial or indoor climates, air quality and urban. In addition to volumes that synthesise the state of knowledge in the subfields, a primary focus will be on important, new and emerging topics in biometeorology that have a bearing on the health of the biosphere and thus human society. An editorial board, the broad role of which will be to elicit and review suggestions for volume themes, will oversee the series.

Hafiz Muhammad Rizwan ·
Muhammad Sohail Sajid
Editors

Ticks in a Changing Climate

Impacts and Emerging Challenges in Tick-Borne Diseases

Springer

Editors
Hafiz Muhammad Rizwan
Department of Pathobiology (Parasitology Section)
University of Veterinary and Animal Sciences
Lahore (Narowal Campus), Pakistan

Muhammad Sohail Sajid
Department of Parasitology
Faculty of Veterinary Science
University of Agriculture
Faisalabad, Pakistan

ISSN 1877-5284 ISSN 2452-1558 (electronic)
Biometeorology
ISBN 978-3-032-12258-2 ISBN 978-3-032-12259-9 (eBook)
https://doi.org/10.1007/978-3-032-12259-9

This Springer imprint is published by the registered company Springer Nature Switzerland AG
The registered company address is: Gewerbestrasse 11, 6330 Cham, Switzerland

Dr. Hafiz Muhammad Rizwan, I am honored to present this book titled Ticks in a Changing Climate: Impacts and Emerging Challenges in Tick-Borne Diseases with heartfelt reverence and deep respect, I dedicate this book to my late father, Mr. Mashooq Ali. His unwavering support, boundless wisdom, and relentless encouragement have been the guiding force behind my academic and professional journey. Though he is no longer with us, his memory continues to inspire and motivate me in every endeavor. May Allah grant my father the highest place in Jannah and shower his soul with eternal peace and blessings. His legacy of kindness, integrity, and dedication to family and community remains a beacon of light for all who knew

him. I pray that his soul finds comfort and that he is rewarded abundantly for all the goodness he brought into this world. Ameen.

—Dr. Hafiz Muhammad Rizwan

Preface

Climate change is no longer a distant threat—it is a pressing reality reshaping ecosystems, disease dynamics, and public health landscapes across the globe. Among the many consequences of this shift is the expanding distribution and growing burden of tick-borne diseases (TBDs), which now pose a serious challenge to both human and animal health in regions that were previously unaffected. This book, *Ticks in a Changing Climate: Impacts and Emerging Challenges in Tick-Borne Diseases*, emerges from the need to understand and respond to these evolving threats through interdisciplinary research and collaborative action.

This volume brings together contributions from experts in entomology, ecology, veterinary and medical sciences, climate research, and public health. It aims to provide a comprehensive view of how climate change influences tick ecology, pathogen transmission, and disease outcomes. Moreover, it emphasizes the socioeconomic and community-level impacts of TBDs, particularly in vulnerable populations where access to healthcare and prevention resources remains limited.

A central theme of this book is the importance of resilience—ecological, public health, and community-based. Readers will find in-depth discussions on adaptive control strategies, emerging diagnostic tools, participatory approaches to disease prevention, and the integration of climate models into tick surveillance. This text supports a One Health perspective, recognizing the interconnectedness of humans, animals, and the environment.

This work is intended for researchers, students, policymakers, and practitioners who are seeking a deeper understanding of TBDs within the context of global environmental change. As editors, we hope this book stimulates further dialogue, fosters innovative research, and contributes to building more resilient communities capable of meeting the challenges of TBDS in a warming world.

Lahore, Pakistan — Hafiz Muhammad Rizwan
Faisalabad, Pakistan — Muhammad Sohail Sajid

About This Book

Ticks in a Changing Climate: Impacts and Emerging Challenges in Tick-Borne Diseases presents an in-depth exploration of how climate change is reshaping the ecology, distribution, and public health significance of ticks and the diseases they transmit. This comprehensive volume brings together the expertise of multidisciplinary researchers to examine the growing threat of tick-borne diseases (TBDs) within the broader context of environmental transformation. As global temperatures rise and ecosystems shift, tick populations are expanding into new geographic regions, elevating the risk of disease emergence in previously unaffected areas. This book addresses these challenges by integrating ecological, epidemiological, and socio-economic perspectives into a unified narrative.

The volume begins by examining how climate change influences tick habitats, distribution, and seasonal activity. It investigates the environmental factors that govern tick development, such as temperature and humidity, and how these changes affect their life cycles and interactions with host species. Understanding these dynamics is essential for anticipating future hotspots of tick activity and associated disease risks. One section of the book focuses on the changing patterns of disease transmission, emphasizing the role of emerging zoonotic threats linked to shifting ecosystems. It evaluates how climate change affects viral, bacterial, and protozoal TBDs, offering evidence from recent case studies and scientific models. These insights highlight how the nature and intensity of disease outbreaks are evolving in response to global warming and ecological disruption.

In addition to the biological aspects, the book pays special attention to the socio-economic consequences of rising TBD burdens. It addresses how limited healthcare access, poverty, and geographic isolation exacerbate vulnerability to TBDs, especially in low-resource and rural communities. The discussion stresses the importance of equitable healthcare policies and increased support for underserved populations. The book also explores a variety of tick control strategies adapted for the challenges posed by climate change. It evaluates both traditional and emerging approaches—from chemical interventions to community-based environmental management—and emphasizes the need for sustainable, integrated vector control methods. The importance of occupational health and safety is underscored, particularly for people working in high-risk environments.

A significant part of the volume highlights the role of community engagement in building resilience against TBDs. It advocates for greater public awareness, health education, and participatory action to reduce exposure and improve early detection. Strengthening community health systems, enhancing surveillance, and fostering collaboration between health and environmental sectors are presented as key pillars of effective disease management. Looking ahead, the book reflects on future challenges in tick research and public health planning. It calls for interdisciplinary and forward-looking research agendas that incorporate climate projections, ecological forecasting, and the One Health approach to address the complex dynamics of ticks, hosts, pathogens, and their shared environments.

With its evidence-based analysis and actionable insights, this book is an essential resource for researchers, practitioners, and decision-makers striving to protect human and animal health in a rapidly changing climate.

Contents

Editors and Contributors

About the Editors

Dr. Hafiz Muhammad Rizwan is an accomplished researcher and academician in the field of Parasitology. He has made significant contributions to the understanding and management of parasitic diseases through his research and publications. Dr. Rizwan's expertise lies in the areas of molecular parasitology, host-parasite interactions, omics-based approaches for diagnosing and managing parasitic diseases, and parasitic control strategies. Dr. Rizwan is currently affiliated with the University of Veterinary and Animal Sciences (UVAS), Lahore (Narowal Campus), Pakistan, where he serves as an Assistant Professor of Parasitology in the Department of Pathobiology. He is actively involved in teaching and mentoring students, sharing his knowledge and passion for the subject. With a strong academic background, Dr. Rizwan holds a Ph.D. in Parasitology. He has published numerous research articles in reputable journals, contributing to the scientific literature on parasitic infections. Dr. Rizwan's dedication and commitment to his field have earned him recognition and respect from his peers. He continues to make valuable contributions to the field of parasitology through his research, teaching, and academic endeavors.

Prof. Dr. Muhammad Sohail Sajid got his early education from the Kamalia and Khanewal and Higher Secondary education from Government Emerson College, Multan, Punjab, Pakistan. He secured silver medal in his Doctor of Veterinary Medicine (DVM) degree from the University of Agriculture, Faisalabad (UAF), Pakistan, in 2002. Since then, he started his professional career as Lecturer in the Department of Parasitology, UAF while continuing in M.Sc. (Hons.) and Ph.D. in Parasitology. He served as Assistant Professor, Associate Professor, Chairman, and Member, Climate Change Chair of the US-Pakistan Center for Advanced Studies in Agriculture and Food Security established through the USAID in the UAF, Pakistan, and finally as Professor tenured. His thematic research area is molecular epidemiology and control of arthropods and arthropod-borne diseases of one health significance. His competitive successes include: Ph.D. fellowship sponsored by the Higher Education Commission (HEC), Islamabad, Pakistan, HEC-funded postdoctoral fellowship from the University of Southern Mississippi (USM), Hattiesburg, MS, USA, USAID-funded project through the HEC Pak-US S&T Cooperation Program (Phase IV), and the Fulbright postdoctoral fellowship at the UC Davis, USA. So far, he has secured 10 grants as PI/Co-PI and mentored >50 M.Phil. and 12 Ph.D. graduate students as chairman/co-supervisor/member. The Ministry of Science and Technology, Islamabad, Pakistan, has awarded him competitive Research Productivity Awards consecutively for the six years (2010–2015). So far, he has published ~90 research papers in ISI-indexed journals having IF of ~100, presented at various national and international conferences, one textbook, two chapters in an international book and 150 abstracts in proceedings. He served as "Section Editor" of the UAF's Official Journal "PJAS" (IF-0.785) from more than a decade and has edited two books published by the CABI, UK and one book published by CRC press, USA. He is the co-author of the first guidelines of the "Institutional Animal Care and Use" at the UAF. His hard work and commitment enabled him to achieve the top ranked academic position of the Tenured Professor and Chairman of Parasitology in the Department of Parasitology, UAF, Pakistan.

Contributors

Haider Abbas Department of Pathobiology (Parasitology Section), KBCMA College of Veterinary and Animal Sciences, Narowal, Sub-Campus UVAS, Lahore, Pakistan

Mudassir Ahmad School of Science and Technology, Nottingham Trent University, England, UK

Faisal Rasheed Anjum Bristol Veterinary School, University of Bristol, Bristol, UK

Muhammad Haziq Bajwa Faculty of Veterinary Science, KBCMA College of Veterinary and Animal Sciences, Narowal, Sub-Campus UVAS, Lahore, Pakistan

Hanène Belkahia Laboratory of Microbiology, National School of Veterinary Medicine of Sidi Thabet, University of Manouba, Manouba, Tunisia

Ali Butt Livestock and Dairy Development Department, Lahore, Pakistan

Ian Daniel Department of Veterinary Pathobiology, College of Veterinary Medicine & Biomedical Sciences, Texas A&M University, College Station, TX, USA

Phelipe Magalhães Duarte Postgraduate Program in Animal Bioscience, Federal Rural University of Pernambuco (UFRPE), Recife, Pernambuco, Brazil

Zahid Fareed Veterinary Research Institute, Lahore, Pakistan

Sadia Ghazanfer Department of Parasitology, Faculty of Veterinary Science, University of Agriculture, Faisalabad, Pakistan

Syed Soban Hassan Faculty of Veterinary Sciences, KBCMA College of Veterinary and Animal Sciences, Narowal, Sub-Campus UVAS, Lahore, Pakistan

Miranda Imeri Agronomy Department, University of Florida, Gainesville, FL, USA

Mahnoor Mahzhar Iqbal School of Environmental Studies, China University of Geosciences, Wuhan, China

Saiful Islam Department of Parasitology, Faculty of Veterinary, Animal and Biomedical Sciences, Sylhet Agricultural University, Sylhet, Bangladesh

Malcolm K. Jones School of Veterinary Science, The University of Queensland, Gatton Queensland, Australia

Muhammad Kamran Faculty of Veterinary Science, KBCMA College of Veterinary and Animal Sciences, Narowal, Sub-Camus UVAS, Lahore, Pakistan

Asma Kausar Veterinary Research Institute, Lahore, Pakistan

Hasti Khalili Pharmaceutical Sciences Research Center, Tehran Medical Sciences, Islamic Azad University, Tehran, Iran

Abdur Rehman Khan Department of Parasitology, University of Veterinary and Animal Sciences, Lahore, Pakistan

Sakandar Khan Section of Parasitology, Department of Pathobiology, KBCMA College of Veterinary and Animal Sciences, Narowal, Sub-Campus UVAS, Lahore, Pakistan

Hira khan Institute of Microbiology, University of Veterinary and Animal Sciences, Lahore, Pakistan;
College of Sciences, Department of Biology, United Arab Emirates University, Abu Dhabi, UAE

Surrya Khanam Department of Zoology, Women University Swabi, Swabi, Pakistan

Mahvish Maqbool Department of Entomology, Texas A & M, College Station, TX, USA

Aiman Maqsood Department of Zoology, Ramal International College, Zafarwal, Punjab, Pakistan

Muhammad Moaaz Faculty of Veterinary Sciences, KBCMA College of Veterinary and Animal Sciences, Narowal, Sub-Campus UVAS, Lahore, Pakistan

Osama B. Mohammed Department of Zoology, College of Science, King Saud University, Riyadh, Saudi Arabia

Bareera Munir College of Earth and Environmental Sciences, University of the Punjab, Lahore, Pakistan

Iqra Munir Faculty of Allied Health Sciences, University Institute of Medical Laboratory Technology, The University of Lahore, Lahore, Pakistan

Maira Munir Department of Parasitology, Faculty of Veterinary Sciences, University of Veterinary and Animal Sciences, Lahore, Pakistan

Nosheen Munir Anesthesiology, CMH Lahore Medical College and Institute of Dentistry, Lahore, Pakistan

Hizqeel Ahmed Muzaffar Faculty of Veterinary Sciences, KBCMA College of Veterinary and Animal Sciences, Narowal, Sub-Camus UVAS, Lahore, Pakistan

Muhammad Noman Naseem School of Veterinary Science, The University of Queensland, Brisbane, Australia

Shumaila Naz Department of Biological Sciences, National University of Medical Sciences, Rawalpindi, Pakistan

Maxwell Nwachukwu Opara Department of Parasitology and Entomology, Faculty of Veterinary Medicine, University of Abuja, Abuja, Nigeria

Muhammad Qasim Grand Valley State University, MI, USA

Mahvish Rajput Department of Parasitology, Faculty of Veterinary Science, University of Agriculture, Faisalabad, Pakistan

Aysha Rana Foot and Mouth Disease Research Centre, Lahore, Pakistan; College of Veterinary Medicine, University of Minnesota, Twin-Cities, USA

Muhammad Imran Rashid Department of Parasitology, University of Veterinary and Animal Sciences, Lahore, Pakistan

Fawad Ali Raza Institute of Soil and Environmental Science, University of Agriculture, Faisalabad, Pakistan

Mohsin Raza Section of Physiology, Department of Basic Sciences, KBCMA College of Veterinary and Animal Sciences, Narowal, Sub-Campus UVAS, Lahore, Pakistan

Muhammad Hussnain Raza Faculty of Veterinary Sciences, KBCMA College of Veterinary and Animal Sciences, Narowal, Sub-Campus UVAS, Lahore, Pakistan

Hafiz Muhammad Rizwan Section of Parasitology, Department of Pathobiology, KBCMA College of Veterinary and Animal Sciences, Narowal, Sub-Campus UVAS, Lahore, Pakistan

Mourad Ben Said Department of Basic Sciences, Higher Institute of Biotechnology of Sidi Thabet, University of Manouba, Manouba, Tunisia; Laboratory of Microbiology, National School of Veterinary Medicine of Sidi Thabet, University of Manouba, Manouba, Tunisia

Muhammad Sohail Sajid Department of Parasitology, Faculty of Veterinary Sciences, University of Agriculture, Faisalabad, Pakistan;
Climate Change Graduate Group, Institute of Soil and Environmental Sciences, University of Agriculture, Faisalabad, Pakistan

Neil Sargison Royal (Dick) School of Veterinary Studies, Easter Bush Veterinary Centre, University of Edinburgh, Roslin, Midlothian, UK

Syeda Hina Shah Department of Biological Sciences, National University of Medical Sciences, Rawalpindi, Pakistan

Olivier Andre Sparagano Department of Infection and Public Health, City University of Hong Kong, Hong Kong, China;
UKMC, Manchester, UK

Maara Raheel Spaul Institute of Microbiology, University of Veterinary and Animal Sciences, Lahore, Pakistan

Urfa Bin Tahir Department of Parasitology, Faculty of Veterinary Science, University of Agriculture, Faisalabad, Pakistan;
School of Natural and Environmental Sciences, Newcastle University, Newcastle Upon Tyne, UK

Sina Salajegheh Tazerji Veterinary Clinical Sciences Department, Science and Research Campus, Islamic Azad University, Tehran, Iran

Muhammad Usman Department of Basic Sciences, KBCMA College of Veterinary and Animal Sciences, Narowal, Sub-Campus UVAS, Lahore, Pakistan

Sahar Younis Department of Parasitology, Faculty of Veterinary Science, University of Agriculture, Faisalabad, Pakistan

Muhammad Younus Department of Pathology, University of Veterinary and Animal Sciences, Lahore, Pakistan

Hina tu Zahra Department of Parasitology, Faculty of Veterinary Science, University of Agriculture, Faisalabad, Pakistan

Chapter 1
Introduction: Ticks in a Changing Climate

Abdur Rehman Khan, Asma Kausar, Muhammad Imran Rashid, Hasti Khalili, Bareera Munir, and Sina Salajegheh Tazerji

Abstract Changes in climate vary greatly with the worldwide spread and population of ticks, thereby intensifying the spread of tick-borne diseases (TBDs). Ticks are obligate hematophagous (actively feed on blood) ectoparasites and act as vectors for many animal and human pathogens. Environmental factors such as temperature, humidity, and precipitation have a great contribution towards the survival, growth, and behavior of ticks. Due to climate change, the world is facing a rise in temperature and a shift in the normal rainfall. This leads to other suitable habitats for various tick species, resulting in colonization of new regions. This increases the risk of human exposure to diseases caused by ticks in places that were formerly devoid of such diseases. Warmer temperatures accelerate the development of seasonal ticks and prolong their periods of activity, thereby increasing the risk of disease transmission. Altered migration patterns and changes in the populations of key tick hosts, including deer and rodents, further intensify climate-based changes to the populations of ticks and pathogens. These shifts have diverse ecological consequences that involve the interplay of ticks, hosts, and pathogens in multiple landscapes. There is an escalating risk posed by changes to tick populations owing to the climate, highlighting the need to manage these shifts from multiple angles.

A. R. Khan · M. I. Rashid (✉)
Department of Parasitology, University of Veterinary and Animal Sciences, Lahore, Pakistan
e-mail: imran.rashid@uvas.edu.pk

A. Kausar
Veterinary Research Institute, Lahore, Pakistan

H. Khalili
Pharmaceutical Sciences Research Center, Tehran Medical Sciences, Islamic Azad University, Tehran, Iran

B. Munir
College of Earth and Environmental Sciences, University of the Punjab, Lahore, Pakistan

S. S. Tazerji
Veterinary Clinical Sciences Department, Science and Research Campus, Islamic Azad University, Tehran, Iran

H. M. Rizwan and M. S. Sajid (eds.), *Ticks in a Changing Climate*, Biometeorology 7,
https://doi.org/10.1007/978-3-032-12259-9_1

Keywords Ecological responses · Habitat shifts · Biodiversity · Ecological interactions · Tick management · Host dynamics

1.1 Introduction to Tick Habitats and Climate Change

Ticks are a type of external parasite that belong to the class Arachnida. Ticks are further divided into soft (Argasidae) and hard ticks (Ixodidae). Hard ticks are more commonly associated with disease transmission due to their prolonged feeding behavior (Peters, 1930). As a vector, ticks transmit several highly significant human and animal diseases (Ma et al., 2024). These include human granulocytic anaplasmosis, Rocky Mountain spotted fever, Lyme disease, human monocytic ehrlichiosis, tularemia, and Colorado tick fever (Alkishe et al., 2021). Some species of ticks take more than five years to complete their life cycle and spend most of their life away from their hosts; their physiology and behavior are greatly influenced by environmental conditions (Apanaskevich et al., 2014).

The ecological implications of these climate-driven changes are profound. Modifications of tick populations and abundance disrupt established host–parasite relationships and drive changes in biodiversity and disease transmission. As ticks spread into new locations, they 'meet' new host species and may outcompete native parasites. This can alter the (ecological) balance of an area. These changes can threaten the health of wildlife animals and livestock, which can affect the stability of the ecosystem and the public health.

Tick populations can serve as indicators of ecosystem health in a region, reflecting the interconnection between ecosystem degradation and human health risk (Rapport, 1999). Ticks prefer ecotones as habitats and require moisture, hosts, and microclimate features such as sandy soil, hardwood trees, and proximity to water sources. These conditions are necessary for their survival and reproduction (Dumas et al., 2024). Macroclimate, such as relative humidity in the surrounding environment, does not influence ticks as much as they are affected by microclimate, such as relative humidity in leaf litter (James et al., 2015). Some tick species have a narrow climate envelope associated with lower elevations, such as American dog ticks (Bishopp & Trembley, 1945), and some tick species, such as *Ixodes (I.) ricinus,* have a large geographical distribution (Macleod, 1936).

Global warming, primarily driven by the greenhouse effect from fossil fuel combustion and deforestation, is intensifying. Climate change leads to abrupt variations in temperature and precipitation which raises anthropogenic concerns because of climate-driven shifts in tick habitats (Milnes & Anderson, 2023). Different stages of tick life cycle, such as egg, larvae, nymphs, and adults are influenced by climate. Climate change alters the temperature, precipitation patterns, and humidity levels which play a significant role in shaping tick habitats. This leads to shifts in the geographical distribution of ticks. Host seeking behavior of ticks is adversely affected by high temperatures, particularly in dry environments (Randolph, 2009). Ticks adapted to dry environments, such as brown dog ticks, are also affected by increasing

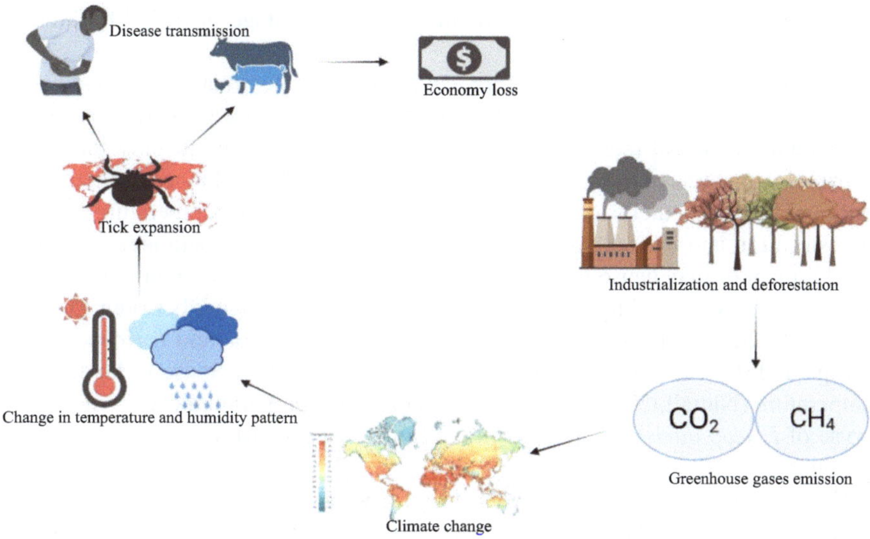

Fig. 1.1 Role of industrialization and deforestation in climate change which is causing tick range expansion, increase prevalence of tick-borne diseases, and causing losses to health systems and the economy

temperature (Yoder et al., 2006). Changes in climate can lead to the expansion or reduction of suitable habitats for tick species, affecting their populations and the potential for transmitting diseases to humans and animals (Chikezie et al., 2024). The expansion of tick habitats owing to climate change can increase the risk of tick-borne diseases (TBDs) by bringing ticks into contact with new host populations and human settlements (Kiragu, 2023). Understanding the dynamics of tick habitats is crucial for effective vector management strategies and disease control efforts, particularly in regions where ticks pose a threat to animal and human health (Fig. 1.1).

1.2 Climate Factors Influencing Tick Distribution

Ticks are present on every continent, highlighting their global distribution and ecological adaptability. Ticks are among the most significant arthropod vectors of zoonotic pathogens. Ticks spend most of their lives away from the host in the external environment and are considered to be greatly affected by climate change (Gilbert, 2021). It has been proposed that climate change affects disease transmission patterns, tick phenology, and geographical distribution. Emerging tick-borne pathogens highlight the importance of monitoring tick distribution. For example, the novel virus TBEV-2871 was isolated from *I. pavlovskyi* in Russia (Tkachev et al., 2017), and the Tyulek virus was discovered in Kyrgyzstan (L'Vov et al., 2014). Ticks are poikilothermic and sensitive to climate change. Each species of tick require specific environmental conditions for their development and survival at each stage of their life (Alkishe et al., 2021).

The primary climate related factors which influence distribution of ticks include temperature, humidity, and precipitation patterns. Ticks are sensitive to extreme temperatures. Annual mean temperature and precipitation significantly impact the distribution patterns of ticks, while temperature has a higher influence on their occurrence compared to other variables (Prakasan et al., 2020). High temperatures can lead to desiccation and mortality of ticks. On the contrary low temperatures can inhibit development or may cause freezing and ultimately death of ticks. Optimal temperature ranges vary depending upon tick species, but general range is between 15–35 °C (Ma et al., 2024). The spread of *I. scapularis* across Canada and *I. ricinus* across Europe and the Great Alpine region is linked to a 2–3 °C increase in land surface temperature (Nuttall, 2022). Furthermore, increasing temperature causes molting of larvae of *I. scapularis* into nymphs in the same year (Belozerov & Naumov, 2002; Ogden et al., 2004).

Humid environment helps prevent desiccation in ticks' desiccation, especially during the immature stages. Humidity levels greater than 80% are considered ideal for most tick species. Precipitation patterns influence habitat suitability. Moreover, ticks favor areas with consistent rainfall and high moisture levels (Bacon et al., 2022). Humidity plays a crucial role in tick habitat suitability, impacting the expansion of populations, particularly for species such as *I. ricinus* in parts of Europe with favorable conditions (Peralbo-Moreno et al., 2022).

Tick activity and reproduction are closely related to seasonal climate patterns. Warmer temperatures and increased precipitation in the spring and summer stimulate host-seeking behavior and egg laying. Cooler and drier conditions in fall and winter cause ticks to enter diapause or quiescence to conserve energy (Nuttall, 2022). As the climate is affected by floods and droughts, the geographical distribution of ticks is also affected (Ogden et al., 2021).

Tick distribution is also affected by vegetation and landscape configuration (Estrada-Peña, 2008). The migration of hosts has a significant impact on tick distribution within a zone, in addition to abiotic factors, such as weather and vegetation (Estrada-Peña, 2003). Ground humidity, diurnal temperature range, and seasonal maxima are primary variables affecting tick distribution (Randolph, 2000; Sonenshine & Roe, 2014). he point to consider here is that tick species such as *I. ricinus* that can feed on almost every land animal are not affected by climate change to an extent as compared to tick species such as the long-legged bat tick *I. vespertilionis,* which are host specific (Gilbert, 2021).

1.3 Ecological Responses to Changing Climate Conditions

Ticks exhibit adaptations that enhance their survival under changing climate conditions. Ticks produce hygroscopic secretions that allow them to absorb moisture from the environment, which is crucial under dry conditions. Additionally, lipid layers help retain water, enabling survival in arid environments (Benoit & Denlinger, 2010). Ticks such as *I. ricinus, Dermacentor (D.) reticulatus*, and *D. marginatus* are expected to benefit from climate change by expanding their ranges into northeastern and central Europe, potentially elevating TBD risks (Cunze et al., 2022). Ticks can modify their behavior which depends on environmental conditions. For instance, during durations of high temperature and low humidity, ticks may reduce their activity and search for shelter in moist microhabitats to avoid desiccation (Nielebeck et al., 2022).

The introduction of non-native tick species such as *Haemaphysalis longicornis* into new regions results in competition with native tick species for hosts and resources. This competition can change dynamics of TBD transmission because different tick species may have varying capacities to transmit pathogens to their hosts (Seo et al., 2024). The introduction of heat-resistive tick species may outcompete native species which ultimately a decline in biodiversity and changes in structure of community. Preserving habitat richness can decrease these risks. This highlights the importance of biodiversity for disease prevention. Ecosystem function and resilience are affected by complex interactions among species (Dagostin et al., 2024). Increased tick populations can lead to higher rates of TBDs in wildlife. This rise may cause a decline in susceptible host species. The decline can disrupt predator–prey relationships. Predators dependent on these hosts may also decline, potentially triggering cascading effects throughout the food web (Perumalsamy et al., 2024; Savage & Moore, 2024).

Climate change influences the geographic distribution potential of *Amblyomma (A.) maculatum* and *D. andersoni* ticks. This is leading to northward range expansion in response to increasing temperatures and altered precipitation patterns. Studies have suggested that climate change could affect the evolutionary processes of pathogens transmitted by *I. scapularis* through the effects of temperature on tick seasonality. Projected seasonal activity patterns for the 2020s and 2050s indicate increased fitness of short-lived, highly pathogenic, and less efficiently transmitted pathogens. Therefore, climate change may affect the frequency and geographic distribution of *I. scapularis*-borne pathogens (Ogden et al., 2008a, 2008b). The nymphal questing peak of *I. ricinus* is expected to shift earlier in the year. This indicates a change in seasonal activity patterns due to influence of climate (Nolzen et al., 2022).

Amblyomma cajannense population in Latin America is facing a decline because of its inability to adapt to changing climate (Oliveira et al., 2017). Monzón et al. (2016) showed that *A. amerioman* ticks found in New York and Oklahoma are genetically distinct from historic range populations in North and South Carolina. Ticks exhibits varied survival and questing responses to varying humidity levels. Lower humidity reduces survival rates, impacting questing behavior differently among tick species and highlighting their ecological sensitivity to climate conditions (Nielebeck et al., 2023).

1.4 Impact of Habitat Shifts on Tick Population Dynamics

Habitat shifts due to climate influence tick population dynamics, as some tick species are expanding their ranges, while some suffer range contractions. Vegetation and host availability regulate tick population dynamics (Estrada-Peña, 2008). The quality and connectivity of habitat patches determine tick dispersal, as hosts can carry ticks into new environments (Estrada-Peña, 2003). Because they may be able to use new hosts, generalized ticks such as *I. ricinus* and *I. scapularis*—the primary tick vectors of the Lyme borreliosis agent *Borrelia burgdorferi*—have a higher chance of being able to expand their ranges due to climate change. They might switch from white-tailed deer (*Odocoileus virginianus*) or red deer (*Cervus elaphus*) to reindeer, *Rangifer tarandus* (Gilbert, 2021).

Likewise, *Hyalomma (H.) marginatum* is also expanding its range in Europe (Fernández-Ruiz & Estrada-Peña, 2021). Climate change causes range contraction in ticks, such as *A. cajennensis* and *A. sculptum* populations (Oliveira et al., 2017). Changing environmental conditions can lead to mismatches between tick life cycles and host availability. These mismatches occur due to the disrupted host-seeking behavior of ticks (Sagurova et al., 2019). Until the beginning of this century, the blacklegged tick (*I. scapularis*) was found throughout the eastern and upper Midwest regions of the United States (Eisen & Eisen, 2018). Since then, it has spread to southern and eastern Canada (Ogden et al., 2008a). As reported by Fowler et al. (2022) and Stafford et al. (2018), the lone star tick (*A. americanum*), which is native to the southeastern United States (Springer et al., 2014), has spread to northern states and southern Canada. Passive surveillance has detected non-native tick species in these regions (Nelder et al., 2019). The rising temperatures will speed up the number of days needed for tick development. This acceleration has contributed to the northward expansion of I. scapularis in Canada. This results in earlier deposition of eggs and larval hatching, more nymphs molting in a single year, and more time for host-seeking operations (McPherson et al., 2017).

Climate change-driven shifts in migratory bird patterns have also led to the expansion of *I. scapularis* (Ogden et al., 2008a). *Amblyomma americanum* has been expanding north to the Canadian border since the 1890s (Springer et al., 2014). *Ixodes affinis* expands northward in the mid-Atlantic US, thriving in mixed pine-hardwood forests with a herbaceous understory and new mammalian hosts (Kidd, 2022). In North America, *A. maculatum* and *D. andersoni* are predicted to expand their ranges northward. *Amblyomma maculatum* moves into the Midwest and New England, and *D. andersoni* potentially reaches Canada (Alkishe & Peterson, 2022).

Tick-friendly conditions—mild warming and reduced precipitation—in the Mediterranean region favor the spread of species like *Rhipicephalus (R.) turanicus* and *H. marginatum* (Estrada-Peña & Venzal, 2007). These two species play a significant role in the spread of diseases that affect humans. The number of acceptable habitats for *R. bursa, R. turanicus,* and *H. marginatum* is expected to sharply increase because of the temperature rise and precipitation reduction. According to Estrada-peña and Venzal (2007), this change would lead to an increase in suitable habitat areas for these three species towards the north. According to Hulme et al. (1999), predicted climate patterns favor alteration in tick fauna and diseases that spread throughout large swaths in Europe. *Ixodes ricinus* (castor bean tick) is projected to expand its range into Scandinavian countries and the Baltic States. It is because of increased habitat suitability caused by climate change (Tytar et al., 2024). As temperature rises, *H. marginatum* is also expected to extend its range northward into central and northern Europe (Hekimoglu et al., 2023). In Norway, *I. ricinus* has expanded its northern distribution limit by approximately 400 km (Jore et al., 2011).

Soft ticks of the genera *Ornithodoros sonrai* and *Ornithodoros marocanus* are adopted to hot and dry climates. They have become more common in Senegal and Northwest Morocco. This is because these regions experience more frequent droughts. Such droughts are manifestation of climate change (Souidi et al., 2014). Estrada-Peña et al. (2006) stated that climate and habitat suitability models in Argentina and Australia predicted increases in the cattle tick *R. Boophilus microplus*, particularly in areas where rainfall is projected to increase. These predictions have negative implications for beef cattle industry. In Zimbabwe, *A. variegatum* is also forced to disperse into places outside of zones with extended dry seasons owing to climate change (Estrada-Peña, 2008).

However, where many species expand their range, some are undergoing range contraction. Similarly, the *A. cajannense* (vectors of spotted fever) species complex in Brazil predicts a contraction in range under future climate scenarios, with positive health implications for spotted fever in the region. (Oliveira et al., 2017).

1.5 Biodiversity Effects and Ecosystem Resilience

Biodiversity plays a crucial role in ecosystem resilience by enhancing the ability of ecosystems to absorb disturbances and maintain functionality despite climate change. Diverse ecosystems can better withstand and adapt to changes and reduce the likelihood of disease outbreaks (Khatsuriya et al., 2024). For neo-tropical tick fauna, tapirs are significant hosts because they inhabit less fragmented biomes with high biodiversity. Therefore, ticks are more abundant in tapirs than in any other vertebrate host (Labruna & Guglielmone, 2009). Changes in habitat can affect the transmission systems of tick-borne infections as well as the communities of terrestrial mammals. The "dilution effect" (Schmidt & Ostfeld, 2001) or the "biodiversity-buffers-disease paradigm" (Salkeld et al., 2013) refers to an inverse link between disease risk and biodiversity. The dilution effect is highly supported by studies on the effects of host distribution on the spread of Lyme disease (Turney et al., 2014).

Increased host diversity can reduce pathogen abundance and human disease risk (Huang et al., 2019), whereas a decline in biodiversity could increase wildlife and human diseases (Civitello et al., 2015). A study found that a high red fox density decreased the density of white-footed mice (*Peromyscus leucopus*), which

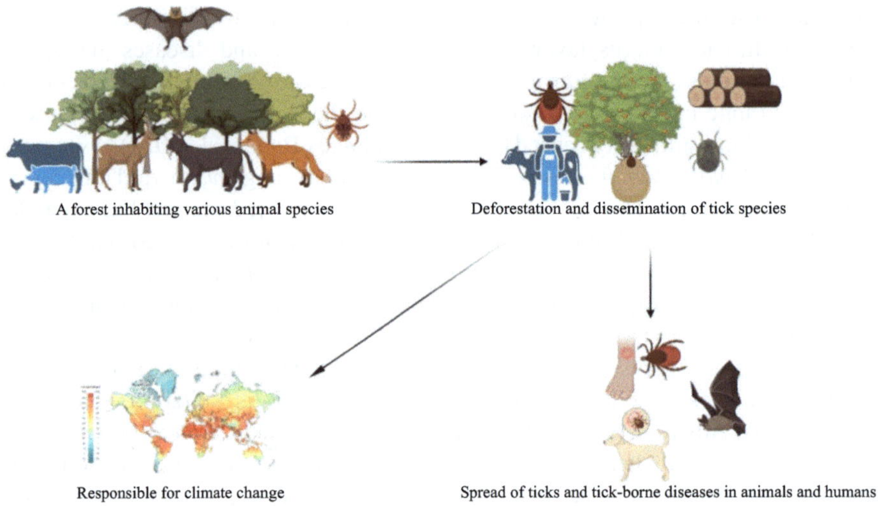

Fig. 1.2 Ecosystem with multiple species leads to less interaction between ticks and competent hosts. Ultimately a smaller proportion of ticks will be infected. In case of loss of biodiversity due to climate change, there will be more interaction between vectors and their hosts leading to tick range expansion and increased prevalence of tick-borne diseases

is a reservoir-competent host for *Borrelia burgdorferi*. Subsequently, the incidence of Lyme borreliosis was low in the northeastern United States (Hofmeester et al., 2017). Role of biodiversity on prevalence of TBDs is illustrated in Fig. 1.2.

The ability of a pathogen or tick species to survive and sustain a viable population despite the effects of climatic change is known as resilience. This could necessitate the ability to adjust to changing environments by genetic evolution or phenotypic adaptability (Gilbert, 2021). Certain species may suffer from the negative effects of climate change, especially in tropical zones where rising temperatures may force certain tick species to colonize new areas and negatively influence habitat fitness. According to research conducted in South Africa, increasing the temperature by 2 °C is expected to make the habitat less suitable for *A. hebraeum*, *R. decoloratus*, *R. appendiculatus*, and *H. truncatum* (Estrada-Pena, 2003). Certain pathogens (e.g., West Nile virus) may have multiple species of reservoirs (birds) and vectors that are locally involved. Variations in the diversity of these reservoirs and vectors due to climate change may lead to functional changes in transmission that will significantly impact the dynamics of the disease (Gould & Fikrig, 2004).

1.6 Ecological Interactions

Predators keep tick populations under check. Changes in predator populations can impact tick densities and disease transmission. Certain birds feed on ticks and contribute to controlling tick populations such as turkeys and robins. The occurrence and abundance of these predators in an area can influence tick densities in that area. For example, regions with higher populations of such birds tend to have lower tick densities (Estrada-Pena et al., 2021). Climate change or habitat alteration can disrupt predator populations. This in turn can disturb the balance of tick populations. A reduction in population of predators can lead to increased tick densities and raise the risk of TBDs (Hoffman & Bradley, 2023). Additionally, if climate change negatively affects predatory species, their declining numbers lead to higher tick populations and increased disease transmission (Lindsay et al., 2022).

Higher temperatures can speed up the development of both ticks and their pathogens. For example, warmer temperatures accelerate the development of Borrelia burgdorferi—the agent of Lyme disease—within ticks, increasing transmission rates (Wallace et al., 2019). Warmer temperatures also extend tick activity seasons, increasing the window for pathogen transmission (Bouchard et al., 2023). Extended warm periods mean ticks remain active for a longer part of the year and this extension raises the risk of TBDs (Randolph, 2004; Eisen & Eisen, 2023).

Climate-driven tick habitat expansion introduces tick-borne pathogens to previously unaffected populations. This expansion can lead to outbreaks of diseases in areas where they were previously rare or not reported (Gray et al., 2023).

1.7 Wildlife Host Dynamics

Wildlife host dynamics play a critical role in the transmission of tick-borne diseases.

1.7.1 Host Species and Tick Transmission

Different tick species have specific host preferences, which influence their distribution and the transmission of pathogens. For example, the black-legged tick (*I. scapularis*) primarily feeds on deer and rodents. However, western black-legged tick (*I. pacificus*) prefers different hosts (Paddock & Yabsley, 2007).

1.7.2 *Host Population Dynamics*

White-footed mice are primary hosts for nymphal ticks. Research has shown that areas with dense populations of these mice experience higher tick densities and greater incidence of Lyme disease (Diuk-Wasser et al., 2006; Ostfeld et al., 2014). Changes in host migration patterns caused by climate change can lead to prevalence of ticks and TBDs into new regions. For instance, expanding deer populations can carry ticks into previously uninhabited areas, introducing tick-borne diseases into new environments (Kilpatrick et al., 2014).

Host Behavior and Disease Risk

Deer browsing in dense vegetation face a higher risk of tick attachment. Changes in such behaviors induced by climate change or land use modifications can change tick distribution and elevate associated disease risks (Salkeld et al., 2013). The activity patterns like movement and nesting behaviors of wildlife hosts also impact tick transmission. Small mammals that are active in the underbrush are more likely to encounter ticks. Thus, increases the risk of pathogen transmission (Ogden et al., 2021).

1.7.3 *Zoonotic Risks*

Zoonotic diseases are transmitted to humans mainly through the bite of infected ticks. For instance, Lyme disease is reported to be mostly spread by bite of infected black-legged ticks (Piesman & Eisen, 2008). The risk of transmission increases in areas with high tick densities and abundant wildlife hosts because of elevated tick bite risks. Wildlife also indirectly contributes to zoonotic risk by serving as pathogen reservoirs for tick transmission. For example, white-footed mice are significant reservoirs for *Borrelia burgdorferi* which is a causative agent of Lyme disease. The mice play a crucial role in maintenance and transmission of the pathogen (Rogovskyy et al., 2024).

1.8 Implications for Human Health and Public Policy

The evolving dynamics of tick populations and associated diseases are influenced by climate change and habitat alterations. These developments pose significant challenges to public health systems and policy planning.

As ticks carrying pathogens expand into new regions, disease prevalence rises. This highlights the urgent need to strengthen public health systems, disease monitoring, and proactive policy development.

1.8.1 Implications for Human Health

Due to climate change, ticks are expanding their range into regions where they were not previously found. This presents an increased possibility for TBDs such as babesiosis, anaplasmosis and Lyme disease. Extended periods of warm weather lengthen tick activity seasons and accelerate pathogen development, increasing transmission rates (Eisen & Eisen, 2023). In Pakistan, the spread of TBDs is believed to be linked to the increased population of ticks which is worsened by climate change and movement of livestock (Khurshid et al., 2021).

Public health initiatives should focus on providing specially targeted education and materials to these high-risk groups (Ostfeld & Keesing, 2013). These interventions aim to reduce exposure risks and improve health outcomes. The emergence and geographical spread of new strains of TBDs in Pakistan is particularly alarming. It demonstrates the need for surveillance and scientific inquiry. These developments could also increase the existing morbidity and mortality rates. The previously noted TBDs example highlights the need for constant research and surveillance (Ahmed et al., 2017). Recent emergence of Powassan virus in areas of the US where it was previously rare highlights the importance of continuous investigation and research (Paddock et al., 2018).

1.8.2 Implications for Public Policy

The establishment of regional tick surveillance networks along with the integration of veterinary and public health offices is imperative for monitoring and control of TBDs in developing countries (Shabbir et al., 2018). These data assist in the formulation of public health intervention, which in this case, could be targeted campaigns for tick eradication and educational programs (Diuk-Wasser et al., 2012). Combining environmental factors with health data serves not only for forecasting possible outbreaks but optimizing resources (Estrada-Pena et al., 2021).

Acaricide resistance research and vaccine development for TBDs are critical globally to mitigate the disease burden (Ibrahim et al., 2017). Farooqi et al. (2020) stressed the need to strengthen the veterinary services and collaboration between the human and animal health care sectors in order to provide timely outbreak detection and response. Furthermore, public health organizations need to outline community targeted guidelines on the prevention, diagnosis, and treatment of TBDs (Crawford et al., 2018). Moreover, policies aimed at improving access to healthcare are significantly more important to reduce the public health burden of TBDs (Eisen & Eisen, 2023).

Sustainable livestock management practices and conservation of natural predators of ticks can contribute to control of tick populations and reduction of disease transmission (Ali et al., 2013). Policies that encourage the preservation of biodiversity and maintenance of healthy ecosystems can also naturally regulate tick populations. Because they support predator species that feed on ticks (Gray et al., 2023).

1.9 Adaptive Strategies for Tick Management

As climate change and shifts in wildlife dynamics impact tick populations and the diseases they transmit, implementing adaptive strategies for tick management is increasingly vital to mitigate risks to human and animal health (Fig. 1.3).

1.9.1 Ecological and Environmental Management

Reducing tick habitats through the management of vegetation around residential and recreational areas is crucial for lowering tick populations. Ticks thrive in areas with tall grasses, shrubs, and leaf litter; therefore, regularly clearing these environments can substantially decrease tick densities. If vegetation is managed around the livestock, then decrease in tick population has been observed and ultimately, TBDs are reduced (Ullah et al., 2023). Strategic landscape design which prevent ticks from

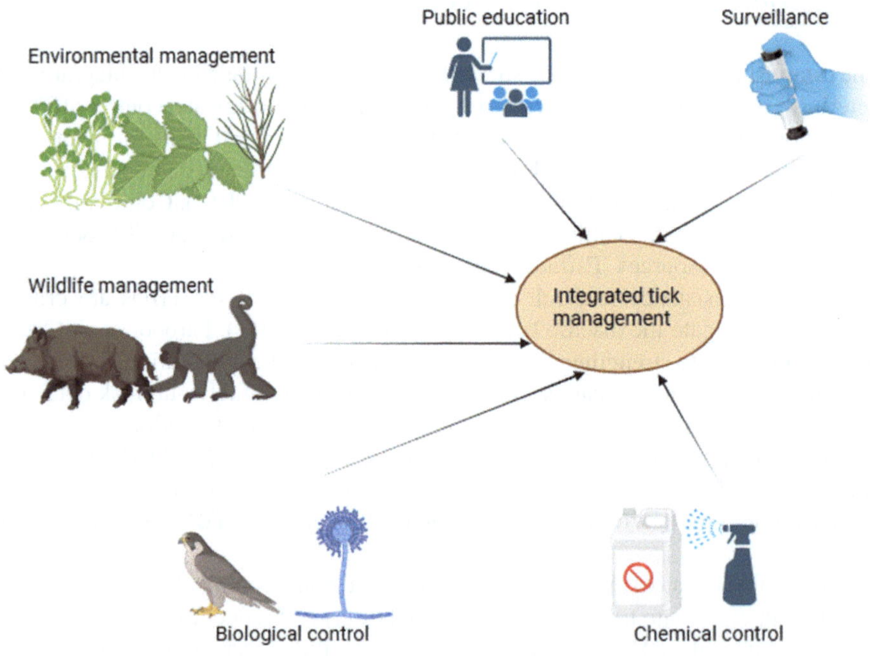

Fig. 1.3 Integrated tick management strategies to control ticks and tick-borne diseases

entering area where people and pets are present can also reduce the threat of TBDs. This strategy is very effective in different regions of the world including Pakistan (Rashid et al., 2019). Tick population has been controlled in many regions of South Asia through this approach (Khalid et al., 2022).

1.9.2 Wildlife Management

Overpopulation of host community leads to more TBDs. Regulated hunting management can be employed to manage the population of deer (Diuk-Wisser et al., 2006; Estrada-Peña et al., 2021). Similar strategies have been implemented in rural areas of Pakistan to manage the interaction of livestock and wildlife, which ultimately reduces the prevalence of TBDs (Qayyum et al., 2010).

1.9.3 Biological Control

Protecting and introducing the predators which prey on ticks such as some bird species can be an effective method to control TBDs. These birds include guinea fowl, chicken, and certain songbird. These bird species consume large populations of tick which is very beneficial in controlling the tick population (Crawford et al., 2018; Hoffman & Bradley, 2023). This technique has been implemented in South Asia and has shown great results to control TBDs (Khan et al., 2021). Use of entomopathogenic fungi also helps in controlling the tick population. These entomopathogenic fungi destroy the different life stages of ticks (Gondim et al., 2004).

1.9.4 Chemical Control

Chemical compounds that are long lasting are available to control ticks and TBDs (Eisen & Paddock, 2021). These products should not be overused as they may lead to development of resistance and have negative environmental consequences (Rizwan et al., 2024). Pesticide monitoring data needs to be collected to assess the effectiveness of these pesticides (Benelli et al., 2022).

1.9.5 Public Health

As public awareness of TBDs has been low, public awareness campaigns are essential that focus specifically on prevention and tick removal techniques. These efforts are expected to improve the public's perception of TBDs and reduce the incidence of

TBDs (Beard et al., 2016). Proper knowledge of ticks as well as their habitat and areas of most activity will help in not getting bitten.

1.9.6 Surveillance and Monitoring

Tracking tick populations, their species distribution, and disease prevalence demands the implementation of tick surveillance programs. These efforts help strategize targeted intervention activities and assist in the identification of high-risk areas (Makwarela et al., 2025). Surveillance of tick and pathogen reservoirs in wildlife helps refine public health strategies and enhances early detection of possible outbreaks (Kilpatrick et al., 2014).

1.9.7 Future Directions in Research and Conservation

As climate change leads to a change in the dynamics of tick populations and the spread of TBDs, future research along with conservation efforts are the need of the hour to address these challenges effectively.

1.9.8 Research Priorities

Future research should be based on the development of climatic models which will predict the effect of changing temperature and humidity on tick distribution, life cycles, and pathogen transmission. These models will predict future risks and will help us to design management strategies (Ogden et al., 2021). In order to determine the localized risks, the investigation of microclimates should be carried out. Prediction accuracy of these models can be enhanced by studying the effects of different habitats on tick populations (Dunn et al., 2019). Genetic modifications of tick-borne pathogens should be studied especially in context of environmental factors (Eisen & Paddock, 2021). Wild animals should be studied as these are reservoir hosts for many tick-borne pathogens. The complex interaction between wildlife, ticks, and pathogens should be studied comprehensively (Dantas-Torres et al., 2021).

In order to replace the use of chemical acaricides, researches are now being conducted to observe the effects of different strains of bacteria, viruses, and fungi for the control of parasites. For example, entomopathogenic fungi such as *Beauveria bassiana* and *Metarhizium anisopliae* have ability to replace chemical acaricides (Kemp et al., 2007). Emerging genetic technologies such as gene drives and CRISPR are innovative approaches to disrupt tick populations in order to reduce pathogen transmission. Although these technologies provide long term solutions but the consequences of these technologies should be evaluated (Harris et al., 2021).

For early detection of ticks and TBDs, advanced surveillance systems should be developed. These surveillance systems should integrate data from environmental sensors and remote sensing technology (Makwarela et al., 2025). In order to improve disease management, new diagnostic tools that offer accurate and rapid detection of TBDs can be promising. It means innovations in molecular diagnostics will have significant results (Schotthoefer & Lane, 2020).

1.9.9 Conservation Strategies

Tick populations can be reduced by restoring the natural habitats which will allow the predators to control tick populations (Ostfeld & Keesing, 2013). Tick populations can be minimized in the green spaces by integrating tick management with land use planning (Hamer et al., 2015). Wildlife management practices for prevention of overpopulation of tick hosts such as deer and rodents is necessary. This involves the control of hunting programs, habitat management, and monitoring of wildlife health (Diuk-Wasser et al., 2006). Predators which control tick populations should be conserved. It can be achieved by maintaining the ecosystems which maintain the predators' populations (Crawford et al., 2018).

Policies that aim to reduce ticks and TBDs should be advocated. This can be achieved by funding researches, conservation programs, and public health initiatives (Beard et al., 2016). Environmental protection along with tick control can be achieved by implementation of policies of habitat conservation, wildlife management, and pesticides use.

References

Ahmed, A., Tanveer, M., Khalid, N., Ahmad, M. U. D., Afzal, M. F., & Nawaz, Z. (2017). Emergence of Crimean-Congo hemorrhagic fever in Pakistan: A public health concern. *Infection, Genetics and Evolution, 49*, 19–25.

Ali, A., Zahid, H., Zeb, I., Siddique, Y. H., Shahnawaz, S., & Khan, M. A. (2013). Tick infestation in livestock: Ecology, economic impact, and integrated management strategies in Pakistan. *Pakistan Journal of Agricultural Sciences, 50*(1), 93–101.

Alkishe, A., & Peterson, A. T. (2022). Climate change influences on the geographic distributional potential of the spotted fever vectors *Amblyomma maculatum* and *Dermacentor andersoni*. *PeerJ, 10*, Article e13279.

Alkishe, A., Raghavan, R. K., & Peterson, A. T. (2021). Likely geographic distributional shifts among medically important tick species and tick-associated diseases under climate change in North America: A review. *Insects, 12*(3), 225.

Anwar, M. M., & Ahmad, N. (2019). Predator-prey dynamics and tick control in Pakistani ecosystems. *Journal of Parasitic Diseases, 43*(3), 456–463.

Apanaskevich, D. A., Oliver, J. H., Sonenshine, D. E., & Roe, R. M. (2014). Life cycles and natural history of ticks. In D. E. Sonenshine & R. M. Roe (Eds.), *Biology of ticks*. Oxford University Press.

Bacon, E. A., Kopsco, H., Gronemeyer, P., Mateus-Pinilla, N., & Smith, R. L. (2022). Effects of climate on the variation in abundance of three tick species in Illinois. *Journal of Medical Entomology, 59*(2), 700–709.

Beard, C. B., Eisen, R. J., Barker, C. M., Garofalo, J. F., Hahn, M., Hayden, M., Monaghan, A., Ogden, N. H., & Schramm, P. J. (2016). Preventing tick-borne diseases: Public health strategies and recommendations. *Journal of Public Health Management and Practice, 22*(5), 470–478.

Belozerov, V. N., & Naumov, R. L. (2002). Nymphal diapause and its photoperiodic control in the tick *Ixodes scapularis* (Acari: Ixodidae). *Folia Parasitologica, 49*(4), 314–318.

Benelli, G., Pavela, R., Canale, A., & Mehlhorn, H. (2022). Tick repellents and acaricides of botanical origin: A green roadmap to control tick-borne diseases? *Parasitology Research, 121*(1), 219–238.

Benoit, J. B., & Denlinger, D. L. (2010). Meeting the challenges of on-host and off-host water balance in blood-feeding arthropods. *Journal of Insect Physiology, 56*(10), 1366–1376.

Bishopp, F. C., & Trembley, H. L. (1945). Distribution and hosts of certain North American ticks. *The Journal of Parasitology, 31*(1), 1–54.

Bouchard, C., Dibernardo, A., Koffi, J. K., Wood, H., & Lindsay, L. R. (2023). The extended season of ticks under climate change. *Nature Climate Change, 13*, 312–318.

Chikezie, F., Opara, K., & Ubulom, P. (2024). Impacts of changing climate on arthropod vectors and diseases transmission. *Nigerian Journal of Environment.*

Civitello, D. J., Cohen, J., Fatima, H., Halstead, N. T., Liriano, J., McMahon, T. A., Oggenfuss, K., Ortega, C. N., & Young, S. (2015). Biodiversity inhibits parasites: Broad evidence for the dilution effect. *Proceedings of the National Academy of Sciences, 112*(28), 8667–8671.

Crawford, C., Jones, A. R., Patel, K., Nguyen, L. M., & Thompson, B. (2018). Effects of tick populations on ecosystem dynamics. *Ecological Applications, 28*(3), 620–632.

Cunze, S., Glock, G., Kochmann, J., & Klimpel, S. (2022). Ticks on the move—Climate change-induced range shifts of three tick species in Europe: Current and future habitat suitability for *Ixodes ricinus* in comparison with *Dermacentor reticulatus* and *Dermacentor marginatus*. *Parasitology Research, 121*(8), 2241–2252.

Dagostin, F., Tagliapietra, V., Marini, G., Ferrari, G., Cervellini, M., Wint, W., Metz, M., Rosà, R., & Rizzoli, A. (2024). High habitat richness reduces the risk of tick-borne encephalitis in Europe: A multi-scale study. *One Health, 18*, Article 100669.

Dantas-Torres, F., Chomel, B. B., & Otranto, D. (2021). Reservoir competence of wildlife for tick-borne pathogens. *Trends in Parasitology, 37*(4), 342–353.

Diuk-Wasser, M. A., VanAcker, M. C., & Fernandez, M. P. (2006). Host population density drives the tick population and pathogen transmission. *Ecological Applications, 16*(2), 122–134.

Diuk-Wasser, M. A., Gatewood, A. G., Cortinas, R., Yaremych-Hamer, S., Tsao, J., Hickling, G., Brown, H., Walker, E., Piesman, J., & Fish, D. (2012). Interactions between climate, tick population dynamics, and pathogen transmission. *Journal of Vector Ecology, 37*(2), 389–396.

Dumas, A., Bouchard, C., Drapeau, P., Lindsay, L. R., Ogden, N. H., & Leighton, P. A. (2024). The risk of contact between visitors and *Borrelia burgdorferi*-infected ticks is associated with fine-scale landscape features in a southeastern Canadian nature park. *BMC Public Health, 24*(1), 1180.

Dunn, R. R., Davies, T. J., Harris, N. C., Gavin, M. C., Sandel, B., & Carlson, C. J. (2019). Microclimate influences on tick populations and pathogen transmission. *Ecology Letters, 22*(12), 2057–2067.

Eisen, R. J., & Eisen, L. (2006). Landscaping for tick prevention: Implications for prevention of tick-borne diseases. *Journal of Vector Ecology, 31*(1), 14–22.

Eisen, R. J., & Eisen, L. (2018). The blacklegged tick, *Ixodes scapularis*: An increasing public health concern. *Trends in Parasitology, 34*(4), 295–309.

Eisen, L., & Eisen, R. J. (2023). Seasonal variations and climate impact on tick-borne diseases. *Journal of Medical Entomology, 60*(3), 512–520.

Eisen, R. J., & Paddock, C. D. (2021). Tick and tick-borne pathogen surveillance as a public health tool in the United States. *Journal of Medical Entomology, 58*(4), 1490–1502.

Ergönül, Ö., Çelikbaş, A., Dokuzoğuz, B., Eren, Ş, Baykam, N., & Esener, H. (2004). Characteristics of patients with Crimean-Congo hemorrhagic fever in a recent outbreak in Turkey and impact of oral ribavirin therapy. *Clinical Infectious Diseases, 39*(2), 284–287.

Estrada-Peña, A. (2003). The relationships between habitat topology, critical scales of connectivity and tick abundance *Ixodes ricinus* in a heterogeneous landscape in northern Spain. *Ecography, 26*(5), 661–671.

Estrada-Peña, A. (2008). Climate, niche, ticks, and models: What they are and how we should interpret them. *Parasitology Research, 103*(1), 87–95.

Estrada-Peña, A., & Venzal, J. M. (2007). Climate niches of tick species in the Mediterranean region: Modeling of occurrence data, distributional constraints, and impact of climate change. *Journal of Medical Entomology, 44*(6), 1130–1138.

Estrada-Peña, A., Corson, M., Venzal, J. M., Mangold, A. J., & Guglielmone, A. A. (2006). Changes in climate and habitat suitability for the cattle tick *Boophilus microplus* in its southern Neotropical distribution range. *Journal of Vector Ecology, 31*(1), 158–167.

Estrada-Peña, A., Mihalca, A. D., & Petney, T. N. (2021). Role of birds in controlling tick populations. *Ticks and Tick-Borne Diseases, 12*(2), 101–107. https://doi.org/10.1016/j.ttbdis.2020.101607

Farooqi, S. H., Ahmad, N., Anwar, M. M., & Khan, M. A. (2020). Strengthening veterinary services for tick-borne disease control in Pakistan. *Tropical Animal Health and Production, 52*(1), 21–29.

Fernández-Ruiz, N., & Estrada-Peña, A. (2021). Towards new horizons: Climate trends in Europe increase the environmental suitability for permanent populations of *Hyalomma marginatum* (Ixodidae). *Pathogens, 10*(2), Article 95.

Fowler, P. D., Nguyentran, S., Quatroche, L., Porter, M. L., Kobbekaduwa, V., Tippin, S., Eisen, R. J., & Tsao, J. I. (2022). Northward expansion of *Amblyomma americanum* (Acari: Ixodidae) into southwestern Michigan. *Journal of Medical Entomology, 59*(5), 1646–1659. https://doi.org/10.1093/jme/tjac089

Gilbert, L. (2021). The impacts of climate change on ticks and tick-borne disease risk. *Annual Review of Entomology, 66*(1), 373–388.

Gondim, M. G. C., Garcia, M. V., & Veríssimo, C. J. (2004). Biological control of ixodid ticks using pathogens. *Journal of Vector Ecology, 29*(1), 1–6.

Gould, L. H., & Fikrig, E. (2004). West Nile virus: A growing concern? *The Journal of Clinical Investigation, 113*(8), 1102–1107.

Gray, J. S., Dantas-Torres, F., Estrada-Peña, A., & Levin, M. (2023). Tick range expansion and the emergence of new tick-borne diseases. In M. M. Cameron, L. H. Molyneux, & R. E. Sinden (Eds.), *Infectious diseases in the new millennium*. Springer.

Hamer, G. L., Goldberg, T. L., Kitron, U. D., & Walker, E. D. (2015). Land use planning strategies for controlling tick populations and reducing disease risk. *EcoHealth, 12*(1), 101–110.

Harris, C., Liu, J., Brown, J., & Narasimhan, S. (2021). Gene editing for tick management: Current status and future prospects. *Frontiers in Veterinary Science, 8*, 123.

Hekimoglu, O., Elverici, C., & Kuyucu, A. C. (2023). Predicting climate-driven distribution shifts in *Hyalomma marginatum* (Ixodidae). *Parasitology, 150*(10), 883–893.

Hoffman, C., & Bradley, K. (2023). Predator reintroduction as a tool for managing tick populations. *Ecology Letters, 26*(1), 78–84.

Hofmeester, T. R., Jansen, P. A., Wijnen, H. J., Coipan, E. C., Fonville, M., Prins, H. H. T., & van Wieren, S. E. (2017). Cascading effects of predator activity on tick-borne disease risk. *Proceedings of the Royal Society b: Biological Sciences, 284*(1859), 20170453.

Huang, C.-I., Kay, S. C., Davis, S., Tufts, D. M., Gaffett, K., Tefft, B., & Diuk-Wasser, M. A. (2019). High burdens of *Ixodes scapularis* larval ticks on white-tailed deer may limit Lyme disease risk in a low biodiversity setting. *Ticks and Tick-Borne Diseases, 10*(2), 258–268.

Hulme, M., Mitchell, J., Ingram, W., Lowe, J., Johns, T., New, M., & Viner, D. (1999). Climate change scenarios for global impacts studies. *Global Environmental Change, 9*(S1), S3–S19.

Ibrahim, N., Ali, A., Khan, M. A., Iqbal, Z., & Abbas, R. Z. (2017). Resistance of *Rhipicephalus* ticks to commonly used acaricides in Pakistan. *Ticks and Tick-Borne Diseases, 8*(5), 747–752.

James, A. M., Burdett, C., McCool, M. J., Fox, A., & Riggs, P. (2015). The geographic distribution and ecological preferences of the American dog tick, *Dermacentor variabilis* (Say), in the USA. *Medical and Veterinary Entomology, 29*(2), 178–188.

Jore, S., Viljugrein, H., Hofshagen, M., Brun-Hansen, H., Kristoffersen, A. B., Nygård, K., & Ytrehus, B. (2011). Multi-source analysis reveals latitudinal and altitudinal shifts in range of *Ixodes ricinus* at its northern distribution limit. *Parasites & Vectors, 4*(1), 84.

Kemp, D. H., Stone, B. F., & Binnington, K. C. (2007). Advances in biological control of ticks. *Trends in Parasitology, 23*(7), 305–312.

Khalid, S., Mahmood, T., & Farooq, U. (2022). Urban tick management and control strategies in South Asia. *Tropical Medicine & Infectious Diseases, 7*(2), 123–130.

Khan, H. A., Akram, W., Shehzad, K., & Khan, M. A. (2021). The role of natural predators in tick management in Pakistan. *Asian Journal of Animal Sciences, 15*(1), 45–52.

Khatsuriya, P. A., Makwana, V. M., & Dodia, P. P. (2024). Biodiversity resilience amidst climate shifts: A global perspective. In S. Idris (Ed.), *Global challenges for the environment and climate change*. IGI Global.

Khurshid, A., Ali, A., Ahmed, H., & Khan, M. A. (2021). Climate change and tick-borne diseases in Pakistan. *Pakistan Journal of Zoology, 53*(3), 1005–1012.

Kidd, L. (2022). Changes. *Veterinary Clinics of North America: Small Animal Practice, 52*(6).

Kilpatrick, H. J., LaBonte, A. M., & Stafford, K. C. (2014). Deer population density and its effects on Lyme disease risk. *Wildlife Society Bulletin, 38*(1), 6–13.

Kiragu, C. (2023). The impact of climate change on animal health and disease patterns. *Journal of Animal Health, 3*(1), 24–33.

Labruna, M. B., & Guglielmone, A. A. (2009). Ticks of new world tapirs. *Medical and Veterinary Entomology, 23*(3), 248–254.

Lindsay, L. R., Ogden, N. H., Leighton, P. A., & Bouchard, C. (2022). Effects of climate change on tick predators and disease spread. *Environmental Research, 214*, Article 113919.

L'Vov, D. K., Al'khovskiĭ, S. V., Shchelkanov, M., Shchetinin, A. M., Deriabin, P. G., Aristova, V. A., & Botikov, A. G. (2014). Taxonomic status of the Tyulek virus (TLKV) (Orthomyxoviridae, Quaranjavirus, Quaranfil group) isolated from the ticks *Argas vulgaris* Filippova, 1961 (Argasidae) from the birds burrow nest biotopes in the Kyrgyzstan. *Voprosy Virusologii, 59*(2), 28–32.

Ma, R., Li, C., Gao, A., Jiang, N., Li, J., Hu, W., & Feng, X. (2024). Tick species diversity and potential distribution alternation of dominant ticks under different climate scenarios in Xinjiang China. *PLOS Neglected Tropical Diseases, 18*(4), Article e0012108.

Macleod, J. (1936). *Ixodes ricinus* in relation to its physical environment: IV. An analysis of the ecological complexes controlling distribution and activities. *Parasitology, 28*(3), 295–319.

Makwarela, T. G., Seoraj-Pillai, N., & Nangammbi, T. C. (2025). Tick control strategies: Critical insights into chemical, biological, physical, and integrated approaches for effective hard tick management. *Veterinary Sciences, 12*(2), 114.

McPherson, M., García-García, A., Cuesta-Valero, F. J., Beltrami, H., Hansen-Ketchum, P., MacDougall, D., & Ogden, N. H. (2017). Expansion of the Lyme disease vector *Ixodes scapularis* in Canada inferred from CMIP5 climate projections. *Environmental Health Perspectives, 125*(5), Article 057008.

Milnes, E. L., & Anderson, N. E. (2023). Chapter 20—Tick and vector-borne disease expansion with climate change. In E. Miller, N. Lamberski, & P. Calle (Eds.), *Fowler's Zoo and Wild Animal Medicine Current Therapy* (Vol. 10, pp. 125–130). W.B. Saunders.

Monzón, J. D., Atkinson, E. G., Henn, B. M., & Benach, J. L. (2016). Population and evolutionary genomics of *Amblyomma americanum*, an expanding arthropod disease vector. *Genome Biology and Evolution, 8*(5), 1351–1360.

Nelder, M. P., Russell, C. B., Clow, K. M., Johnson, S., Weese, J. S., Cronin, K., & Patel, S. N. (2019). Occurrence and distribution of *Amblyomma americanum* as determined by passive surveillance in Ontario, Canada (1999–2016). *Ticks and Tick-Borne Diseases, 10*(1), 146–155.

Nielebeck, C., Kim, S. H., Dedmon, L., Pangilinan, M., Quan, J., Ota, W., & Monzón, J. D. (2022). A novel laboratory method to simulate climatic stress with successful application to experiments with medically relevant ticks. *PLoS ONE, 17*(9), Article e0275314.

Nielebeck, C., Kim, S., Pepe, A., Himes, L., Miller, Z., Zummo, S., Tang, M., & Monzón, J. D. (2023). Climatic stress decreases tick survival but increases rate of host-seeking behavior. *Ecosphere, 14*(4), Article e4369.

Nolzen, H., Brugger, K., Reichold, A., Brock, J., Lange, M., & Thulke, H.-H. (2022). Model-based extrapolation of ecological systems under future climate scenarios: The example of *Ixodes ricinus* ticks. *PLoS ONE, 17*(4), Article e0267196.

Nuttall, P. A. (2022). Climate change impacts on ticks and tick-borne infections. *Biologia, 77*(6), 1503–1512.

Ogden, N., Lindsay, L., Beauchamp, G., Charron, D., Maarouf, A., O'Callaghan, C., & Barker, I. (2004). Investigation of relationships between temperature and developmental rates of tick *Ixodes scapularis* (Acari: Ixodidae) in the laboratory and field. *Journal of Medical Entomology, 41*(4), 622–633.

Ogden, N. H., Bigras-Poulin, M., Hanincová, K., Maarouf, A., O'Callaghan, C. J., & Kurtenbach, K. (2008a). Projected effects of climate change on tick phenology and fitness of pathogens transmitted by the North American tick *Ixodes scapularis. Journal of Theoretical Biology, 254*(3), 621–632.

Ogden, N. H., Lindsay, L. R., Hanincová, K., Barker, I. K., Bigras-Poulin, M., Charron, D. F., & Thompson, R. A. (2008b). Role of migratory birds in introduction and range expansion of *Ixodes scapularis* ticks and of *Borrelia burgdorferi* and *Anaplasma phagocytophilum* in Canada. *Applied and Environmental Microbiology, 74*(6), 1780–1790.

Ogden, N. H., Ben Beard, C., Ginsberg, H. S., & Tsao, J. I. (2021). Possible effects of climate change on ixodid ticks and the pathogens they transmit: Predictions and observations. *Journal of Medical Entomology, 58*(4), 1536–1545.

Oliveira, S. V. d., Romero-Alvarez, D., Martins, T. F., Santos, J. P. d., Labruna, M. B., Gazeta, G. S., & Gurgel-Gonçalves, R. (2017). *Amblyomma* ticks and future climate: Range contraction due to climate warming. *Acta Tropica, 176*, 340–348.

Ostfeld, R. S., & Keesing, F. (2013). Biodiversity and the dilution effect: Implications for disease management. *Ecology Letters, 16*(5), 601–609.

Paddock, C. D., & Yabsley, M. J. (2007). Ecological aspects of tick-borne diseases in the United States. *Emerging Infectious Diseases, 13*(3), 306–311.

Paddock, C. D., Kuehnert, M. J., Eng, J., & Zaki, S. R. (2018). Powassan virus disease in the United States, 2006–2016. *Emerging Infectious Diseases, 24*(4), 804–812.

Peralbo-Moreno, A., Baz-Flores, S., Cuadrado-Matías, R., Barroso, P., Triguero-Ocaña, R., Jiménez-Ruiz, S., & Ruiz-Fons, F. (2022). Environmental factors driving fine-scale ixodid tick abundance patterns. *Science of the Total Environment, 853*, Article 158633.

Perumalsamy, N., Sharma, R., Subramanian, M., & Nagarajan, S. A. (2024). Hard ticks as vectors: The emerging threat of tick-borne diseases in India. *Pathogens, 13*(7).

Peters, H. S. (1930). Ectoparasites and bird-banding. *Bird-Banding, 1*(2), 51–60.

Piesman, J., & Eisen, R. J. (2008). Prevention of tick-borne diseases. *Vector-Borne and Zoonotic Diseases, 8*(1), 19–28.

Prakasan, K., Aiswarya, M., & Aswathi, R. (2020). Climate—the principal factor influencing tick plethora and tick-borne zoonoses: A review. *UP Journal of Zoology.*

Qayyum, M., Khan, M. S., & Zaman, M. A. (2010). Management of ticks in livestock in Pakistan. *Pakistan Journal of Zoology, 42*(1), 35–41.

Randolph, S. E. (2000). Ticks and tick-borne disease systems in space and from space. In *Advances in Parasitology.* Academic Press.

Randolph, S. E. (2009). Chapter 6: Epidemiological consequences of the ecological physiology of ticks. In *Advances in Insect Physiology.* Academic Press.

Rapport, D. J. (1999). Epidemiology and ecosystem health: Natural bridges. *Ecosystem Health, 5*(3), 174–180.

Rashid, M., Ghosh, S., & Rajput, Z. I. (2019). Epidemiology and control strategies of tick-borne diseases in Pakistan. *Asian Journal of Animal and Veterinary Advances, 14*(3), 135–142.

Rizwan, H. M., Naeem, M. A., Younus, M., Sajid, M. S., & Chen, X. (2024). *Anti-parasitic drug resistance in veterinary practice*. CABI.

Rogovskyy, A. S., Pliasas, V. C., Buhrer, R., Lewy, K., Wiener, D. J., Jung, Y., Bova, J., Rogovska, Y., Kim, S. J., & Jeon, E. (2024). Do white-footed mice, the main reservoir of the Lyme disease pathogen in the United States, clinically respond to the borrelial tenancy? *Infection and Immunity, 92*(12), Article e0038224.

Sagurova, I., Ludwig, A., Ogden, N. H., Pelcat, Y., Dueymes, G., & Gachon, P. (2019). Predicted northward expansion of the geographic range of the tick vector *Amblyomma americanum* in North America under future climate conditions. *Environmental Health Perspectives, 127*(10), Article 107014.

Salkeld, D. J., Lane, R. S., & Niebylski, M. L. (2013). Landscape composition, habitat type, and the prevalence of tick-borne pathogens in wildlife. *Journal of Vector Ecology, 38*(1), 119–129.

Savage, J. D. T., & Moore, C. M. (2024). How do host population dynamics impact Lyme disease risk dynamics in theoretical models? *PLoS ONE, 19*(5), Article e0302874.

Schmidt, K. A., & Ostfeld, R. S. (2001). Biodiversity and the dilution effect in disease ecology. *Ecology, 82*(3), 609–619.

Schotthoefer, A. M., & Lane, R. S. (2020). Advances in tick-borne disease diagnostics. *Clinical Microbiology Reviews, 33*(4), e00065-e119.

Seo, J., Kim, G., Lim, J.-A., Song, S., Yoo, D.-S., Cho, H.-S., & Oh, Y. (2024). Tick diversity and pathogen transmission in Daejeon, Korea: Implications from companion animals and walking trails. *Veterinary Sciences, 11*(2), 85.

Shabbir, M. Z., Rashid, M., Nawaz, M., Khan, M. T., Aslam, M. A., Ghafoor, A., & Saeed, M. (2018). Molecular detection and epidemiology of tick-borne diseases in Pakistan. *Journal of Parasitic Diseases, 42*(4), 641–649.

Sonenshine, D. E., & Roe, R. M. (2014). *Biology of ticks volume 2* (Vol. 2). Oxford University Press, USA.

Souidi, Y., Boudebouch, N., Ezikouri, S., Belghyti, D., Trape, J. F., & Sarih, M. H. (2014). Borrelia crocidurae in *Ornithodoros* ticks from northwestern Morocco: A range extension in relation to climatic change? *Journal of Vector Ecology, 39*(2), 316–320.

Springer, Y. P., Eisen, L., Beati, L., James, A. M., & Eisen, R. J. (2014). Spatial distribution of counties in the continental United States with records of occurrence of *Amblyomma americanum* (Ixodida: Ixodidae). *Journal of Medical Entomology, 51*(2), 342–351.

Stafford, K. C., III., Molaei, G., Little, E. A., Paddock, C. D., Karpathy, S. E., & Labonte, A. M. (2018). Distribution and establishment of the lone star tick in Connecticut and implications for range expansion and public health. *Journal of Medical Entomology, 55*(6), 1561–1568.

Tkachev, S. E., Chicherina, G. S., Golovljova, I., Belokopytova, P. S., Tikunov, A. Y., Zadora, O. V., & Tikunova, N. V. (2017). New genetic lineage within the Siberian subtype of tick-borne encephalitis virus found in Western Siberia, Russia. *Infection, Genetics and Evolution, 56*, 36–43.

Turney, S., Gonzalez, A., & Millien, V. (2014). The negative relationship between mammal host diversity and Lyme disease incidence strengthens through time. *Ecology, 95*(12), 3244–3250.

Tytar, V., Kozynenko, I., Pupins, M., Škute, A., Čeirāns, A., & Nekrasova, O. (2024). Species distribution modeling of the castor bean tick, *Ixodes ricinus* (Linnaeus, 1758), under current and future climates, with a special focus on Ukraine and Latvia. *bioRxiv*.

Ullah, R., Rehman, S., & Khan, A. (2023). Recent advances in tick management strategies in Pakistan. *Journal of Parasitic Diseases, 47*(1), 23–30.

Wallace, D., Ratti, V., Kodali, A., Winter, J. M., Ayres, M. P., Chipman, J. W., Aoki, C. F., Osterberg, E. C., Silvanic, C., Partridge, T. F., & Webb, M. J. (2019). Effect of rising temperature on Lyme disease: *Ixodes scapularis* population dynamics and *Borrelia burgdorferi* transmission and prevalence. *Canadian Journal of Infectious Diseases & Medical Microbiology, 2019*, 9817930. https://doi.org/10.1155/2019/9817930

Yoder, J. A., Benoit, J. B., Rellinger, E. J., & Tank, J. L. (2006). Developmental profiles in tick water balance with a focus on the new Rocky Mountain spotted fever vector *Rhipicephalus Sanguineus*. *Medical and Veterinary Entomology, 20*(4), 365–372.

Chapter 2
Tick Life Cycle Modifications in Response to Changing Climates

Hafiz Muhammad Rizwan, Muhammad Moaaz, Muhammad Younus, Muhammad Noman Naseem, and Neil Sargison

Abstract Ticks are blood feeding ectoparasites that undergo a complex life cycle consisting of four stages: egg, larva, nymph, and adult. With the exception of egg stage, each stage requires a blood meal to progress. Completion of the life cycle is strongly dependent on environmental conditions. The life of a tick typically varies between several months and three years, depending on the species and ecological factors. The environmental conditions that influence tick development, survival and seasonal activity are temperature and humidity. Warmer climates tend to speed up the transition between life stages and prolong the period of tick activity; thereby potentially increasing population densities. Cold weather reduces the survival rate of some tick species. Some tick species contain glycoproteins which provide protection from freezing conditions, while certain species possess adaptive traits, such as quiescence and diapause, which help counter environmental stress. These traits are exhibited to different extents by different tick populations, leading to differing seasonal activities across regions. Knowledge of such developmental approaches and environmental dependencies is critical to account for the impacts of habitat and climate changes on tick abundance and ability to transmit diseases; hence informing control strategies.

H. M. Rizwan (✉)
Section of Parasitology, Department of Pathobiology, KBCMA College of Veterinary and Animal Sciences, Narowal, Sub-Campus UVAS, Lahore, Pakistan
e-mail: hm.rizwan@uvas.edu.pk

M. Moaaz
Faculty of Veterinary Sciences, KBCMA College of Veterinary and Animal Sciences, Narowal, Sub-Campus UVAS, Lahore, Pakistan

M. Younus
Department of Pathology, University of Veterinary and Animal Sciences, Lahore, Pakistan

M. N. Naseem
School of Veterinary Science, The University of Queensland, Brisbane, Australia

N. Sargison
Royal (Dick) School of Veterinary Studies, Easter Bush Veterinary Centre, University of Edinburgh, Roslin, Midlothian, UK

H. M. Rizwan and M. S. Sajid (eds.), *Ticks in a Changing Climate*, Biometeorology 7,
https://doi.org/10.1007/978-3-032-12259-9_2

Keywords Environmental triggers · Temperature effects · Host availability · Season variability · Geographic variations

2.1 Introduction

Over the past few decades, global climate change has become an increasing concern due to the continuous emission of greenhouse gases. Human activities have significantly contributed to this phenomenon, accelerating environmental changes that threaten ecosystems worldwide (Nuttall, 2022). Change mitigation strategies are insufficient on a global scale, resulting in unrestricted climate change that could permanently damage the world. The changes in environment are unfavorable to all life forms within the ecosystem, including to arthropods such as ticks that play the role of intermediary hosts in many zoonotic diseases (Esposito et al., 2023).

For decades, scientists have studied the correlation of climate change and vector-related diseases. Early studies on the impact of climate change tended to center around human to human transmission via mosquitoes, for example causing malaria and dengue fevers (Reiter, 2001). Research rapidly shifted towards assessing the effect of climate changes on other vector-borne organisms such as ticks that are known to carry numerous pathogens of veterinary and zoonotic significance (Caminade et al., 2019). The extent to which ticks can evolve in response to climate shifts is paramount to their geographic range, how active they are in different seasons and their general numbers. Climate has an impact on the activity patterns, habitat, and expected hosts of ticks; thus affecting their survival and reproduction (Nuttall, 2022; Tsoumani et al., 2023). For understanding and tackling the risks of tick-related issues, a unified 'One Health' approach in considering human, animal, and environmental health is necessary (Machtinger et al., 2024).

For different tick species, the duration of a complex life cycle (comprising of an egg, larva, nymph, and adult) ranges from three months to three years. The black-legged tick, also known as deer tick, is a species that takes two years to complete its life cycle (Mehlhorn, 2015). While males die after mating, females die after laying eggs. The tick life cycle depends on a variety of environmental variables, especially temperature and humidity (Ogden et al., 2021). The development of ticks can be accelerated by warmer temperatures which can increase the risk of disease transmission. Species such as *Ixodes ricinus* are known to survive freezing weather conditions, giving them a better chance at reproduction in seasonally cold climates. Ticks are only temporarily parasitic, spending a part of their life history, depending on whether they are one, two, or three host species; therefore are highly susceptible to variation in environmental conditions (Yano et al., 1987; Eisen et al., 2024).

Temperature fluctuations significantly influence the rate of tick development, particularly during critical periods such as the transition from engorged female to egg-laying female (pre-oviposition period), egg development (pre-eclosion period), and the transformation of engorged larvae and nymphs into questing nymphs and adults (Yano et al., 1987). Generally, the relationship between temperature and tick development is nonlinear; as temperatures decrease, the duration of each developmental stage increases (Ogden et al., 2004). While extreme cold environments can be detrimental to tick populations, some species can survive in subzero temperatures by utilising antifreeze glycoproteins in their hemolymph (Ogden et al., 2021). Studies have shown that ticks exposed to temperatures below −5 °C experience high mortality rates (Ogden et al., 2004). However, certain woodland habitats offer microclimates that help in protection against freezing temperatures. For example, snowpack allows for soil insulation covering in which the insulating of snow can facilitate to ensure the presence of ticks even when the surrounding altitude goes below −20 °C for long stretches (Burtis et al., 2016; Templer et al., 2012).

Ticks have two major survival mechanisms, quiescence and diapause, in response to extreme environmental conditions. Quiescence refers to the reduction in metabolic processes to minimum levels, progressing to dormancy in extremely hot, cold, or dry situations. Ticks start to move normally again once the environmental conditions improve. In contrast, diapause is a genetically programmed developmental arrest to enable the ticks to survive during prolonged adverse conditions (Belozerov, 2008). There are two types of diapause that influence the life cycle of tick: behavioral and developmental diapause (Cabrera & Labruna, 2009). The existence of diapause or lack of it is not consistent in all species of ticks and eco-geographical populations. Within the genus *Amblyomma*, some species have been shown to have diapauses while others do not. In the same way, populations of *Ixodes scapularis* have been shown to exhibit variability in diapause expression, whereby some populations do not enter diapause at certain points of their life cycle (Belozerov & Naumov, 2002).

Changes in climate, habitat structure, and host dynamics are affecting the population and distribution of ticks. As the world gets warmer, ticks are moving into new, previously unfavorable areas, thereby increasing the risk of tick-borne diseases (TBDs) being spread. These movements, to a large extent, require tick spread along with migrating hosts (Tonk-Rugen et al., 2025). Many wildlife hosts, including deer and birds, are instrumental in moving ticks from one region to another and enabling them to establish in new regions. The complex interactions of climate change with tick evolution and host movement necessitate the application of comprehensive surveillance and predictive modeling in order to estimate the prospective disease burden (Deshpande et al., 2024). In the context of designing workable interventions that mitigate the veterinary and public health burden of TBDs, it is crucial to understand the underlying mechanisms that govern tick population dynamics.

2.2 Basic Components of the Tick Life Cycle

Different species of ticks, on average, can live from three months to three years. Ticks go through a life cycle consisting of four stages: namely egg, larva, nymph and adult. Larvae, nymphs and adults require a blood meal from a host in order for the tick to undergo the necessary progression to the next stage (Fig. 2.1). At any point if the tick is unable to take a blood meal, they will not complete their life history (Mehlhorn,

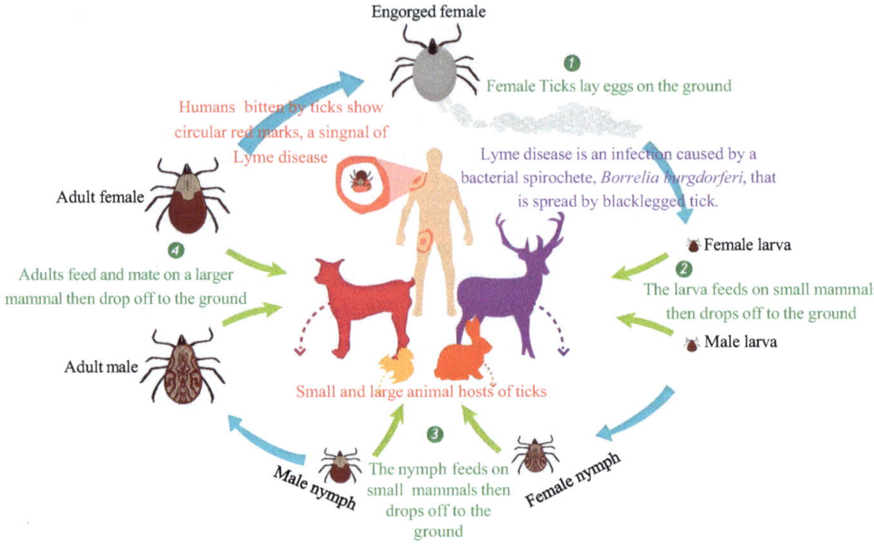

Fig. 2.1 Life cycle stages of ticks and transmission of diseases

2015). Different species and stages of ticks exhibit a degree of host preference. For example, members of the *Rhipicephalus* and *Hyalomma* genera predominantly infest cattle, posing a significant risk to their health (Hussain et al., 2023).

As with other arthropods, the life cycle of ticks starts with eggs. Following the mating process, an adult female tick will deposit thousands of eggs into protected environmental niches such as leaf litter, soil, or animal burrows (Nicholson et al., 2019). Temperature and humidity strongly influence the hatching of the eggs, which usually occur within 60 days. The process usually speeds up with higher temperatures, but extreme cold can slow it down. Upon hatching, small six-legged larvae, sometimes referred to as 'seed ticks' emerge and search for their first meal of blood (Ajayi et al., 2024).

Unlike their later forms, larval ticks possess only six legs. Because of their small size, sometimes less than 1 mm long, they prefer to feed on small hosts such as mice, birds, reptiles, and small mammals (Mehlhorn, 2015). After a larva attaches to a host, it begins to feed on blood, from which it gains nutrients critical for the next stage of development. After a few days, the larvae stop feeding and detach from the host so that they can enter dormancy. As the temperature rises, the larvae go through the molting process into nymphs (Nuss et al., 2017). One host ticks stay on their host, and two or three host ticks which leave their hosts and fall into the environment. Then, the time spent in the environment varies with tick species (Leal et al., 2020).

Nymphs are generally less than 2 mm in length and have a full arachnid body structure, with four pairs of limbs. Their small size when unfed can make them difficult to identify (Mehlhorn, 2015). Adult ticks are much larger than larvae and nymphs, and easily identifiable. Three host ticks seek for hosts by waiting on tall grass or shrubs, sometimes referred to as 'questing'; while one and two host species are already present on the fur of animals. Questing ticks will extend their front legs to grab the host after sensing heat, carbon dioxide, and movement (Nicholson et al., 2019). An adult tick will grab a host and feed on its blood for up to about 10 days. Males usually die after mating, but females must complete feeding and drop to the environment before they can lay eggs. Engorged females retreat from their host and shelter themselves while laying thousands of eggs (Eisen et al., 2024).

Factors such as climate, habitat conditions, and host availability all determine how long life cycle stages take and how successful they are. Various species of ticks have different lifespans and rates of development, and some of these species can live for a long time without blood meals (Nuttall, 2022). Understanding these life cycle components is critical for designing appropriate tick control methods to reduce the incidence of TBDs within human and animal populations.

2.3 Environmental Triggers and Developmental Responses

Tick activities and development are affected by climatic and environmental variables; temperature and humidity inn particular. These variables are the most important for the growth, survival and population change of tick species. Among these, temperature is a critical environmental driver of tick development (Rebaudo & Rabhi, 2018), biological processes (Gillooly et al., 2001), and population growth (Savage et al., 2004). In different regions, the peak abundance of *I. scapularis* at different post-egg life stages depends on a combination of temperature, humidity, and photoperiod (Ogdenet al., 2018).

Ticks, unlike most other parasites, do not rely exclusively on vertebrate hosts for a large part of their life cycle. As such, they have to cope with daily, seasonal and annual changes in temperature and humidity. Such shifts in environmental conditions affect their survival, activity patterns, and questing behavior (Couret et al., 2022). Furthermore, ticks have intricate symbiotic, or pathogenic relations with microbes, which are also influenced by environmental conditions. Such relations may assist arthropod hosts in managing environmental stress, especially in reducing the impact of heat shock while attached to a vertebrate host (Lemoine et al., 2020). It is crucial to understand how these environmental factors modify the chances of survival and behavioral patterns of different tick species for estimating their potential distribution and the risk of tick-borne pathogens in the prediction of climate change.

The transmission, survival, and development processes of tick-borne pathogens such as *Borrelia burgdorferi,* which causes Lyme disease, are largely influenced by temperature. This bacterium is highly sensitive to external temperature (Tilly et al., 2008). Changes in temperature from 23 °C to 37 °C, (which occurs when a tick moves from the environment to a mammalian host) result in *B. burgdorferi* transcriptome changes (Ojaimi et al., 2002). These transcriptional alterations are critical for the microorganism's adaptation, colonisation in ticks, transmission to a host, and survival in various environments (Radolf et al., 2012; Revel et al., 2002).

Nonetheless, there is evidence indicating that temperature alone is not sufficient to explain the survival of *Borrelia* in a vertebrate or invertebrate host (Mulay et al., 2009). Rather, it underscores the manner in which climatic changes may modify disease transmission dynamics. Higher temperatures and low humidity cause ticks to remain under leaf litter which restrict their ability to seek avian or mammalian hosts. However, some host animals such as lizards, reside in the leaf litter, where they can behave as tick reservoirs. These lizards to not transmit *B. burgdorferi* infections, enabling an inverse relationship between tick populations, infection rates, and the incidence of Lyme disease in some areas (Tijsse-Klasen et al., 2010).

Humidity has a large impact on the survival of ticks, their host-seeking behavior and their overall range. For example, *Ixodes scapularis* is highly sensitive to humidity changes; potentially more so than to temperature variations (Ginsberg et al., 2017). *Ixodes* spp blacklegged ticks spend more than 95% of their life cycle off-host; hence are prone to dehydration. This is particularly true for smaller larvae and nymph stages. These stages are exposed to high amount of environmental selection pressures due to their diminutive size combined with lower levels of mobility (Fleshman et al., 2021). In order to live, ticks depend on places where humidity is near 100% as occurs underneath leaf litter in areas of dense vegetation, including woodlands and forests (Berger et al., 2014). Because ticks lose water through their skin while breathing, they need to periodically migrate above the leaf litter to moisten themselves. Their survival in these conditions depends on the timing and amount of previous blood meals providing them with nutritional reserves to support motile environmental life stages (Crooks & Randolph, 2006).

The water acquisition capacity of ticks is fundamental for their hydration and survival. In humid environments, ticks generally acquire water passively (Fielden & Lighton, 1996). As the humidity increases and reaches a certain level, known as the critical equilibrium humidity, water vapor starts to be actively absorbed, which permits water uptake. Ticks are able to actively absorb water in a humidity range of 75–96% relative humidity (Rodgers et al., 2007). The synergistic impact of temperature and relative humidity on tick survival and pathogen transmission is indicative of the multifaceted environmental interactions which determine tick populations (Fig. 2.2). Understanding these impacts will be important in order to estimate tick distribution in the coming years, which will be needed to develop sustainable control measures for ticks and TBDS.

2.4 Temperature Effects on Tick Life Cycle Dynamics

The epidemiology of vector-borne diseases is highly sensitive to temperature fluctuations and their variations (Estrada-Peña, 2009; Ogden et al., 2007). Ticks have long life cycles and go through multiple developmental stages that are each dependent on temperature (Estrada-Peña et al., 2020). Consequently, tick abundance and activity is greatly affected by temperature (Walker et al., 2001). The traditional seasonal occurrence of tick populations, such as *I. Ricinus* and *I. Scapularis*, is well known. Cold

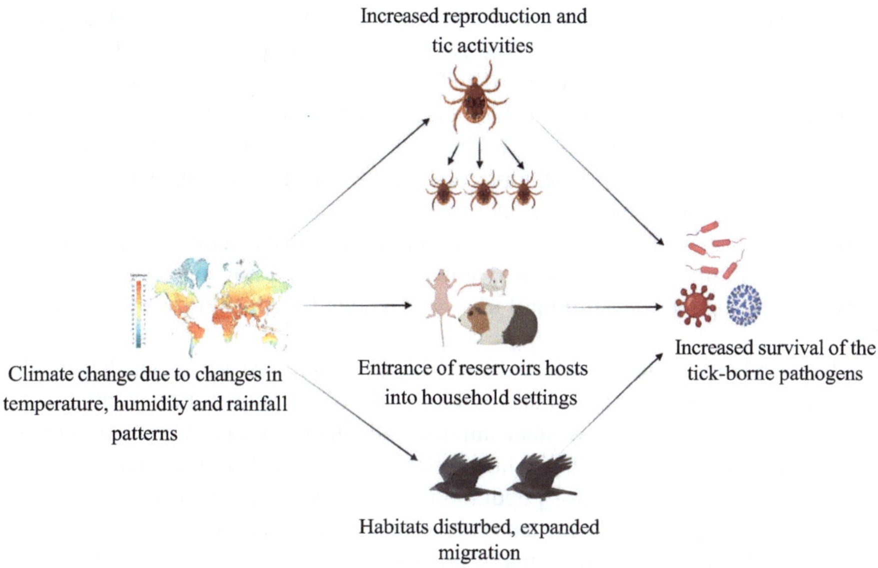

Fig. 2.2 Effect of environmental changes like temperature, humidity, and rainfall on the survival and spread of ticks and tick-borne diseases

weather permits interstadial development in addition to changing questing activity (Ogden et al., 2005; Randolph et al., 2002). In Europe and North America, there are seasonalm peaks in annual tick abundance, primarily during the summer months (Estrada-Peña et al., 2004; Saleh et al., 2021). However, the timing of these peaks varies between different life stages. For instance, in spring and autumn, *I. ricinus* nymphs and adults reach peak questing activity, whereas larvae may exhibit a single mid-year peak (Gray, 1991; Gray et al., 2016).

The transmission of TBDs, such as tick-borne encephalitis, Lyme borreliosis, louping ill, and tick pyaemia, is closely linked to temperature-driven tick life cycle stages (Rocha et al., 2022). Pathogen transmission occurs when larvae and nymphs acquire infections from hosts and subsequently transmit the disease to susceptible hosts as they mature and take additional blood meals as nymphs or adults (de la Fuente et al., 2017; Rocha et al., 2022). Since tick populations are highly dependent on climatic conditions, the temporal synchronisation of life cycle stages plays a key role in disease outbreak patterns (Gray et al., 2009). Disease transmission is more likely to persist in environments where the active life cycle stages overlap synchronously, particularly for pathogens that require a brief period of host infectivity (Hancock et al., 2011). The transmission cycle of ticks may be disrupted, if life stages become desynchronised due to climate shifts and this may lead to fluctuations in disease prevalence (Gray et al., 2009).

The development of ixodid ticks is dependent on temperature, usually showing an upward shift of growth with an increase in temperature, although this is sometimes non-linear (Randolph et al., 2002; Troughton & Levin, 2007). However, dry predictions based on temperature alone often do not work because of diapause; whereby development ceases in response to unfavorable environmental conditions (Gray et al., 2016). The beginning and end of a period of a programed dormancy is at least partially defined by changes in temperature and photoperiod. Diapause regulates tick phenology rather than allowing for a simple linear relationship between development and temperature, therefore, understanding diapause is critical. This in turn has an effect on the geographical distribution of ticks; whereby thay may migrate to other more favorable regions in response to climatic change driven environmental restrictions. For example, *I. scapularis* may undergo diapause as an engorged nymph and does not develop into an adult until triggered by day length (Belozerov & Naumov, 2002).

The abundance of ticks is affected not only by temperature per se, but also by other climatic factors such as rainfall and extreme cold temperatures. For example, early season droughts and severe winters can influence the mortality of questing nymphs and ticks that are not attached with hosts (Berger et al., 2014; Jones & Kitron, 2000). In addition, alterations in host populations that could be associated with climatic change, for example, in rodents that serve as hosts for immature ticks, or in deer that serve as hosts for adults, have a major influence on tick populations (Ostfeld et al., 2006). Although *I. ricinus* is tolerant of cold weather, that these ticks can only live for about 24 h if winter temperatures in their environemtal niche reach −14 °C to −18.9 °C. Unfed nymphs and diapausing larvae die following exposure to −10 °C for a month's duration (Dautel & Knülle, 1997). If summer temperatures were not high enough to permit full maturation, these ticks may die at higher winter temperatures of lower duration (Voyiatzaki et al., 2022).

Changes in climate over the last several decades have markedly changed the range of distribution of ticks. In Sweden, noticeable climate changes in the late 1980s manifested in the northward migration of *I. ricinus* (Tälleklint & Jaenson, 1998). Prior to this, the geographic distribution of *I. ricinus* did not extend north of 61°N. However, as temperatures have warmed, these ticks are now found along the entire coastline of the Baltic Sea up to 66°N and have also migrated into several northern river valleys and lakes (Jaenson et al., 1994). The northern shift of range expansion of tick populations is directly associated with rising temperature that affects the chances of survival, activity, and developmental processes of ticks (Gray et al., 2009). Warmer temperatures decrease chances of overwintering mortality and speed up the development of ticks, which results in gaining increased density of ticks and expansion of regions of risk for diseases. This phenomenon draws attention to the necessity of developing climate-oriented tick management methods to minimise the risk of introduction of TBD public health hazards to new areas.

2.5 Humidity and Moisture Requirements

Ticks have a life cycle composed of a series of short phases of parasitic feeding, interspersed with a prolonged non-feeding phase including no intake of food or water. These extended periods in the environment place considerable stress on their water balance systems (Buczek, 1999). When feeding, ticks ingest large quantities of blood containing water and ions. To maintain osmotic and ionic balance and concentrate nutrients, they must efficiently eliminate excess water and ions (Kaufman & Phillips, 1973a). In ixodid ticks, this function is primarily performed by the salivary glands, which return excess water and specific ions back to the host while the tick remains attached (Kaufman & Phillips, 1973a). In contrast, argasid ticks involving multiple blood meals and feed more rapidly, utilise their coxal organs for the elimination of surplus water and ions, enabling them to maintain homeostasis during feeding. This efficient regulation of water and ion balance is crucial for their survival (Lazzari et al, 2021).

While engorging ticks must eliminate excess water, non-feeding ticks face the opposite challenge of conserving water for extended periods that may last weeks or even months. For example, in *Amblyomma hebraeum*, only about 2% of its three-year life cycle in South Africa's Eastern Cape Province is spent on a host. The remaining time is spent in harsh, dry environments, where maintaining water balance is critical for survival. In such conditions, ticks must avoid dehydration, as excessive water loss leads to death (Norval, 1977).

To counteract water loss, ticks have evolved various adaptations that help them to conserve water during their prolonged non-parasitic phases. A waxy lipid layer of epicuticular lipids on the tick's integument and specialised closure mechanisms on the spiracles confer an ability to absorb water vapor from unsaturated air. Altered behavior by seeking humid microhabitats directly influence tick distribution, survival, activity, and role in disease transmission (Estrada-Peña et al., 2013; Wrońska et al., 2023). Unlike feeding ticks, which regulate water balance through excess fluid elimination, unfed ticks must conserve water through specialised excretory processes that enable the impact of excretion and metabolic water production on tick hydration to be minimal. In ixodid ticks, waste elimination, including the excretion of guanine and other purines as nitrogenous by-products, occurs within a short timeframe, typically up to four weeks after moulting. Ticks limit excretion to this period to prevent losing water during the long non feeding time. Unfed ticks depend on physical adaptations such as their waxy epicuticle for waterproofing and physiological mechanisms such as closure of spiracles for survival during dry conditions. By matching their physiological activities with the ambient conditions, ticks efficiently save water until a new host is found. These adaptations enable ticks to withstand long periods of host-seeking while remaining hydrated in arid regions (Hamdy, 1977). Understanding these structural and physiological mechanisms may aid in the development of sustainable control strategies.

Instead of relying on metabolic water production, ticks depend on water preservation tactics during extended periods in which they are not feeding. Poorly soluble excrement, such as guanine, slowly builds up in the rectal pouch and Malpighian tubules, helping ticks to conserve water for longer time periods (Alasmari & Wall, 2021). Moreover, unfed ticks have diminished levels of respiratory gas exchange, which saves water. Previous research found that for extended non-parasitic active phases, ticks also reduce their respiration which helps to increase the duration that they can go without feeding. These physiological adaptations allow ticks to waterproof their bodies to a degree where they can spend long periods without hosts (Earls & Oyen, 2024). These adaptations are crucial for estimating tick survival, disease transmission risk, and geographic expansion in a changing climate (Fig. 2.2).

2.6 Host Availability and Blood Feeding Pattern

As obligate hematophageous species, tick feeding behavior and host availability is vital for their general biology (Nicholson et al., 2019). With climate change shifting host populations, tick habitats, and seasonal activity, some aspects of their feeding strategies and host-seeking behavior are also changing. The availability of suitable hosts is affected by warmer temperatures and changes in humidity, which impacts tick feeding dynamics, attachment duration, and the rate of successful blood feeding (Ogden et al., 2004).

Ticks adopt a variety of ways in which to feed on blood, but they will eventually resort to pool feeding, which is cutting a hole in the host's skin and waiting for blood to flow. To receive blood through this method, a tick must first pierce the host's skin to the level of the superficial dermal blood vessels. The mouthparts of different species of ticks can result in different degrees of host tissue damage (Binnington & Kemp, 1980). Hard ticks are classified as slow feeders; whereby lesions develop gradually as a haematoma forms. Soft ticks are rapid feeders that can complete their meal within minutes. Their method of feeding enables blood leakage, without firm connection to the host (Binnington & Kemp, 1980). When soft ticks probe for blood, tiny blood vessels are severed leading to the destruction of host cells, causing uncontrollable blood loss. They can change their feeding cycle lengths and attachment behavior due to climate-driven changes regarding the movement and availability of the host (Estrada-Peña et al., 2012). For example, warmer conditions can enable ticks to gather nourishment more quickly as a result of elevated metabolic activity (Nielebeck et al., 2023).

The act of feeding on a host's blood incites a variety of defense mechanisms, such as immune response, hemostatic response, and blood coagulation (Ribeiro, 1987). In addition to these immune responses, the increase in the blood volume at the feeding site also activates other host hemostatic defences, including the coagulation cascade and platelet aggregation (Law et al., 1992). In an attempt to counteract, the host's immune responses, ticks possess salivary proteins that are potent disruptors of platelet aggregation, while simultaneously being weak interfering agents of blood coagulation. Because of this, blood flow is guaranteed during feeding (Ribeiro, 1987). Salivary proteins that disrupt blood clotting act at multiple stages of platelet aggregation: pre-activation and post-activation. Such mechanisms of these proteins help the tick to effectively feed without triggering rapid clot formation and allow them to finish taking blood before the host's defenses activate (Šimo et al., 2017).

Changing climate is impacting the availability of hosts and blood feeding behavior, which ticks rely on for survival and reproduction. These shifts include: migration of hosts and their availability (ticks are exposed to new hosts because of rising temperatures, which pose a greater risk of spreading infections), increased length of activity seasons (tick activity increases with temperature, resulting to more meals being consumed annually), modified feeding times (warmer weather could increase metabolic processes, allowing ticks to feed and let go more quickly), and increased transmission of diseases (changes in feeding behavior can make it easier to pass along diseases to humans and animals). There is an increasing concern with climate change because the continued increase in temperature will create new ranges for ticks to survive in; making them more likely to carry a wider range of tick-borne infections (Bouchard et al., 2019; Ogden et al., 2021; Voyiatzaki et al., 2022). To mitigate adverse outcomes, there is need for better understanding of the implications of climate change on the blood feeding behavior of ticks (Fig. 2.3).

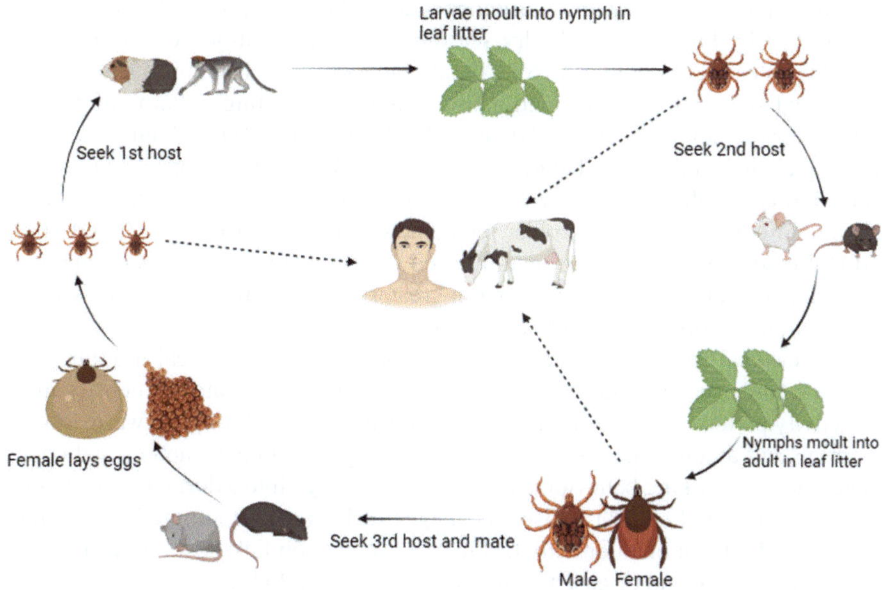

Fig. 2.3 Host availability and blood feeding pattern affect the tick life cycle dynamics

2.7 Season Variability in Tick Life Cycle

The seasonal changes in weather patterns define how ticks develop, leading to changes in reproduction and seeking of hosts. Temperature, humidity, and the photoperiod significantly impact ticks' capability to complete various developmental cycles (Vail & Smith, 2002). These environmental elements control the intensity and timing of tick infestations, which affects their distribution and the associated disease transmission risk (Busi et al., 2024). Change in global climate are modifying seasonal extremes, which is causing tick populations to experience changes in their life cycle timing including prolonged periods of activity, which increases the risk of TBDs. Additionally, the duration and success of each life stage varies profoundly with seasonal temperature and humidity periods (Perret et al., 2000).

After their last blood meal, female ticks can deposit thousands of eggs in the environment. Humidity and temperature control the rate of egg hatching. Higher temperatures speed up the hatching process, but dry conditions can lead to desiccation as well as death (Sutherst et al., 2006). Changes in seasons that arise because of climate change could change the time periods for hatching, resulting in ticks infesting earlier. Once the larvae have hatched, they actively seek their blood meal that is needed for molting into nymphs (Brunner et al., 2011). Nymphal and larval survival is enhanced by humidity, while dry seasons cause dehydration, leading to higher mortality rates (Eisen et al., 2016). Climate induced shifts leading to shortened or lengthened seasonal activity could cause changes in molting patterns of nymphs which impacts tick populations dynamically (Gray et al., 2009). Adult ticks show seasonal changes in seeking behavior, which are affected by photoperiod as well as temperature. Heightened activity in searching for a host is brought about by warm and humid conditions, while colder temperatures induce dormancy (Gilbert et al., 2014). Regions with climate change-induced warming are experiencing longer tick reproductive cycles, which is resulting in greater burden of ticks and increased risk of TBD transmission (Bouchard et al., 2009).

The seasonal variation impacts host availability, which is critical in influencing the tick feeding success and life cycle completion. In the moderate tick density areas, the activity pattern of ticks is bimodal, showing an increase during the warm months of spring and autumn (Burtis et al., 2023). During hot summer months and colder winter months, seeking behavior is less active as ticks go into a dormant state to save energy (Ogden et al., 2021). Seasonal changes such as wildlife migration, livestock grazing, and human outdoor activities have an impact on tick-host interactions and affect tick survival and disease transmission (Jori et al., 2021).

Changes in seasonal patterns due to climate change may shift tick life cycles. Some alterations include warmer temperatures triggering earlier spring activity and more questing, increased tick-host interactions extending tick activity into late autumn and winter, and expansion of their geographical distribution to previously unsuitable regions (Lindsay et al., 1998; Nuttall, 2022). These changes that are suspected to occur due to climate change are important for public health and veterinary medicine, as prolonged tick activity leads to a higher risk of TBDs.

2.8 Geographic Variations in Life Cycle Modifications

Geographical variations can modify the life cycle of ticks, as they drive changes in developmental stages, seasonal activity, and pathogen transmission. There are distinct regional life cycle patterns of ticks such as *I. scapularis* that influence the efficiency of pathogen transmission (Eisen & Eisen, 2018). Understanding of geographic differences are highly important if we are to anticipate changes in tick population dynamics and evaluate the impact of climate change on TBDs.

Ticks exhibit seasonal activity patterns which vary significantly with geography in relation to nymphal and larval questing behaviors, which in turn affect the cycles of disease transmission. In the Northeast region of the United States, *Ixodes* spp. nymphs are most active in the spring and larvae in the late summer (Goethert et al., 2023; Burtis et al., 2023). This assists in the transmission of *B. burgdorferi* as infected nymphs feed on small mammals in spring and uninfected larvae also feed on the same hosts in late summer to pass on the infection (Voordouw, 2015). However, recent modeling studies suggest that asynchronous activity of nymphs and larvae might also increase transmission efficiency, as infected rodents may die before new larvae emerge, possibly disrupting the cycle (Ogden et al., 2007). The seasonal pattern of nymph and larval tick activity in the Southeast region is opposite, with larvae emerging in the early year and nymphs appearing in late autumn (Clark et al., 1998). This opposite activity pattern leads to lowered rates of *B. burgdorferi*, which is believed to be the reason for fewer Lyme disease cases in the area (Schwartz et al., 2017). In longer warmer climates, the higher disease spread by other tick-borne pathogens is expected (Bouchard et al., 2019; Estrada-Peña et al., 2012).

Nymphs that feed after the summer solstice may enter developmental diapause, triggered by declining day length, delaying their transition into adults until after winter (Belozerov & Naumov, 2002; Ogden et al., 2004). The geographic consistency of these patterns is uncertain, as diapause mechanisms may differ between populations. Some southern tick populations may exhibit temperature-independent diapause, while northern populations rely on photoperiod cues (Belozerov, 2008).

Understanding how tick seasonality varies across regions requires long-term field studies that monitor tick activity, host interactions, and environmental conditions. Studies analysing north–south variations in tick activity suggest that temperature plays a major role in shaping life cycle modifications. For example, in the Upper Midwest United States, the seasonal activity of larvae has been observed during the spring and late summer, indicating a possible impact of temperature on their developmental stages and questing behavior (Gatewood et al., 2009).

2.9 Implications for Disease Transmission and Control

The scope of climate change greatly modifies the range, lifecycle, and population numbers of ticks which, in turn, affects the transmission of TBDs (Gray et al., 2009). Increases in average temperatures, shifts in the amount of rainfall, and changes in the timing of the seasons, has enhanced the range and duration of activity of ticks, as well as their geographical and developmental rates (Ogden et al., 2021). These changes pose serious challenges with regards to controlling ticks and the diseases which they cause. An understanding of how changes in the tick life cycle resulting from climate change are related to the transmission of the pathogen is critical for formulating effective control measures. Increased tick activity heightens the chances of infecting livestock, wildlife, and humans with pathogens (Brites-Neto et al., 2015). Subsequently, higher tick survival rates translate into higher tick infestations in warmer seasons. In addition, greater humidity in some places leads to enhanced tick egg and larval development and survival rates, aiding in increased tick populations (Eisen et al., 2016; Ogden et al., 2008).

With the aid of new technologies, conventional control measures need to be adapted to changing environmental conditions. The traditional methods employed in tick control such as using chemical acaricides, habitat modification, and biocontrol, need to be modified because of the growing dangers as a result of global warming (Rizwan & Sajid, 2023). Because these ticks have an increased geographic extent and differ seasonally, better monitoring systems are needed to track tick populations and their associated disease risks. Predictive modeling of climate, remote sensing, and GIS can pinpoint regions of concern that would require active management control (Baldwin et al., 2022). Certain habitat manipulation practices such as restricting grazing, rotational grazing, alteration of plant cover can decrease tick density by modifying the microclimatic conditions which are highly conducive to tick survival (Rizwan & Sajid, 2023). Furthermore, rather than applying acaricides indiscriminately, applying them at critical periods of activity can make treatment more effective and less environmentally damaging (Makwarela et al., 2025).

Although climate change is enabling ticks to reproduce faster which may increase the rate of selection for acaricide resistance development (Rojas-Cabeza et al., 2025), chemical acaricides are still the major method used for tick control. Abuse of acaricide application often results in the development of resistant tick populations. To mitigate this, integrated pest management (IPM) approaches should include discontinuous application of acaricides with different mechanisms of action to delay the development of resistance, application during periods of maximum tick activity determined through surveillance data, and use of both chemical and biological controls in order to minimize the use of chemical acaricide (Makwarela et al., 2025; Rizwan & Sajid, 2023).

Climate change impacts on tick populations require greater investment in biological control strategies. The implementation of biocontrol approaches, such as employing predatory arthropods like fungi, nematodes, and parasitoid wasps, provide environmentally friendly approaches to substitute chemical acaricides. Research in fungal tick pathogens such as *Metarhizium anisopliae* and *Beauveria bassiana* have shown efficacy in lowering tick populations in some environments (Siddique et al., 2023). Moreover, as the environment becomes less favorable for conventional tick control methods, oral or injectable systemic acaricides may exert more favorable effects on tick control. Although, as evidence suggests, the primary limitation to adopting these strategies is the high expense and risks of accumulating chemical residues (George et al., 2004; Makwarela et al., 2025). Tackling ticks through anti-tick vaccines is a highly promising strategy considering the increase of tick infestations driven by climate change. Vaccines against tick salivary proteins or gut antigens can limit feeding and subsequent pathogen transmission. Attention is needed on the study of host genetic resistance to ticks, particularly as selective breeding of tick-resistant livestock has shown to greatly lower the need for acaricides (Abbass et al., 2023; Antunes & Domingos, 2023). By adapting control strategies to climate-driven modifications in tick life cycles, it is possible to mitigate the increasing burden of TBDs while ensuring the sustainability of livestock production and public health protection.

References

]Dautel, H., & Knülle, W. (1997). Cold hardiness, supercooling ability and causes of low-temperature mortality in the soft tick, *Argas reflexus*, and the hard tick, *Ixodes ricinus* (Acari: Ixodea) from Central Europe. *Journal of Insect Physiology, 43*(9), 843–854. https://doi.org/10.1016/s0022-1910(97)00025-5

Abbas, M. N., Jmel, M. A., Mekki, I., Dijkgraaf, I., & Kotsyfakis, M. (2023). Recent advances in tick antigen discovery and anti-tick vaccine development. *International Journal of Molecular Sciences, 24*(5), 4969. https://doi.org/10.3390/ijms24054969

Ajayi, O. M., Oyen, K. J., Davies, B., Finch, G., Piller, B. D., Harmeyer, A. A., Wendeln, K., Perretta, C., Rosendale, A. J., & Benoit, J. B. (2024). Egg hatching success is influenced by the time of thermal stress in four hard tick species. *Journal of Medical Entomology, 61*(1), 110–120. https://doi.org/10.1093/jme/tjad142

Alasmari, S., & Wall, R. (2021). Metabolic rate and resource depletion in the tick *Ixodes ricinus* in response to temperature. *Experimental and Applied Acarology, 83*(1), 81–93. https://doi.org/10.1007/s10493-020-00568-1

Antunes, S., & Domingos, A. (2023). Tick vaccines and concealed versus exposed antigens. *Pathogens, 12*(3), 374. https://doi.org/10.3390/pathogens12030374

Baldwin, H., Landesman, W. J., Borgmann-Winter, B., & Allen, D. (2022). A geographic information system approach to map tick exposure risk at a scale for public health intervention. *Journal of Medical Entomology, 59*(1), 162–172. https://doi.org/10.1093/jme/tjab169

Belozerov, V. N., & Naumov, R. L. (2002). Nymphal diapause and its photoperiodic control in the tick *Ixodes scapularis* (Acari: Ixodidae). *Folia Parasitologica, 49*(4), 314–318. https://doi.org/10.14411/fp.2002.058

Belozerov, N. V. (2008). Diapause and quiescence as two main kinds of dormancy and their significance in life cycles of mites and ticks (Chelicerata: Arachnida: Acari). Part 1. *Acariformes, 16*(2), 79–130.

Berger, K. A., Ginsberg, H. S., Dugas, K. D., Hamel, L. H., & Mather, T. N. (2014). Adverse moisture events predict seasonal abundance of Lyme disease vector ticks (*Ixodes scapularis*). *Parasites and Vectors, 7*, 181. https://doi.org/10.1186/1756-3305-7-181

Binnington, K. C., & Kemp, D. H. (1980). Role of tick salivary glands in feeding and disease transmission. *Advances in Parasitology, 18*, 315–339. https://doi.org/10.1016/s0065-308x(08)60403-0

Bouchard, C., Dibernardo, A., Koffi, J., Wood, H., Leighton, P. A., & Lindsay, L. R. (2019). Increased risk of tick-borne diseases with climate and environmental changes. *Canada Communicable Disease Report, 45*(4), 83–89. https://doi.org/10.14745/ccdr.v45i04a02

Brites-Neto, J., Duarte, K. M., & Martins, T. F. (2015). Tick-borne infections in human and animal populations worldwide. *Veterinary World, 8*(3), 301–315. https://doi.org/10.14202/vetworld.2015

Brunner, J. L., Cheney, L., Keesing, F., Killilea, M., Logiudice, K., Previtali, A., & Ostfeld, R. S. (2011). Molting success of *Ixodes scapularis* varies among individual blood meal hosts and species. *Journal of Medical Entomology, 48*(4), 860–866. https://doi.org/10.1603/me10256

Buczek, A. (1999). Physiology of ticks (Ixodida) in the non-parasitic phase of the life cycle. *Wiadomości Parazytologiczne, 45*(2), 151–159.

Burtis, J. C., Ostfeld, R. S., Yavitt, J. B., & Fahey, T. J. (2016). The relationship between soil arthropods and the overwinter survival of *Ixodes scapularis* (Acari: Ixodidae) under manipulated snow cover. *Journal of Medical Entomology, 53*(1), 225–229. https://doi.org/10.1093/jme/tjv151

Burtis, J. C., Bjork, J., Johnson, T. L., Schiffman, E., Neitzel, D., & Eisen, R. J. (2023). Seasonal activity patterns of host-seeking *Ixodes scapularis* (Acari: Ixodidae) in Minnesota, 2015–2017. *Journal of Medical Entomology, 60*(4), 769–777. https://doi.org/10.1093/jme/tjad048

Busi, A., Martínez-Sánchez, E. T., Alvarez-Londoño, J., Rivera-Páez, F. A., Ramírez-Chaves, H. E., Fontúrbel, F. E., & Castaño-Villa, G. J. (2024). Environmental and ecological factors affecting tick infestation in wild birds of the Americas. *Parasitology Research, 123*, 254. https://doi.org/10.1007/s00436-024-08246-6

Cabrera, R. R., & Labruna, M. B. (2009). Influence of photoperiod and temperature on the larval behavioral diapause of *Amblyomma cajennense* (Acari: Ixodidae). *Journal of Medical Entomology, 46*(6), 1303–1309. https://doi.org/10.1603/033.046.0608

Caminade, C., McIntyre, K. M., & Jones, A. E. (2019). Impact of recent and future climate change on vector-borne diseases. *Annals of the New York Academy of Sciences, 1436*(1), 157–173. https://doi.org/10.1111/nyas.13950

Clark, K. L., Oliver, J. H., Jr., McKechnie, D. B., & Williams, D. C. (1998). Distribution, abundance, and seasonal activities of ticks collected from rodents and vegetation in South Carolina. *Journal of Vector Ecology, 23*(1), 89–105.

Couret, J., Schofield, S., & Narasimhan, S. (2022). The environment, the tick, and the pathogen—It is an ensemble. *Frontiers in Cellular and Infection Microbiology, 12*, 1049646. https://doi.org/10.3389/fcimb.2022.1049646

Crooks, E., & Randolph, S. E. (2006). Walking by *Ixodes ricinus* ticks: Intrinsic and extrinsic factors determine the attraction of moisture or host odour. *Journal of Experimental Biology, 209*(Pt 11), 2138–2142. https://doi.org/10.1242/jeb.02238

de la Fuente, J., Antunes, S., Bonnet, S., Cabezas-Cruz, A., Domingos, A. G., Estrada-Peña, A., Johnson, N., Kocan, K. M., Mansfield, K. L., Nijhof, A. M., Papa, A., Rudenko, N., Villar, M., Alberdi, P., Torina, A., Ayllón, N., Vancová, M., Golovchenko, M., Grubhoffer, L., & Rego, R. O. M. (2017). Tick-pathogen interactions and vector competence: Identification of molecular drivers for tick-borne diseases. *Frontiers in Cellular and Infection Microbiology, 7*, 114. https://doi.org/10.3389/fcimb.2017.00114

Deshpande, G., Beetch, J. E., Heller, J. G., Naqvi, O. H., & Kuhn, K. G. (2024). Assessing the influence of climate change and environmental factors on the top tick-borne diseases in the United States: A systematic review. *Microorganisms, 12*(1), 50. https://doi.org/10.3390/microorganisms12010050

Earls, K. N., & Oyen, K. J. (2024). Metabolic rate does not scale with body size or activity in some tick species. *Experimental and Applied Acarology, 93*, 869–885. https://doi.org/10.1007/s10493-024-00958-9

Eisen, R. J., & Eisen, L. (2018). The blacklegged tick, *Ixodes scapularis*: An increasing public health concern. *Trends in Parasitology, 34*(4), 295–309. https://doi.org/10.1016/j.pt.2017.12.006
Eisen, R. J., & Eisen, L. (2024). Evaluation of the association between climate warming and the spread and proliferation of *Ixodes scapularis* in northern states in the Eastern United States. *Ticks and Tick-Borne Diseases, 15*(1), Article 102286. https://doi.org/10.1016/j.ttbdis.2023.102286
Eisen, R. J., Eisen, L., Ogden, N. H., & Beard, C. B. (2016). Linkages of weather and climate with *Ixodes scapularis* and *Ixodes pacificus* (Acari: Ixodidae), enzootic transmission of *Borrelia burgdorferi*, and Lyme disease in North America. *Journal of Medical Entomology, 53*(2), 250–261. https://doi.org/10.1093/jme/tjv199
Esposito, M. M., Turku, S., Lehrfield, L., & Shoman, A. (2023). The impact of human activities on zoonotic infection transmissions. *Animals, 13*(10), 1646. https://doi.org/10.3390/ani13101646
Estrada-Peña, A. (2009). Tick-borne pathogens, transmission rates and climate change. *Frontiers in Bioscience (Landmark Ed.), 14*(7), 2674–2687. https://doi.org/10.2741/3405
Estrada-Peña, A., & Fernández-Ruiz, N. (2020). A retrospective assessment of temperature trends in Northern Europe reveals a deep impact on the life cycle of *Ixodes ricinus* (Acari: Ixodidae). *Pathogens, 9*(5), 345. https://doi.org/10.3390/pathogens9050345
Estrada-Peña, A., Martinez, J. M., Sanchez Acedo, C., Quilez, J., & Del Cacho, E. (2004). Phenology of the tick, *Ixodes ricinus*, in its southern distribution range (central Spain). *Medical and Veterinary Entomology, 18*(4), 387–397. https://doi.org/10.1111/j.0269-283X.2004.00523.x
Estrada-Peña, A., Ayllón, N., & de la Fuente, J. (2012). Impact of climate trends on tick-borne pathogen transmission. *Frontiers in Physiology, 3*, 64. https://doi.org/10.3389/fphys.2012.00064
Estrada-Peña, A., Gray, J. S., Kahl, O., Lane, R. S., & Nijhof, A. M. (2013). Research on the ecology of ticks and tick-borne pathogens—Methodological principles and caveats. *Frontiers in Cellular and Infection Microbiology, 3*, 29. https://doi.org/10.3389/fcimb.2013.00029
Fielden, L. J., & Lighton, J. R. B. (1996). Effects of water stress and relative humidity on ventilation in the tick *Dermacentor andersoni* (Acari: Ixodidae). *Physiological Zoology, 69*(3), 599–617. https://doi.org/10.1086/physzool.69.3.30164218
Fleshman, A. C., Graham, C. B., Maes, S. E., Foster, E., & Eisen, R. J. (2021). Reported county-level distribution of Lyme disease spirochetes, *Borrelia burgdorferi* sensu stricto and *Borrelia mayonii* (Spirochaetales: Spirochaetaceae), in host-seeking *Ixodes scapularis* and *Ixodes pacificus* ticks (Acari: Ixodidae) in the contiguous United States. *Journal of Medical Entomology, 58*(3), 1219–1233. https://doi.org/10.1093/jme/tjaa283
Gatewood, A. G., Liebman, K. A., Vourc'h, G., Bunikis, J., Hamer, S. A., Cortinas, R., Melton, F., Cislo, P., Kitron, U., Tsao, J., Barbour, A. G., Fish, D., & Diuk-Wasser, M. A. (2009). Climate and tick seasonality are predictors of *Borrelia burgdorferi* genotype distribution. *Applied and Environmental Microbiology, 75*(8), 2476–2483. https://doi.org/10.1128/AEM.02633-08
George, J. E., Pound, J. M., & Davey, R. B. (2004). Chemical control of ticks on cattle and the resistance of these parasites to acaricides. *Parasitology, 129*, S353–S366. https://doi.org/10.1017/S0031182003004682
Gilbert, L., Aungier, J., & Tomkins, J. L. (2014). Climate of origin affects tick (*Ixodes ricinus*) host-seeking behavior in response to temperature: Implications for resilience to climate change? *Ecology and Evolution, 4*(7), 1186–1198. https://doi.org/10.1002/ece3.1014
Gillooly, J. F., Brown, J. H., West, G. B., Savage, V. M., & Charnov, E. L. (2001). Effects of size and temperature on metabolic rate. *Science, 293*(5538), 2248–2251. https://doi.org/10.1126/science.1061967
Ginsberg, H. S., Albert, M., Acevedo, L., Dyer, M. C., Arsnoe, I. M., Tsao, J. I., Mather, T. N., & LeBrun, R. A. (2017). Environmental factors affecting survival of immature *Ixodes scapularis* and implications for geographical distribution of Lyme disease: The climate/behavior hypothesis. *PLoS ONE, 12*(1), Article e0168723. https://doi.org/10.1371/journal.pone.0168723
Ginsberg, H. S., Hickling, G. J., Burke, R. L., Ogden, N. H., Beati, L., LeBrun, R. A., Arsnoe, I. M., Gerhold, R., Han, S., Jackson, K., Maestas, L., Moody, T., Pang, G., Ross, B., Rulison, E. L., & Tsao, J. I. (2021). Why Lyme disease is common in the northern US, but rare in the south:

The roles of host choice, host-seeking behavior, and tick density. *PLoS Biology, 19*(1), Article e3001066. https://doi.org/10.1371/journal.pbio.3001066

Goethert, H. K., Mather, T. N., O'Callahan, A., & Telford, S. R. III. (2023). Host-utilization differences between larval and nymphal deer ticks in northeastern U.S. sites enzootic for *Borrelia burgdorferi* sensu stricto. *Ticks and Tick-Borne Diseases, 14*(6), 102230. https://doi.org/10.1016/j.ttbdis.2023.102230

Gray, J. S., Dautel, H., Estrada-Peña, A., Kahl, O., & Lindgren, E. (2009). Effects of climate change on ticks and tick-borne diseases in Europe. *Interdisciplinary Perspectives on Infectious Diseases, 2009*, Article 593232. https://doi.org/10.1155/2009/593232

Gray, J. S., Kahl, O., Lane, R. S., Levin, M. L., & Tsao, J. I. (2016). Diapause in ticks of the medically important *Ixodes ricinus* species complex. *Ticks and Tick-Borne Diseases, 7*(5), 992–1003. https://doi.org/10.1016/j.ttbdis.2016.05.006

Hamdy, B. H. (1977). Biochemical and physiological studies of certain ticks (Ixodea). Excretion during ixodid feeding. *Journal of Medical Entomology, 14*(1), 15–18. https://doi.org/10.1093/jmedent/14.1.15

Hancock, P. A., Brackley, R., & Palmer, S. C. (2011). Modelling the effect of temperature variation on the seasonal dynamics of *Ixodes ricinus* tick populations. *International Journal of Parasitology, 41*(5), 513–522. https://doi.org/10.1016/j.ijpara.2010.12.012

Hussain, N., Shabbir, R. M. K., Ahmed, H., Afzal, M. S., Ullah, S., Ali, A., Irum, S., Naqvi, S. K., Yin, J., & Cao, J. (2023). Prevalence of different tick species on livestock and associated equines and canine from different agro-ecological zones of Pakistan. *Frontiers in Veterinary Science, 9*, 1089999. https://doi.org/10.3389/fvets.2022.1089999

Jaenson, T. G. T., Tälleklint, L., Lundqvist, L., Olsen, B., Chirico, J., & Mejlon, H. (1994). Geographical distribution, host associations, and vector roles of ticks (Acari: Ixodidae, Argasidae) in Sweden. *Journal of Medical Entomology, 31*(2), 240–256. https://doi.org/10.1093/jmedent/31.2.240

Jones, C. J., & Kitron, U. D. (2000). Populations of *Ixodes scapularis* (Acari: Ixodidae) are modulated by drought at a Lyme disease focus in Illinois. *Journal of Medical Entomology, 37*(3), 408–415. https://doi.org/10.1603/0022-2585(2000)037[0408:POISAI]2.0.CO;2

Jori, F., Hernandez-Jover, M., Magouras, I., Dürr, S., & Brookes, V. J. (2021). Wildlife-livestock interactions in animal production systems: What are the biosecurity and health implications? *Animal Frontiers, 11*(5), 8–19. https://doi.org/10.1093/af/vfab045

Kaufman, W. R., & Phillips, J. E. (1973a). Ion and water-balance in ixodid tick *Dermacentor andersoni*. 1. Routes of ion and water excretion. *Journal of Experimental Biology, 58*, 523–536.

Kaufman, W. R., & Phillips, J. E. (1973b). Ion and water-balance in ixodid tick *Dermacentor andersoni*. 2. Mechanism and control of salivary secretion. *Journal of Experimental Biology, 58*, 537–547.

Law, J. H., Ribeiro, J. M., & Wells, M. A. (1992). Biochemical insights derived from insect diversity. *Annual Review of Biochemistry, 61*, 87–111. https://doi.org/10.1146/annurev.bi.61.070192.000511

Lazzari, C. R., Fauquet, A., Lahondère, C., Araújo, R. N., & Pereira, M. H. (2021). Soft ticks perform evaporative cooling during blood-feeding. *Journal of Insect Physiology, 130*, Article 104197.

Leal, B., Zamora, E., Fuentes, A., Thomas, D. B., & Dearth, R. K. (2020). Questing by tick larvae (Acari: Ixodidae): A review of the influences that affect off-host survival. *Annals of the Entomological Society of America, 113*(6), 425–438. https://doi.org/10.1093/aesa/saaa013

Lemoine, M. M., Engl, T., & Kaltenpoth, M. (2020). Microbial symbionts expanding or constraining abiotic niche space in insects. *Current Opinion in Insect Science, 39*, 14–20. https://doi.org/10.1016/j.cois.2020.01.003

Lindsay, L. R., Barker, I. K., Surgeoner, G. A., McEwen, S. A., Gillespie, T. J., & Addison, E. M. (1998). Survival and development of the different life stages of *Ixodes scapularis* (Acari: Ixodidae) held within four habitats on Long Point, Ontario, Canada. *Journal of Medical Entomology, 35*, 189–199.

Machtinger, E. T., Poh, K. C., Pesapane, R., & Tufts, D. M. (2024). An integrative framework for tick management: The need to connect wildlife science, One Health, and interdisciplinary perspectives. *Current Opinion in Insect Science, 61*, Article 101131. https://doi.org/10.1016/j.cois.2023.101131

Makwarela, T. G., Seoraj-Pillai, N., & Nangammbi, T. C. (2025). Tick control strategies: Critical insights into chemical, biological, physical, and integrated approaches for effective hard tick management. *Veterinary Sciences, 12*(2), 114. https://doi.org/10.3390/vetsci12020114

Mehlhorn, H. (2015). Ticks. In H. Mehlhorn (Ed.), *Encyclopedia of parasitology*. Springer, Berlin, Heidelberg. https://doi.org/10.1007/978-3-642-27769-6_3181-2

Mulay, V. B., Caimano, M. J., Iyer, R., Dunham-Ems, S., Liveris, D., Petzke, M. M., Schwartz, I., & Radolf, J. D. (2009). *Borrelia burgdorferi* bba74 is expressed exclusively during tick feeding and is regulated by both arthropod- and mammalian host-specific signals. *Journal of Bacteriology, 191*(8), 2783–2794. https://doi.org/10.1128/JB.01802-08

Nicholson, W. L., Sonenshine, D. E., Noden, B. H., & Brown, R. N. (2019). Ticks (Ixodida). In G. R. Mullen & L. A. Durden (Eds.), *Medical and Veterinary Entomology* (3rd ed., pp. 603–672). Academic Press.

Nielebeck, C., Kim, S. H., Pepe, A., Himes, L., Miller, Z., Zummo, S., Tang, M., & Monzón, J. D. (2023). Climatic stress decreases tick survival but increases rate of host-seeking behavior. *Ecosphere, 14*(1), Article e4369. https://doi.org/10.1002/ecs2.4369

Norval, R. A. I. (1977). Studies on the ecology of the tick *Amblyomma hebraeum* Koch in the Eastern Cape Province of South Africa. II. Survival and development. *The Journal of Parasitology, 63*(4), 740–747.

Nuss, A. B., Mathew, M. G., & Gulia-Nuss, M. (2017). Rearing *Ixodes scapularis*, the black-legged tick: Feeding immature stages on mice. *Journal of Visualized Experiments, 123*, 55286. https://doi.org/10.3791/55286

Nuttall, P. A. (2022). Climate change impacts on ticks and tick-borne infections. *Biologia, 77*, 1503–1512. https://doi.org/10.1007/s11756-021-00927-2

Ogden, N. H., Lindsay, L. R., Beauchamp, G., Charron, D., Maarouf, A., O'Callaghan, C. J., Waltner-Toews, D., & Barker, I. K. (2004). Investigation of relationships between temperature and developmental rates of tick *Ixodes scapularis* (Acari: Ixodidae) in the laboratory and field. *Journal of Medical Entomology, 41*(4), 622–633. https://doi.org/10.1603/0022-2585-41.4.622

Ogden, N. H., Bigras-Poulin, M., O'Callaghan, C. J., Barker, I. K., Lindsay, L. R., Maarouf, A., Smoyer-Tomic, K. E., Waltner-Toews, D., & Charron, D. (2005). A dynamic population model to investigate effects of climate on geographic range and seasonality of the tick *Ixodes scapularis. International Journal of Parasitology, 35*(4), 375–389. https://doi.org/10.1016/j.ijpara.2004.12.013

Ogden, N. H., Bigras-Poulin, M., O'Callaghan, C. J., Barker, I. K., Kurtenbach, K., Lindsay, L. R., & Charron, D. F. (2007). Vector seasonality, host infection dynamics, and fitness of pathogens transmitted by the tick *Ixodes scapularis. Parasitology, 134*(Pt 2), 209–227. https://doi.org/10.1017/S0031182006001417

Ogden, N. H., Bigras-Poulin, M., Hanincova, K., Maarouf, A., O'Callaghan, C. J., & Kurtenbach, K. (2008). Projected effects of climate change on tick phenology and fitness of pathogens transmitted by the North American tick *Ixodes scapularis. Journal of Theoretical Biology, 254*, 621–632. https://doi.org/10.1016/j.jtbi.2008.06.020

Ogden, N. H., Pang, G., Ginsberg, H. S., Hickling, G. J., Burke, R. L., Beati, L., & Tsao, J. I. (2018). Evidence for geographic variation in life-cycle processes affecting phenology of the Lyme disease vector *Ixodes scapularis* (Acari: Ixodidae) in the United States. *Journal of Medical Entomology, 55*(6), 1386–1401. https://doi.org/10.1093/jme/tjy104

Ogden, N. H., Beard, C. B., Ginsberg, H. S., & Tsao, J. I. (2021). Possible effects of climate change on ixodid ticks and the pathogens they transmit: Predictions and observations. *Journal of Medical Entomology, 58*(4), 1536–1545. https://doi.org/10.1093/jme/tjaa220

Ojaimi, C., Brooks, C., Akins, D., Casjens, S., Rosa, P., Elias, A., Barbour, A., Jasinskas, A., Benach, J., Katonah, L., Radolf, J., Caimano, M., Skare, J., Swingle, K., Sims, S., & Schwartz,

I. (2002). *Borrelia burgdorferi* gene expression profiling with membrane-based arrays. *Methods in Enzymology, 358*, 165–177. https://doi.org/10.1016/s0076-6879(02)58088-5
Ostfeld, R. S., Canham, C. D., Oggenfuss, K., Winchcombe, R. J., & Keesing, F. (2006). Climate, deer, rodents, and acorns as determinants of variation in Lyme-disease risk. *PLoS Biology, 4*(6), Article e145. https://doi.org/10.1371/journal.pbio.0040145
Perret, J. L., Guigoz, E., Rais, O., & Gern, L. (2000). Influence of saturation deficit and temperature on *Ixodes ricinus* tick questing activity in a Lyme borreliosis-endemic area (Switzerland). *Parasitology Research, 86*(7), 554–557. https://doi.org/10.1007/s004360000209
Pound, J. M., Hilburn, L. R., & George, J. E. (1993). Implications of selection and hybridization studies on the mode of inheritance of photoperiodically induced developmental diapause in laboratory strains of the lone star tick (Acari: Ixodidae). *Journal of Medical Entomology, 30*(1), 100–106. https://doi.org/10.1093/jmedent/30.1.100
Radolf, J. D., Caimano, M. J., Stevenson, B., & Hu, L. T. (2012). Of ticks, mice and men: Understanding the dual-host lifestyle of Lyme disease spirochaetes. *Nature Reviews Microbiology, 10*(2), 87–99. https://doi.org/10.1038/nrmicro2714
Randolph, S. (2009). Epidemiological consequences of the ecological physiology of ticks. *Advances in Insect Physiology, 37*, 297–339.
Randolph, S. E., Green, R. M., Hoodless, A. N., & Peacey, M. F. (2002). An empirical quantitative framework for the seasonal population dynamics of the tick *Ixodes ricinus*. *International Journal of Parasitology, 32*(8), 979–989. https://doi.org/10.1016/s0020-7519(02)00030-9
Rebaudo, F., & Rabhi, V. B. (2018). Modeling temperature-dependent development rate and phenology in insects: Review of major developments, challenges, and future directions. *Entomologia Experimentalis Et Applicata, 166*(8), 607–617.
Reiter, P. (2001). Climate change and mosquito-borne disease. *Environmental Health Perspectives, 109*(Suppl 1), 141–161. https://doi.org/10.1289/ehp.01109s1141
Revel, A. T., Talaat, A. M., & Norgard, M. V. (2002). DNA microarray analysis of differential gene expression in *Borrelia burgdorferi*, the Lyme disease spirochete. *Proceedings of the National Academy of Sciences of the United States of America, 99*(3), 1562–1567. https://doi.org/10.1073/pnas.032667699
Ribeiro, J. M. C. (1987). Role of saliva in blood-feeding by arthropods. *Annual Review of Entomology, 32*, 463–478.
Rizwan, H. M., & Sajid, M. S. (2023). *Parasitism and parasitic control in animals: Strategies for the developing world*. CABI, Wallingford, UK. https://doi.org/10.1079/9781800621893.0000
Rocha, S. C., Velásquez, C. V., Aquib, A., Al-Nazal, A., & Parveen, N. (2022). Transmission cycle of tick-borne infections and co-infections, animal models and diseases. *Pathogens, 11*(11), 1309. https://doi.org/10.3390/pathogens11111309
Rochlin, I., & Toledo, A. (2020). Emerging tick-borne pathogens of public health importance: A mini-review. *Journal of Medical Microbiology, 69*(6), 781–791. https://doi.org/10.1099/jmm.0.001206
Rodgers, S. E., Miller, N. J., & Mather, T. N. (2007). Seasonal variation in nymphal blacklegged tick abundance in southern New England forests. *Journal of Medical Entomology, 44*(5), 898–900. https://doi.org/10.1093/jmedent/44.5.898
Rojas-Cabeza, J. F., Moreno-Cordova, E. N., Ayala-Zavala, J. F., Ochoa-Teran, A., Sonenshine, D. E., Valenzuela, J. G., & Sotelo-Mundo, R. R. (2025). A review of acaricides and their resistance mechanisms in hard ticks and control alternatives with synergistic agents. *Acta Tropica, 261*, Article 107519. https://doi.org/10.1016/j.actatropica.2024.107519
Saleh, M. N., Allen, K. E., Lineberry, M. W., Little, S. E., & Reichard, M. V. (2021). Ticks infesting dogs and cats in North America: Biology, geographic distribution, and pathogen transmission. *Veterinary Parasitology, 294*, Article 109392. https://doi.org/10.1016/j.vetpar.2021.109392
Sauer, J. R., & Hair, J. A. (1971). Water balance in the lone star tick (Acarina: Ixodidae): The effects of relative humidity and temperature on weight changes and total water content. *Journal of Medical Entomology, 8*(5), 479–485. https://doi.org/10.1093/jmedent/8.5.479

Savage, V. M., Gillooly, J. F., Woodruff, W. H., West, G. B., Allen, A. P., Enquist, B. J., & Brown, J. H. (2004). The predominance of quarter-power scaling in biology. *Functional Ecology, 18*, 257–282.

Schwartz, A. M., Hinckley, A. F., Mead, P. S., Hook, S. A., & Kugeler, K. J. (2017). Surveillance for Lyme disease—United States, 2008–2015. *MMWR Surveillance Summaries, 66*(SS-22), 1–12. https://doi.org/10.15585/mmwr.ss6622a1

Siddique, R. M., Iqbal, A., Nisa, F. U., Arijo, A. G., & Liaqat, H. A. (2023). Parasite control strategies: Biological materials. In H. M. Rizwan & M. S. Sajid (Eds.), *Parasitism and parasitic control in animals: Strategies for the developing world* (pp. 183–200). CAB International.

Šimo, L., Kazimirova, M., Richardson, J., & Bonnet, S. I. (2017). The essential role of tick salivary glands and saliva in tick feeding and pathogen transmission. *Frontiers in Cellular and Infection Microbiology, 7*, 281. https://doi.org/10.3389/fcimb.2017.00281

Sutherst, R. W., & Bourne, A. S. (2006). The effect of desiccation and low temperature on the viability of eggs and emerging larvae of the tick, *Rhipicephalus (Boophilus) microplus* (Canestrini) (Ixodidae). *International Journal of Parasitology, 36*(2), 193–200. https://doi.org/10.1016/j.ijpara.2005.09.007

Tälleklint, L., & Jaenson, T. G. T. (1998). Increasing geographical distribution and density of *Ixodes ricinus* (Acari: Ixodidae) in Central and Northern Sweden. *Journal of Medical Entomology, 35*(4), 521–526. https://doi.org/10.1093/jmedent/35.4.521

Templer, P. H., Schiller, A. F., Fuller, N. W., Socci, A. M., Campbell, J. L., Drake, J. E., & Kunz, T. H. (2012). Impact of a reduced winter snowpack on litter arthropod abundance and diversity in a northern hardwood forest ecosystem. *Biology and Fertility of Soils, 48*, 413–424.

Tijsse-Klasen, E., Fonville, M., Reimerink, J. H., Spitzen-van der Sluijs, A., & Sprong, H. (2010). Role of sand lizards in the ecology of Lyme and other tick-borne diseases in the Netherlands. *Parasites & Vectors, 3*, 42. https://doi.org/10.1186/1756-3305-3-42

Tilly, K., Rosa, P. A., & Stewart, P. E. (2008). Biology of infection with *Borrelia burgdorferi*. *Infectious Disease Clinics of North America, 22*(2), 217–234. https://doi.org/10.1016/j.idc.2007.12.013

Tonk-Rügen, M., Kratou, M., & Cabezas-Cruz, A. (2025). A warming world, a growing threat: The spread of ticks and emerging tick-borne diseases. *Pathogens, 14*(3), 213. https://doi.org/10.3390/pathogens14030213

Troughton, D. R., & Levin, M. L. (2007). Life cycles of seven ixodid tick species (Acari: Ixodidae) under standardized laboratory conditions. *Journal of Medical Entomology, 44*(5), 732–740. https://doi.org/10.1603/0022-2585(2007)44[732:lcosit]2.0.co;2

Tsoumani, M. E., Papailia, S. I., Papageorgiou, E. G., & Voyiatzaki, C. (2023). Climate change impacts on the prevalence of tick-borne diseases in Europe. *Environmental Sciences Proceedings, 26*(1), 18. https://doi.org/10.3390/environsciproc2023026018

Vail, S. G., & Smith, G. (2002). Vertical movement and posture of blacklegged tick (Acari: Ixodidae) nymphs as a function of temperature and relative humidity in laboratory experiments. *Journal of Medical Entomology, 39*, 842–846. https://doi.org/10.1603/0022-2585-39.6.842

Voordouw, M. J. (2015). Co-feeding transmission in Lyme disease pathogens. *Parasitology, 142*(2), 290–302. https://doi.org/10.1017/S0031182014001486

Voyiatzaki, C., Papailia, S. I., Venetikou, M. S., Pouris, J., Tsoumani, M. E., & Papageorgiou, E. G. (2022). Climate changes exacerbate the spread of *Ixodes ricinus* and the occurrence of Lyme borreliosis and tick-borne encephalitis in Europe—How climate models are used as a risk assessment approach for tick-borne diseases. *International Journal of Environmental Research and Public Health, 19*(11), 6516. https://doi.org/10.3390/ijerph19116516

Walker, A. R., Alberdi, M. P., Urquhart, K. A., & Rose, H. (2001). Risk factors in habitats of the tick *Ixodes ricinus* influencing human exposure to *Ehrlichia phagocytophila* bacteria. *Medical and Veterinary Entomology, 15*(1), 40–49. https://doi.org/10.1046/j.1365-2915.2001.00271.x

Wrońska, A. K., Kaczmarek, A., Boguś, M. I., & Kuna, A. (2023). Lipids as a key element of insect defense systems. *Frontiers in Genetics, 14*, 1183659. https://doi.org/10.3389/fgene.2023.1183659

Yano, Y., Shiraishi, S., & Uchida, T. A. (1987). Effects of temperature on development and growth in the tick *Haemaphysalis Longicornis. Experimental and Applied Acarology, 3*(1), 73–78. https://doi.org/10.1007/BF01200415

Chapter 3
Spatial Dynamics of Tick-Borne Diseases Under Climate Change

Mudassir Ahmad, Hizqeel Ahmed Muzaffar, Sakandar Khan, Sadia Ghazanfer, Fawad Ali Raza, and Phelipe Magalhães Duarte

Abstract Climatic variables such as temperature, humidity, and precipitation significantly influence the distribution of TBDs. Changes in the seasonal activity of ticks, tick genotype, tick development rate, and geographical distribution are important indicators of the impact of climate change on ticks and TBDs. Ecological factors such as deforestation and habitat fragmentation also lead to climate change which ultimately influences the distribution patterns of ticks. Ticks also cause zoonotic TBDs which highlights the importance of understanding the epidemiology and distribution patterns of ticks. Climate-based models have been developed to observe the effect of climate on different TBDs along with recording the geographical distribution of ticks. Different research has indicated that the incidence of Lyme disease has been increasing in different regions of the world such as Europe and North America. Socioeconomic factors play an important role in regulating the transmission of TBDs. Important socioeconomic factors include deforestation, educational index, unemployment, and recreation. Differentiating the effect of climate on the

M. Ahmad
School of Science and Technology, Nottingham Trent University, England, UK

H. A. Muzaffar
Faculty of Veterinary Sciences, KBCMA College of Veterinary and Animal Sciences, Narowal, Sub-Campus UVAS, Lahore, Pakistan

S. Khan (✉)
Section of Parasitology, Department of Pathobiology, KBCMA College of Veterinary and Animal Sciences, Narowal, Sub-Campus UVAS, Lahore, Pakistan
e-mail: sakandar.khan@uvas.edu.pk

S. Ghazanfer
Department of Parasitology, Faculty of Veterinary Science, University of Agriculture, Faisalabad, Pakistan

F. A. Raza
Institute of Soil and Environmental Science, University of Agriculture, Faisalabad, Pakistan

P. M. Duarte
Postgraduate Program in Animal Bioscience, Federal Rural University of Pernambuco (UFRPE), Recife, Pernambuco, Brazil

H. M. Rizwan and M. S. Sajid (eds.), *Ticks in a Changing Climate*, Biometeorology 7,
https://doi.org/10.1007/978-3-032-12259-9_3

distribution of ticks from other factors such as the host population will help us to design the control strategies for TBDs more effectively.

Keywords Ticks · Tick-borne diseases · Climate change · Spatial epidemiology · Ecological Factors · Spatiotemporal dynamics

3.1 Introduction

Pathogens transmitted directly from the source typically have the characteristic of spreading rapidly. In contrast, disease-carrying organisms that feed deliberately, have multistage life cycles, and reproduce slowly generally lack this rapid transmission capability (Aldahi, 2017). Though, they are serious threat to public health, and ticks are one of those organisms. Ticks are small blood feeding parasites, belongs to family Acari, responsible for causing serval diseases in humans and animals (Perveen et al., 2021). The increasing number of tick populations has elevated the incidence of diseases of public health concern. There can be multiple reasons for the outbreaks of tick- borne diseases, climate change is most crucial among them (Gilbert, 2021). The complex interplay of environmental and ecological factors greatly influence the spatial distribution of vector borne diseases including tick-borne diseases (TBDs). The increase in distance of pathogen source and host drastically decrease the probability of disease spread, which highlights the significance of spatial patterns of disease spread and associated risks (Wang et al., 2016).

Responses to various abiotic and biotic factors are shifting. The epidemiology of TBDs is being impacted due to shift in tick distribution patterns and populations which necessitates the rigorous surveillance, monitoring and potential role by interdisciplinary approaches to deal with emerging public health concerns (Rowan et al., 2023). Moreover, the synergy of host population such as deer's and rodents are also crucial for the understanding of transmission dynamics of TBDs. Rodents also play role for the transmission of tick borne pathogens. Hence availability of the food source of the populations carrying ticks increase tick density and prevalence of infection (Ostfeld et al., 2018; Tardy et al., 2021). The eco-epidemiology of TBDs also profoundly affected by change in usage of land which increases the exposure of humans and animals to the higher densities of tick populations carrying several pathogens (Zhao et al., 2021).

Through various studies, it has been observed that climate variables including temperature, precipitation and humidity play signification role for the distribution of tick-borne pathogens. A study conducted in North America analyzed difference in geographical distribution of anaplasma and ehrlichia that were influenced by climatic variables and forest covers (Wimberly et al., 2008). Similarly, studies in Europe also showed change in ecological balance and transmission season of *Ixodes (I.) ricinus*, vector for tick-borne encephalitis and Lyme disease (Voyiatzaki et al., 2022). To access the future risks associated with TBDs under climate change conditions, predictive modeling is essential. Combining environmental climatic variables and

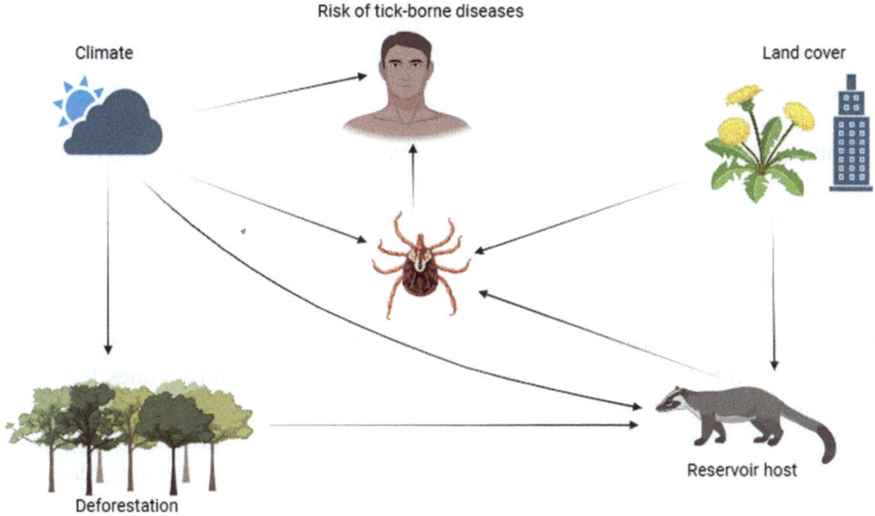

Fig. 3.1 Relation of different factors affecting spatial dynamics of tick-borne diseases

these models predict geographical distribution of tick population and incidence of tick-borne infections. A particular study revealed that considering climate change, ecological modeling has been used to access and predict the spread of *I. ricinus* in eastern and northern parts of Europe (Alkishe et al., 2017). In the similar way, such approaches have also been used to identify disease hotspots (Lee et al., 2022).

Although several advancements have been introduced for the assessment of TBDs, there is need to understand the impact of climate change on TBD distribution. Long term empirical studies are needed to understand the association between disease patterns and climatic variations. This chapter will explore the complex interplay environmental, ecological and biological factors that greatly influence the spatial dynamics of TBDs considering rapidly changing climate. Figure 3.1 explains the relation of different factors on the spatial dynamics of tick-borne diseases.

3.2 Climate Change and Geographical Distribution of Ticks

Though climate change is impacting the ticks and TBDs across the world, however, change has not been recorded in many regions. This is may be due to the lack of long term studies on the TBDs, lack understanding of complex interaction of ticks and their environment, and absence of data on TBDs (de la Fuente et al., 2023). Contrary, in other countries any socioeconomic change reflects towards the possible cause of that change. For instance, in 1999, Russia recorded the highest number of cases of tick-bone encephalitis followed by decline in that number. This peak was recorded as a result of legislative change allowing increased contact of city dwellers with ticks during the usage of undeveloped land for housing (Korenberg et al., 2009). Similarly, USA witnessed an increased population of *I. scapularis* in areas previously inhabited by deer population (Hoen et al., 2009).

The change in seasonal activity of the ticks, geographical distribution, and tick genotype and development rate are the indicators of impact of climate change on ticks and tick-borne infections (Ogden et al., 2021). Long term studies are required to appropriately understand the impact of climate change on ticks, TBDs transmission and disease dynamics. As ticks are exposed to environment during their life cycle, they respond to various biotic and abiotic factors. Environmental factors including moisture, humidity and temperature possess great significance for the survival of ticks (Ogden et al., 2016). Since, increasing global warming is affecting the TBD dynamics and tick survival, the tick developmental and tick biting rates have been increased crucially (Ostfeld & Brunner, 2015).

A study conducted in United States and Canada observed range of stability patterns of eight species of ticks of medical and veterinary importance. These species were I. scapularis, *I. cookie, I. pacificus, Amblyomma (A.) maculatum, A. americanum, Rhipicephalus (R.) singuineus, Dermacentor (D.) oriabilis,* and *D. andersoni*. The study concluded that ticks in North America have become sensitive to fluctuations in ambient abiotic factors. The pattern showing range of stability among most of the species was found dominant particularly increasing towards North America. Moreover, the potential expansion in range was observed in South Dakota, Northern Minnesota, and some parts of northern and western Canada (Alkishe et al., 2021). The most evident impact of change can be noticed in Canada where tick-borne human infections in 1990, when rate of tick-bone human infections were very low. The expansion of *I. scapularis* from USA to Canada have resulted in spread of Lyme borreliosis in many regions of Canada (Guillot et al., 2019). However, expansion in distribution of local tick species were also noticed that was mainly due to increased temperature of the region due to global warming providing favorable environment for tick activity and survivability (Dantas-Torres, 2015).

Another study was conducted to investigate the potential geographical distribution of *I. ricinus*. The results revealed that tick population expansion in North Africa, Europe and Middle East have been rapidly increasing with changing climate and global warming. Considering the current patterns, it is believed that Northern Europe will be the most affected by TBDs in future (Alkishe et al., 2017). The potential geographical distribution of *R. microplus*, vector of bovine anaplasmosis and babesiosis was examined focusing on two periods of 2050 and 2070. It has been anticipated that there will be change in suitability patterns of *R. microplus* worldwide. Moreover, this suitability pattern in the regions of cattle exposure to this specie will increase to $\geq$30% in 2050 (Marques et al., 2020). In South Korea, the impact of climate change on incidence of TBDs was observed at 5 km resolution. A rise in index based on environmental indicators signifies the implementation of robust prevention strategies to control tick-borne infections (Lee & Chung et al., 2023).

It also have been noted that the prevalence of tick borne encephalitis, Lyme disease and babesiosis and anaplasmosis in animals is increasing with the expansion of tick populations in Scandinavian countries. The main cause behind these changes are warmer temperature (Hvidsten et al., 2020). Similarly, the populations of *A. americanum* and *D. varibabilis* are expanding their range in North America. Moreover, the prevalence of *R. microplus* is also growing during to warmer temperature in North America (Minigan et al., 2018). Additionally, warmer climate is a driving factor for the illegal movement of livestock from Africa to the other regions carrying a risk of introducing new tick species (Githaka et al., 2021). These studies show that environmental factors such as moisture, humidity, and temperature play a crucial role in the spread of ticks and TBDs. These factors contribute to the increased survivability and expanded geographic range of tick populations. Therefore, it is essential to plan strategically and implement effective control measures to manage TBDs.

3.3 Ecological Factors Shaping Disease Transmission Patterns

It is estimated the global temperature increased by 1 °C during 1859–1900. With the increased human activities facilitating elevated global temperature, it has reached to 2 °C by twentieth century. Although the climate changing phenomena is not same over the globe as temperature over land is increasing rapidly as compared to over the oceans (Joshi, 2008). Ticks do spend major part of their life off the host, hence exposed to different environmental conditions. For instance, the warmer temperature is highly favorable for the survival of *I. ricinus*, showing increased oviposition, higher egg development rates, and increased rates of interstadial development (Estrada-Peña et al., 2004). Through studies it has been observed that increased population *I. ricinus* is associated with warmer temperatures. Moreover, the warmer temperature and relative humidity facilitate the saturation deficit which adds a lot to the drying power of air. Hence low temperature lead to inhibition of tick activity and increased tick mortality. Additionally, studies revealed that *I. ricinus* cannot survive temperature below -15 °C. On the other hand, temperature above 30°C cause tick mortality even at higher humidity level $\geq$80% (Oshiro et al., 2021).

The major factors that are responsible for changing the epidemiology of TBDs include agricultural changes such as deforestation, and the extent of interaction with the ecosystem of the disease (Estrada-Peña & de la Fuente, 2014). As ticks are susceptible to environmental conditions, ecological factors have a particular impact on the epidemiology of TBDs (Pongsiri et al., 2009). Climate change and habitat fragmentation increase the risk of TBDs (Wimberly et al., 2008). Three types of environmental change have been identified by researchers as the main drivers of *I. ricinus*'s continued range expansion in Europe: (1) climate change, which helps the species spread geographically into higher latitudes (like Scandinavia) and altitudes (like central Europe); (2) factors associated with shifting tick host distributions, like ecological shifts or habitat connectivity; and (3) anthropogenic changes, like adjustments to forest management techniques, tourism, or leisure activities (Boulanger

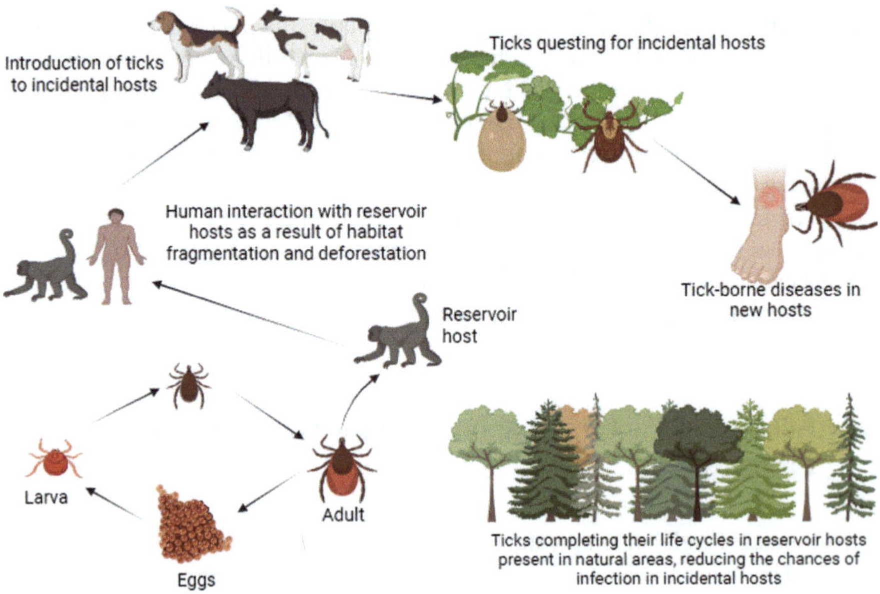

Fig. 3.2 Effect of deforestation on disease transmission patterns of ticks

et al., 2024). Figure 3.2 gives an overview of the effect of deforestation, one of the ecological factors affecting disease transmission.

3.4 Anthropic Actions

The destruction of natural habitats due to anthropogenic actions can lead to increased contact between wild animals and humans, which may heighten the occurrence of emerging and re-emerging diseases (Volpato et al., 2020). Among these anthropogenic actions, we can mention deforestation, trafficking and consumption of wild animals, agriculture and livestock, and mining. Additionally, activities such as ecotourism and outdoor sports can increase exposure to these animals and their parasites.

3.4.1 Deforestation

The increase in urbanization is a factor related to the rise in deforestation and the loss of natural habitats, leading to a higher number of emerging and re-emerging disease cases (Wu et al., 2017). The deforestation of these natural areas triggers a severe imbalance in ecosystems (Bloomfield et al., 2020). The loss of these areas poses a public health risk and drawing a parallel to the impact of deforestation on the rise of Tick-Borne Diseases, each 1 km^2 of deforested Amazon rainforest may be equivalent to 27 new cases of malaria (Chaves et al., 2018).

3.4.2 *Trafficking and Consumption of Wild Animals*

The trafficking of animals poses a risk of species extinction, environmental imbalance, and exposure of these animals to other ecosystems and humans. Additionally, these animals carry their parasites and microorganisms, which may lead to the adaptation of these pathogens to humans. It is important to highlight that many of these animals are traded for consumption purposes (Chow et al., 2014). In some communities in Asia, Africa, and Latin America, between 2004 and 2010, more than 53.5% of households were supplied with wild animals and plants (Rowland et al., 2017).

3.4.3 *Agriculture*

Agriculture has historically been the main driver of global deforestation (Lindquist et al., 2012). In Brazil, for example, the estimated production of cereals, legumes, and oilseeds for 2025 is projected to reach 328.4 million tons, a 12.2% increase compared to 2024 (292.7 million tons). The harvested area is expected to be 81.0 million hectares, representing a 2.5% increase compared to 2024, with an expansion of 2.0 million hectares (Anonymous, 2025). While population growth demands greater food production, it is essential that this expansion be rational, sustainable, and mindful of its environmental impact.

In Turkey, a predictive model has been developed to map the suitability of habitats for the tick vector of Crimean-Congo Hemorrhagic Fever, using satellite-based climate data and high-resolution vegetation characteristics from Landsat imagery covering the entire country. It was found that the highest-risk areas were correlated with regions of high climatic suitability for the tick, along with a high rate of fragmentation of agricultural lands interspersed with forest and shrub vegetation (Estrada-Peña et al., 2007).

3.4.4 *Ecotourism*

Ecotourists are exposed to a range of physical and infectious health risks that are both similar to and different from other types of travel experiences. In addition to the common risks associated with travel in developing countries, significant threats to ecotourists include vector-borne and zoonotic diseases, lack of access to emergency healthcare, and physical injuries related to intense outdoor activities in diverse environments. Ecotourists may be at risk of severe injuries caused by wildlife, and combined with the threat of zoonotic diseases, they should be discouraged from having any contact with animals during their travels. Travel health specialists are in an excellent position to educate travelers about how they may negatively impact destination communities and the wildlife they seek to visit. Ethical travel should be emphasized (Muehlenbein et al., 2019).

3.4.5 Extreme Sports

Engaging in sports and participating in competitions in extreme locations such as jungles, mountains, deserts, lakes, rivers, caves, and canyons can expose humans to vector-borne insects. Outbreaks of infections have been linked to activities such as caving, canoeing, kayaking, rafting, triathlons, and multisport races. The lack of immediate medical care can complicate and worsen the severity of these diseases (Igreja et al., 2017). Among them, we can mention ehrlichiosis (Standaert et al., 1995), Lyme disease (Smith et al., 2001) and Crimean-Congo Hemorrhagic Fever (CCHF) (Maltezou et al., 2010).

3.5 Modeling Approaches for Predicting Disease Spread Under Climate Change

The growing rate of emerging zoonotic pathogens has highlighted the urgency to understand the epidemiology and geographic distribution of ticks and TBDs with relevancy to favorable habitat and climate change (McDonald et al., 2019). Tick-borne diseases has direct impact on veterinary and human wellbeing causing significant damage to livelihoods and global economies. It is of utmost importance to recognize and prioritize surveillance strategies to control ticks and TBDs (Zanet et al., 2020).

Previously several species distribution models have been used to predict the species distribution of ticks and TBDs. Max ENT model has been used to predict the habitat suitable for ticks. It also have been used to predict the range of tick population (Ma et al., 2023). In Europe, several studies have shown its application to rule out the distribution of primary vector of tick-borne encephalitis and Lyme disease (Kopsco et al., 2022). Another model which was climate driven cohort based population model was used to predict the variation in tick density and any change in tick activity under different climatic conditions. For this, high resolution climate model and data was used to access the seasonal tick population distribution, particularly *I. ricinus* (Nolzen et al., 2022). Climate based models predict the distribution of tick population under climate change. These models incorporates certain specific variables including prolonged transmission season, elevated proliferation rates, environmental equilibrium, climate induced migration of vectors, availability of reservoir host, and human and animal population. These models helps to understand the relationship between climate change and epidemiological patterns of TBDs (Voyiatzaki et al., 2022).

Spatial regression framework models has also been used to analyze the influence of landscape and climatic variables on the disease dynamics and its incidence. For instance, a study conducted in Minnesota, America found that presence of forests and temperature plays crucial role for the spread of tick-borne Lyme disease (Robinson et al., 2015). A study conducted in UK used stage structured host vector model to analyze future risks of disease spread under climate change conditions. This model incorporates the complex interplay between populations of host and vector, varying land usage and environmental variables (Ewing et al., 2021). Dynamic population models incorporate disease progression over large spatial regions. These models used phenology and temperature driven population models to analyze and predict varying vector population size considering different climatic conditions (Dinghra et al., 2013). Mechanistic models includes traditional frameworks based on movement of the host, impacts of seasonal climate and transmission routes. These models predict population growth of vectors and disease spread based on reproductive activity (Zhang et al., 2021).

Temporal and spatial distribution of TBDs vary greatly with changing climatic conditions. Increase in temperature and patterns are driving factors for increase in range of ticks and tick-borne pathogens. Rapidly changing climate forecast shift in disease distribution, tick survivability, and activity. For instance, several regions have shown increased incidence of TBDs due to rise in temperature (Robinson et al., 2015). The modeling approaches are significant to forecast disease dynamics and its prevalence focusing on tick population distribution considering ecological, climatic and environmental factors. The model based predictions of disease dynamics necessitates incorporation of advanced modeling approaches to evaluate more appropriate future risks and inform disease control strategies to ensure One Health (Adegoke et al., 2024).

3.6 Case Studies; Regional Variations in Disease Distribution

The impact of rapidly changing climate on the prevalence of TBDs have been observed across the world. Several tick species, mainly *I. ricinus* is now more prevalent in Europe and this expansion is growing towards North vertically (Medlock et al., 2013). The burden of Lyme diseases and tick-borne encephalitis that is linked with increasing tick population has been increased in United Kingdom and Nordic countries (Gilbert et al., 2009, Leibovici et al., 2021a). Moreover, seasonal durations effected by increasing global temperature are driving these changes which highlights the urgency of regional surveillance and potential interventions in Europe (Semenza & Paz, 2021).

Research have shown that climate change can provide favorable situations by increasing tick activity season and facilitating prolonged tick survivability (Ogden et al., 2020). These significant shifts in species distribution pattern increases the incidences of Rocky Mountain spotted fever and Lyme diseases in North America due to warmer winters and extended autumns.

Over the years, the distribution of *I. ricinus* has expanded 400 km upwards, and northwards in Norway. Climate change is associated with its elevation. Global warming has led to 2–3 weeks of shortening of the winter season which has resulted in an earlier and warmer spring in Norway which synchronizes the emergence of this vector (Mysterud et al., 2017). Similar effects of climate change have also been observed in Scotland (Li et al., 2016). The emergence of TBDs has been associated with an increased spatial distribution of the vector in northern altitudes in Norway which has been associated with climate change (Mysterud et al., 2018). Changes in land use leading to habitat conversion are responsible for alteration in host communities leading to the emergence of Lyme disease in the United States (Wood & Lafferty, 2013).

In a similar vein, *I. scapularis*, another generalized feeder, is thought to be found in Canada and nearly the whole eastern part of the United States (Eisen et al., 2019). The furthest reaches of this species' latitudinal and altitudinal distribution are probably where climate change will have the biggest direct effects. Furthermore, it is evident that species with broad geographic ranges and climate envelopes have adapted to local host groups and climates (Gilbert, 2021). Severe fever with thrombocytopenia syndrome (SFTS) in Japan has been found to cause mortalities among the affected people. This disease has seasonal variations and cases of SFTS have been reported in Japan from May to August which are the months of farming and mountain climbing in Japan (Yamaji et al., 2018).

3.7 Health Implications

Climate change is a key driving factor influencing the distribution of Lyme disease, contributing to an increase in both the number of human cases and their geographical spread (Léger et al., 2013; Medlock et al., 2013). Other TBDs such as human babesiosis, anaplasmosis, and tick-borne encephalitis are also increasing worldwide. Along with many other causes, climate change is one of the most important factors that is responsible for the transmission of TBDs to humans (Ostfeld & Brunner, 2015). As the temperature increases and warmer climate occurs, humans become vulnerable to the attack of *I. ricinus* (James et al., 2013). The lone star tick, which is common in the United States is now expanding towards the Canadian border (Springer et al., 2014).

Tick-borne encephalitis is now increasing in Europe (Gilbert, 2021). *I. ricinus* is now expanded to the Czech Republic (Danielová et al., 2008), Sweden (Lindgren et al., 2000), and Slovakia (Lukan et al., 2010). Climate factors such as rapid spring warming are also responsible for the increased prevalence of tick-borne encephalitis (Hartemink & Takken, 2016). Until 1990, TBDs were rare and limited in Canada (Bouchard et al., 2019). The conditions changed when the *I. scapularis* was transmitted from the United States to Canada (Guillot et al., 2020) because of climate change (Nuttall, 2022; Sagurova et al., 2019). A warmer environment is acting as a key factor for this expansion (Jaenson et al., 2012a, 2012b). For tick-borne encephalitis, a warmer environment increases the virus load in affected ticks, ultimately increasing the prevalence in humans (Daniel et al., 2018).

Tick-borne diseases have also increased in domesticated and wild animals (Esteve-Gassent et al., 2016) because of climate change (Dantas-Torres, 2015). Climate change also increases the prevalence of *R. microplus* which is a great threat to the cattle industry (Giles et al., 2014). In Africa, climate change is an important contributing factor in the illegal and cross-border movement of livestock which is responsible for increased transmission to new regions (Githaka et al., 2021). Climate change facilitates the establishment of exotic tick species in new locations. Examples include *Hyalomma marginatum marginatum* which is now established in France and *Haemaphysalis longicornis* which was a native of Asia but is now established in the United States (Raghavan et al., 2019; Stachurski & Vial, 2018). As tick expansion is predicted in Northern regions, tick prevalence is expected to decline in the Southern regions. For example, *I. scapularis* is expected to expand to the Northern region as compared to the Southern region because of the temperature changes (Ginsberg et al., 2014). Climate change is affecting the distribution of soft ticks also (Estrada-Peña et al., 2013).

3.8 Socioeconomic Factors and Vulnerability

The educational index is an important socio-economic factor that is responsible for the transmission of TBDs (Zhang et al., 2022). Forests are the regions where ticks are highly prevalent. Humans may have to carry out the harvesting of wild food for human consumption because of unemployment and poverty (Šumilo et al., 2008). Socio-economic factors are important factors playing a role in the transmission of TBDs along with climatic factors (Parham et al., 2015). People who spend most of their time outside of their homes such as tourists are more vulnerable to TBDs (Klingelhöfer et al., 2022) as recreational places play a key role in the transmission of tick species ultimately leading to the transmission of TBDs (Slunge et al., 2019).

Socio-economic transition occur as a result of TBDs because TBDs cause unemployment which in turn increases poverty (Randolph, 2008). Livestock farming is a source of income for many small-scale farmers (Namgyal et al., 2021). Tick-borne diseases affect the livelihoods of these small-scale farmers. Another considerable concern regarding the poor communities is the lack of knowledge regarding TBDs among these communities (Katswara & Mukaratirwa, 2021), which acts as a driving factor for the increased transmission of TBDs in these communities (Valente et al.,

2015). Furthermore, limited epidemiological data is available from these communities which makes it difficult the implementation of control strategies (Simuunza et al., 2011). As the human population increases, people have to migrate to the cities to earn their livelihoods which disturbs the natural habitats ultimately increasing the risk of TBDs (Pfäffle et al., 2013). Human encroachment on wildlife habitats and deforestation has been proposed as a reason for the continuous increase in the incidence of TBDs (Nchu et al., 2020). If they can't effectively connect with the public, experts on TBDs might not be able to update them on scientific developments.

The ultimate goal of the medical, veterinary, and scientific areas' efforts to address new or developing TBDs is to benefit the general public. Many TBD experts, however, can shy away from public speaking, particularly if they feel that the subject is politicized and that the intended audience is not interested in listening to them (Munz et al., 2024). Patients may have misconceptions about diseases spread by ticks as a result of their exposure to television, social media, and online research. Occasionally, local or regional efforts to improve public education on tick bite prevention may exist in highly populous or financially supported areas (Eisen & Stafford III, 2021). However, these methods are often ineffective in reaching disadvantaged groups, where distribution through technology is not feasible. It is important to assess the general public's understanding of mitigating tick bites, resistance to insecticide use, ability to recognize tick-borne infection signs, and the necessity of veterinarian or medical consultation (Sakamoto, 2018).

Directing the public to researchers who focus on tick identification is also crucial (Sakamoto, 2018). Individuals with less education are more likely to perform a tick check than those with greater education. Less educated people might be more likely to choose occupations that need them to spend more time outside, such as landscaping or yard care, which will require them to perform tick inspections more regularly. People with more education and income can choose more expensive options, such as landscaping and lawn maintenance. Similarly, those with lower salaries are more likely to use insect repellent on themselves than people with greater incomes. Those who have treated their yards may think that personal bug repellent is unnecessary. Alternatively, people who don't typically apply insect repellent could apply it to their yards (Niesobecki et al., 2019). Socioeconomic factors that play an important role in the transmission of tick-borne diseases have been listed in in table 3.1.

3.9 Challenges and Opportunities for Future Research

Since the dairy and meat business is growing quickly, it is important to conduct nationwide epidemiological surveys to determine the economic effect and geographic distribution of TBDs. This will help direct future research and control efforts (Jabbar et al., 2015). Public health officials have recently expressed serious concerns about the TBDs rapid geographic spread throughout the world. In enzootic cycles, ticks and wild animals serve as hosts for tick-borne infections, which can then spread epizootically to other mammals, such as humans and livestock (Rocha et al., 2022).

Table 3.1 Socioeconomic factors playing an important role in the transmission of tick-borne diseases

Socioeconomic factor	Reference
The educational index is an important socioeconomic factor that is responsible for the transmission of TBDs	Zhang et al., 2022
Forests are the regions where ticks are highly prevalent. Humans may have to carry out the harvesting of wild food for human consumption because of unemployment and poverty	Šumilo et al., 2008
Socio-economic factors are important factors playing a role in the transmission of TBDs along with climatic factors	Parham et al., 2015
People who spend most of their time outside of their homes such as tourists are more vulnerable to TBDs	Klingelhöfer et al., 2022
Recreational places play a key role in the transmission of tick species ultimately leading to the transmission of TBDs	Slunge et al., 2019
Socio-economic transition occur as a result of TBDs because TBDs cause unemployment which in turn increases poverty	Randolph, 2008
Another considerable concern regarding the poor communities is the lack of knowledge regarding TBDs among these communities	Katswara & Mukaratirwa, 2021
Limited epidemiological data is available from poor communities which makes it difficult the implementation of control strategies	Simuunza et al., 2011
As the human population increases, people have to migrate to the cities to earn their livelihoods which disturbs the natural habitats ultimately increasing the risk of TBDs	Pfäffle et al., 2013
Human encroachment on wildlife habitats and deforestation has been proposed as a reason for the continuous increase in the incidence of TBDs	Nchu et al., 2020
Many TBD experts, however, can shy away from public speaking, particularly if they feel that the subject is politicized and that the intended audience is not interested in listening to them	Munz et al., 2024
Patients may have misconceptions about diseases spread by ticks as a result of their exposure to television, social media, and online research	Eisen & Stafford III, 2021
Individuals with less education are more likely to perform a tick check than those with greater education. Less educated people might be more likely to choose occupations that need them to spend more time outside, such as landscaping or yard care, which will require them to perform tick inspections more regularly. People with more education and income can choose more expensive options, such as landscaping and lawn maintenance. Similarly, those with lower salaries are more likely to use insect repellent on themselves than people with greater incomes. Those who have treated their yards may think that personal bug repellent is unnecessary. Alternatively, people who don't typically apply insect repellent could apply it to their yards	Niesobecki et al., 2019

Even in nearby places, the variation of environment, vector, and host characteristics limits prediction models of future hazards posed by TBDs, and current diagnostic tests and treatments for TBDs are insufficient despite growing public health concerns (Yamaji et al., 2018). Being able to differentiate the effects of climate change from those of other factors that contribute to TBD change, such as shifts in deer, rodent, and predator populations or in human behavior, is essential for improving the approach to disease control and mitigation. Controlling deer densities, for instance, could be seen as a mitigation approach if an increase in disease is caused by an increase in deer densities (Gilbert et al., 2012).

A different aspect of the disease cycle must be addressed, such as reducing human exposure via awareness campaigns or enhancing post-infection therapy through practitioner education, if climate change is the main driver of the rise in disease. Although predictive models that forecast climate change-related changes in tick populations and the risk of TBDs are improving, it is critical to concentrate on estimates of prediction uncertainty in the absence of validation data sets, which cannot be obtained until time has passed (de Oliveira et al., 2017). The idea of climate change resilience through tick and host adaptation to local conditions, as well as the degree to which ticks and pathogens can use new hosts outside of the existing climate envelope, should also be included in future studies (Gilbert, 2021). Future studies should examine the particular, local abiotic factors impacted by regional or global climate change, as well as how these factors affect tick behavior, development, and population dynamics and tick-borne pathogen dynamics. Researchers can forecast how the risk of human exposure to these pathogens will alter as the climate continues to change by studying how the vector and pathogen are affected by climate change. Finally, it will be possible to predict epidemiological patterns due to changes in human behavior and climate-driven changes (Haque et al., 2024).

Separating the climatic impacts from temporal variability unrelated to climate change is another significant obstacle to comprehending the relative significance of directional climate change in TBD dynamics (Ostfeld & Brunner, 2015). Last but not least, attempts to forecast the risk and spread of TBDs must make the assumption that the same mechanisms that currently govern ticks and TBDs will continue to function in the future. To put it another way, it is simply assumed that ticks, their hosts, and the diseases they transmit are not changing quickly in response to climate change (Gilbert et al., 2014). Over the past ten years, several temperate and peri-Arctic areas of the Northern Hemisphere have seen the trends for TBDs that have been described, involving an altitudinal and latitudinal movement of *Ixodes* ticks and the diseases they spread. With more hospitable temperatures and milder winters, climate change has a direct effect on the habitats of *Ixodes* ticks. If we don't mitigate and adapt to climate change, these regions will probably see similar trends in the future. It is important to take into account both natural and man-made elements that contribute to ticks spreading into new areas. Tick vectors are also transported into new areas by wild animal hosts, and afforestation as well as land and wildlife management are significant potential drivers (Caminade et al., 2019; Medlock et al., 2013).

References

Adegoke, B. O., Odugbose, T., & Adeyemi, C. (2024). Data analytics for predicting disease outbreaks: A review ofmodels and tools. *International Journal of Life Science Research Updates, 2*(2), 1–9.

Aldahi, A. (2017). *Studies on microbes including potential human pathogens from insects and other invertebrates* (Doctoral dissertation, University of Sheffield).

Alkishe, A. A., Peterson, A. T., & Samy, A. M. (2017). Climate change influences on the potential geographic distribution of the disease vector tick *Ixodes ricinus*. *PLoS ONE, 12*(12), Article e0189092.

Alkishe, A., Raghavan, R. K., & Peterson, A. T. (2021). Likely geographic distributional shifts among medically important tick species and tick-associated diseases under climate change in North America: A review. *Insects, 12*(3), 225.

Anonymous (2025). *In April, IBGE forecasts a harvest of 328.4 million tons for 2025.* https://agenciadenoticias.ibge.gov.br/agencia-sala-de-imprensa/2013-agencia-de-noticias/releases/43400-em-abril-ibge-preve-safra-de-328-4-milhoes-de-toneladas-para-2025

Bloomfield, L. S., McIntosh, T. L., & Lambin, E. F. (2020). Habitat fragmentation, livelihood behaviors, and contact between people and nonhuman primates in Africa. *Landscape Ecology, 35*(4), 985–1000.

Bouchard, C., Dibernardo, A., Koffi, J., Wood, H., Leighton, P. A., & Lindsay, L. R. (2019). Increased risk of tick-borne diseases with climate and environmental changes. *Canada Communicable Disease Report, 45*(4), 83–89.

Boulanger, N., Aran, D., Maul, A., Camara, B. I., Barthel, C., Zaffino, M., Lett, M. C., Schnitzler, A., & Bauda, P. (2024). Multiple factors affecting Ixodes ricinus ticks and associated pathogens in European temperate ecosystems (northeastern France). *Scientific Reports, 14*(1), 9391.

Caminade, C., McIntyre, K. M., & Jones, A. E. (2019). Impact of recent and future climate change on vector-borne diseases. *Annals of the New York Academy of Sciences, 1436*(1), 157–173.

Chaves, L. S. M., Conn, J. E., López, R. V. M., & Sallum, M. A. M. (2018). Abundance of impacted forest patches less than 5 km2 is a key driver of the incidence of malaria in Amazonian Brazil. *Scientific Reports, 8*(1), 7077.

Chow, A. T., Cheung, S., & Yip, P. K. (2014). Wildlife markets in south China. *Human-Wildlife Interactions, 8*(1), 108–112.

Daniel, M., Danielová, V., Fialová, A., Malý, M., Kříž, B., & Nuttall, P. A. (2018). Increased relative risk of tick-borne encephalitis in warmer weather. *Frontiers in Cellular and Infection Microbiology, 8*, 90.

Danielová, V., Schwarzová, L., Materna, J., Daniel, M., Metelka, L., Holubová, J., & Kříž, B. (2008). Tick-borne encephalitis virus expansion to higher altitudes correlated with climate warming. *International Journal of Medical Microbiology, 298*(1), 68–72.

Dantas-Torres, F. (2015). Climate change, biodiversity, ticks and tick-borne diseases: The butterfly effect. *International Journal for Parasitology: Parasites and Wildlife, 4*(3), 452–461.

de la Fuente, J., Estrada-Peña, A., Rafael, M., Almazán, C., Bermúdez, S., Abdelbaset, A. E., Kasaija, P. D., Kabi, F., Akande, F. A., Ajagbe, D. O., & Dahal, A. (2023). Perception of ticks and tick-borne diseases worldwide. *Pathogens, 12*(10), 1258.

de Oliveira, S. V., Romero-Alvarez, D., Martins, T. F., Dos Santos, J. P., Labruna, M. B., Gazeta, G. S., Escobar, L. E., & Gurgel-Gonçalves, R. (2017). *Amblyomma* ticks and future climate: Range contraction due to climate warming. *Acta Tropica, 176*, 340–348.

Dhingra, R., Jimenez, V., Chang, H., Gambhir, M., Fu, J., Liu, Y., & Remais, J. (2013). Spatially-explicit simulation modeling of ecological response to climate change: Methodological considerations in predicting shifting population dynamics of infectious disease vectors. *ISPRS International Journal of Geo-Information, 2*(3), 645–664.

Diuk-Wasser, M., VanAcker, M., & Fernandez, M. (2020). Impact of land use changes and habitat fragmentation on the eco-epidemiology of tick-borne diseases. *Journal of Medical Entomology, 58*(4), 1546–1564.

Eisen, L., & Stafford, K. C., III. (2021). Barriers to effective tick management and tick-bite prevention in the United States (Acari: Ixodidae). *Journal of Medical Entomology, 58*(4), 1588–1600.

Eisen, R. J., Eisen, L., Graham, C. B., Hojgaard, A., Mead, P. S., Kersch, G., Karpathy, S., Paddock, C. D., Savage, H., Herwaldt, B., & Bradbury, R. (2019). Surveillance for Ixodes scapularis and pathogens found in this tick species in the United States. *Centers for Disease Control and Prevention, 1*, 34.

Esteve-Gassent, M. D., Castro-Arellano, I., Feria-Arroyo, T. P., Patino, R., Li, A. Y., Medina, R. F., de León, A. A. P., & Rodríguez-Vivas, R. I. (2016). Translating ecology, physiology, biochemistry, and population genetics research to meet the challenge of tick and tick-borne diseases in North America. *Archives of Insect Biochemistry and Physiology, 92*(1), 38–64.

Estrada-Peña, A., & de la Fuente, J. (2014). The ecology of ticks and epidemiology of tick-borne viral diseases. *Antiviral Research, 108*, 104–128.

Estrada-Peña, A., Martinez, J. M., Sanchez Acedo, C., Quilez, J., & Del Cacho, E. (2004). Phenology of the tick, Ixodes ricinus, in its southern distribution range (central Spain). *Medical and Veterinary Entomology, 18*(4), 387–397.

Estrada-Peña, A., Gray, J. S., Kahl, O., Lane, R. S., & Nijhof, A. M. (2013). Research on the ecology of ticks and tick-borne pathogens—methodological principles and caveats. *Frontiers in Cellular and Infection Microbiology, 3*, 29.

Estrada-Pena, A., Zatansever, Z., Gargili, A., Aktas, M., Uzun, R., Ergonul, O., & Jongejan, F. (2007). Modeling the spatial distribution of Crimean-Congo hemorrhagic fever outbreaks in Turkey. *Vector-Borne and Zoonotic Diseases, 7*(4), 667–678.

Ewing, D., Purse, B., Cobbold, C., & White, S. (2021). A novel approach for predicting risk of vector-borne disease establishment in marginal temperate environments under climate change: West Nile virus in the UK. *Journal of the Royal Society Interface, 18*(179), 20210049.

Gilbert, L. (2009). Altitudinal patterns of tick and host abundance: A potential role for climate change in regulating tick-borne diseases? *Oecologia, 162*(1), 217–225.

Gilbert, L. (2021). The impacts of climate change on ticks and tick-borne disease risk. *Annual Review of Entomology, 66*(1), 373–388.

Gilbert, L., Maffey, G. L., Ramsay, S. L., & Hester, A. J. (2012). The effect of deer management on the abundance of *Ixodes ricinus* in Scotland. *Ecological Applications, 22*(2), 658–667.

Gilbert, L., Aungier, J., & Tomkins, J. L. (2014). Climate of origin affects tick (*Ixodes ricinus*) host-seeking behavior in response to temperature: Implications for resilience to climate change? *Ecology and Evolution, 4*(7), 1186–1198.

Giles, J. R., Peterson, A. T., Busch, J. D., Olafson, P. U., Scoles, G. A., Davey, R. B., Pound, J. M., Kammlah, D. M., Lohmeyer, K. H., & Wagner, D. M. (2014). Invasive potential of cattle fever ticks in the southern United States. *Parasites & Vectors, 7*, 189.

Ginsberg, H. S., Rulison, E. L., Azevedo, A., Pang, G. C., Kuczaj, I. M., Tsao, J. I., & LeBrun, R. A. (2014). Comparison of survival patterns of northern and southern genotypes of the North American tick *Ixodes scapularis* (Acari: Ixodidae) under northern and southern conditions. *Parasites & Vectors, 7*, 394.

Githaka, N., Kanduma, E., & Bishop, R. (2021). Role of climate and other factors in determining the dynamics of tick and tick-transmitted pathogen populations and distribution in western, central and eastern Africa. In *Climate, ticks and disease.* Wallingford, UK: CABI.

Guillot, C., Badcock, J., Clow, K., Cram, J., Dergousoff, S., Dibernardo, A., Evason, M., Fraser, E., Galanis, E., Gasmi, S., & German, G. J. (2020). Sentinel surveillance of Lyme disease risk in Canada, 2019: Results from the first year of the Canadian Lyme Sentinel Network (CaLSeN). *Canada Communicable Disease Report, 46*(10), 354–361.

Haque, S., Mengersen, K., Barr, I., Wang, L., Yang, W., Vardoulakis, S., Bambrick, H., & Hu, W. (2024). Towards development of functional climate-driven early warning systems for climate-sensitive infectious disease: statistical models and recommendations. *Environmental Research*, 118568.

Hartemink, N., & Takken, W. (2016). Trends in tick population dynamics and pathogen transmission in emerging tick-borne pathogens in Europe: An introduction. *Experimental and Applied Acarology, 68*(3), 269–278.

Hoen, A. G., Margos, G., Bent, S. J., Diuk-Wasser, M. A., Barbour, A., Kurtenbach, K., & Fish, D. (2009). Phylogeography of *Borrelia burgdorferi* in the eastern United States reflects multiple independent Lyme disease emergence events. *Proceedings of the National Academy of Sciences, 106*(35), 15013–15018.

Hvidsten, D., Frafjord, K., Gray, J. S., Henningsson, A. J., Jenkins, A., Kristiansen, B. E., Lager, M., Rognerud, B., Slåtsve, A. M., Stordal, F., & Stuen, S. (2020). The distribution limit of the common tick, *Ixodes ricinus*, and some associated pathogens in north-western Europe. *Ticks and Tick-Borne Diseases, 11*(4), Article 101388.

Igreja, R. P. (2017). Infectious diseases and extreme sports. *Extreme Sports Medicine*, 35–40.

Jabbar, A., Abbas, T., Sandhu, Z. U. D., Saddiqi, H. A., Qamar, M. F., & Gasser, R. B. (2015). Tick-borne diseases of bovines in Pakistan: Major scope for future research and improved control. *Parasites and Vectors, 8*, 283.

Jaenson, T. G. T., Hjertqvist, M., Bergström, T., & Lundkvist, Å. (2012a). Why is tick-borne encephalitis increasing? A review of the key factors causing the increasing incidence of human TBE in Sweden. *Parasites and Vectors, 5*(1), 184.

Jaenson, T. G. T., Jaenson, D. G. E., Eisen, L., Petersson, E., & Lindgren, E. (2012b). Changes in the geographical distribution and abundance of the tick *Ixodes ricinus* during the past 30 years in Sweden. *Parasites and Vectors, 5*, 8.

James, M. C., Bowman, A. S., Forbes, K. J., Lewis, F., McLeod, J. E., & Gilbert, L. (2013). Environmental determinants of *Ixodes ricinus* ticks and the incidence of *Borrelia burgdorferi* sensu lato, the agent of Lyme borreliosis Scotland. *Parasitology, 140*(2), 237–246.

Joshi, M. M., Gregory, J. M., Webb, M. J., Sexton, D. M., & Johns, T. C. (2008). Mechanisms for the land/sea warming contrast exhibited by simulations of climate change. *Climate Dynamics, 30*, 455–465.

Katswara, T., & Mukaratirwa, S. (2021). Knowledge, attitudes and practices on African tick bite fever of rural livestock communities living in a livestock-wildlife interface area in the Eastern Cape Province of South Africa. *BMC Infectious Diseases, 21*(1), 497.

Klingelhöfer, D., Braun, M., Brüggmann, D., & Groneberg, D. A. (2022). Ticks in medical and parasitological research: Globally emerging risks require appropriate scientific awareness and action. *Travel Medicine and Infectious Disease, 50*, Article 102468.

Kopsco, H., Smith, R., & Halsey, S. (2022). A scoping review of species distribution modeling methods for tick vectors. *Frontiers in Ecology and Evolution, 10*, Article 893016.

Korenberg, E. I. (2009). Recent epidemiology of tick-borne encephalitis: An effect of climate change? *Advances in Virus Research, 74*, 123–144.

Lee, J., & Chung, S. (2022). The threat of climate change on tick-borne infections: Rising trend of infections and geographical distribution of climate risk factors associated with ticks. *The Journal of Infectious Diseases*. Advance online publication.

Lee, J. S., & Chung, S. Y. (2023). The threat of climate change on tick-borne infections: Rising trend of infections and geographic distribution of climate risk factors associated with ticks. *The Journal of Infectious Diseases, 227*(2), 295–303.

Léger, E., Vourc'h, G., Vial, L., Chevillon, C., & McCoy, K. D. (2013). Changing distributions of ticks: Causes and consequences. *Experimental and Applied Acarology, 59*(1), 219–244.

Leibovici, D., Bylund, H., Björkman, C., Tokarevich, N., Thierfelder, T., Evengård, B., & Quegan, S. (2021). Associating land cover changes with patterns of incidences of climate-sensitive infections: An example on tick-borne diseases in the Nordic area. *International Journal of Environmental Research and Public Health, 18*(20), 10963.

Li, S., Gilbert, L., Harrison, P. A., & Rounsevell, M. D. (2016). Modelling the seasonality of Lyme disease risk and the potential impacts of a warming climate within the heterogeneous landscapes of Scotland. *Journal of the Royal Society Interface, 13*(116), 20160140.

Lindgren, E., Tälleklint, L., & Polfeldt, T. (2000). Impact of climatic change on the northern latitude limit and population density of the disease-transmitting European tick *Ixodes ricinus*. *Environmental Health Perspectives, 108*(2), 119–123.

Lindquist, E. J., D'Annunzio, R., Gerrand, A., MacDicken, K., Achard, F., Beuchle, R., & Stibig, H. J. (2012). *Global forest land-use change 1990–2005*.

Lukan, M., Bullová, E., & Petko, B. (2010). Climate warming and tick-borne encephalitis Slovakia. *Emerging Infectious Diseases, 16*(3), 524–526.

Ma, R., Li, C., Tian, H., Zhang, Y., Feng, X., Li, J., & Hu, W. (2023). The current distribution of tick species in Inner Mongolia and inferring potential suitability areas for dominant tick species based on the MaxEnt model. *Parasites and Vectors, 16*(1), 286.

MacDonald, A. J., O'Neill, C., Yoshimizu, M. H., Padgett, K. A., & Larsen, A. E. (2019). Tracking seasonal activity of the western blacklegged tick across California. *Journal of Applied Ecology, 56*(11), 2562–2573.

Maltezou, H. C., Andonova, L., Andraghetti, R., Bouloy, M., Ergonul, O., Jongejan, F., & Zeller, H. (2010). *Crimean-Congo hemorrhagic fever in Europe: current situation calls for preparedness*.

Marques, R., Krüger, R. F., Peterson, A. T., de Melo, L. F., Vicenzi, N., & Jiménez-García, D. (2020). Climate change implications for the distribution of the babesiosis and anaplasmosis tick vector, *Rhipicephalus (Boophilus) microplus*. *Veterinary Research, 51*(1), 94.

Medlock, J. M., Hansford, K. M., Bormane, A., Derdakova, M., Estrada-Peña, A., George, J. C., Golovljova, I., Jaenson, T. G., Jensen, J. K., Jensen, P. M., & Van Bortel, W. (2013). Driving forces for changes in geographical distribution of Ixodes ricinus ticks in Europe. *Parasites and Vectors, 6*, 1–11.

Minigan, J. N., Hager, H. A., Peregrine, A. S., & Newman, J. A. (2018). Current and potential future distribution of the American dog tick (*Dermacentor variabilis*, Say) in North America. *Ticks and Tick-Borne Diseases, 9*(2), 354–362.

Muehlenbein, M. P. (2019). *Ecotourism*. Elsevier.

Munz, S. M., McKenna-Buchanan, T., & Wright, A. M. (Eds.). (2024). *The Routledge handbook of public speaking research and theory*. Routledge.

Mysterud, A., Jore, S., Østerås, O., & Viljugrein, H. (2017). Emergence of tick-borne diseases at northern latitudes in Europe: A comparative approach. *Scientific Reports, 7*(1), 16316.

Mysterud, A., Stigum, V. M., Seland, I. V., Herland, A., Easterday, W. R., Jore, S., Østerås, O., & Viljugrein, H. (2018). Tick abundance, pathogen prevalence, and disease incidence in two contrasting regions at the northern distribution range of Europe. *Parasites and Vectors, 11*, 1–11.

Namgyal, J., Tenzin, T., Checkley, S., Lysyk, T. J., Rinchen, S., Gurung, R. B., Dorjee, S., Couloigner, I., & Cork, S. C. (2021). A knowledge, attitudes, and practices study on ticks and tick-borne diseases in cattle among farmers in a selected area of eastern Bhutan. *PLoS ONE, 16*(2), Article e0247302.

Nchu, F., Nyangiwe, N., Muhanguzi, D., Nzalawahe, J., Nagagi, Y. P., Msalya, G., Joseph, N. A., Kimaro, E. G., Mollel, M., Temba, V., & Harouna, D. V. (2020). Development of a practical framework for sustainable surveillance and control of ticks and tick-borne diseases in Africa. *Veterinary World, 13*(9), 1910–1921.

Niesobecki, S., Hansen, A., Rutz, H., Mehta, S., Feldman, K., Meek, J., Niccolai, L., Hook, S., & Hinckley, A. (2019). Knowledge, attitudes, and behaviors regarding tick-borne disease prevention in endemic areas. *Ticks and Tick-Borne Diseases, 10*(6), Article 101264.

Nolzen, H., Brugger, K., Reichold, A., Brock, J., Lange, M., & Thulke, H. (2022). Model-based extrapolation of ecological systems under future climate scenarios: The example of *Ixodes ricinus* ticks. *PLoS ONE, 17*(5), Article e0267196.

Nuttall, P. A. (2022). Climate change impacts on ticks and tick-borne infections. *Biologia, 77*(6), 1503–1512.

Ogden, N. H., & Lindsay, L. R. (2016). Effects of climate and climate change on vectors and vector-borne diseases: Ticks are different. *Trends in Parasitology, 32*(8), 646–656. https://doi.org/10.1016/j.pt.2016.04.015

Ogden, N., Beard, C., Ginsberg, H., & Tsao, J. (2020). Possible effects of climate change on ixodid ticks and the pathogens they transmit: Predictions and observations. *Journal of Medical Entomology, 58*(4), 1536–1545.

Ogden, N. H., Ben Beard, C., Ginsberg, H. S., & Tsao, J. I. (2021). Possible effects of climate change on ixodid ticks and the pathogens they transmit: Predictions and observations. *Journal of Medical Entomology, 58*(4), 1536–1545.

Oshiro, L. M., da Silva Rodrigues, V., Garcia, M. V., de Oliveira Souza Higa, L., Suzin, A., Barros, J. C., & Andreotti, R. (2021). Effect of low temperature and relative humidity on reproduction and survival of the tick Rhipicephalus microplus. *Experimental and Applied Acarology, 83*, 95–106.

Ostfeld, R. S., & Brunner, J. L. (2015). Climate change and *Ixodes* tick-borne diseases of humans. *Philosophical Transactions of the Royal Society b: Biological Sciences, 370*(1665), 20140051.

Ostfeld, R., Levi, T., Keesing, F., Oggenfuss, K., & Canham, C. (2018). Tick-borne disease risk in a forest food web. *Ecology, 99*(7), 1562–1573.

Parham, P. E., Waldock, J., Christophides, G. K., Hemming, D., Agusto, F., Evans, K. J., Fefferman, N., Gaff, H., Gumel, A., LaDeau, S., & Lenhart, S. (2015). Climate, environmental and socio-economic change: Weighing up the balance in vector-borne disease transmission. *Philosophical Transactions of the Royal Society b: Biological Sciences, 370*(1665), 20130551.

Perveen, N., Muzaffar, S. B., & Al-Deeb, M. A. (2021). Ticks and tick-borne diseases of livestock in the Middle East and North Africa: A review. *Insects, 12*(1), 83.

Pfäffle, M., Littwin, N., Muders, S. V., & Petney, T. N. (2013). The ecology of tick-borne diseases. *International Journal for Parasitology, 43*(12–13), 1059–1077.

Pongsiri, M. J., Roman, J., Ezenwa, V. O., Goldberg, T. L., Koren, H. S., Newbold, S. C., Ostfeld, R. S., Pattanayak, S. K., & Salkeld, D. J. (2009). Biodiversity loss affects global disease ecology. *BioScience, 59*(11), 945–954.

Raghavan, R. K., Barker, S. C., Cobos, M. E., Barker, D., Teo, E. J. M., Foley, D. H., Nakao, R., Lawrence, K., Heath, A. C. G., & Peterson, A. T. (2019). Potential spatial distribution of the newly introduced long-horned tick, *Haemaphysalis longicornis*, in North America. *Scientific Reports, 9*, 498.

Randolph, S. E. (2008). Tick-borne disease systems. *Revue Scientifique Et Technique (International Office of Epizootics), 27*(2), 499–512.

Robinson, S., Neitzel, D., Moen, R., Craft, M., Hamilton, K., Johnson, L., Mulla, D., Munderloh, U., Redig, P., Smith, K., Turner, C., Umber, J., & Pelican, K. (2015). Disease risk in a dynamic environment: The spread of tick-borne pathogens in Minnesota, USA. *EcoHealth, 12*(1), 152–163.

Rocha, S. C., Velásquez, C. V., Aquib, A., Al-Nazal, A., & Parveen, N. (2022). Transmission cycle of tick-borne infections and co-infections, animal models and diseases. *Pathogens, 11*(11), 1309.

Rowan, S., Mohseni, N., Chang, M., Burger, H., Peters, M., & Mir, S. (2023). From tick to test: A comprehensive review of tick-borne disease diagnostics and surveillance methods in the United States. *Life, 13*(10), 2048.

Rowland, D., Ickowitz, A. M. Y., Powell, B., Nasi, R., & Sunderland, T. (2017). Forest foods and healthy diets: Quantifying the contributions. *Environmental Conservation, 44*(2), 102–114.

Sagurova, I., Ludwig, A., Ogden, N. H., Pelcat, Y., Dueymes, G., & Gachon, P. (2019). Predicted northward expansion of the geographic range of the tick vector *Amblyomma americanum* in North America under future climate conditions. *Environmental Health Perspectives, 127*(10), Article 107014.

Sakamoto, J. M. (2018). Progress, challenges, and the role of public engagement to improve tick-borne disease literacy. *Current Opinion in Insect Science, 28*, 81–89.

Semenza, J. C., & Paz, S. (2021). Climate change and infectious disease in Europe: Impact, projection and adaptation. *The Lancet Regional Health–Europe, 9*.

Simuunza, M., Weir, W., Courcier, E., Tait, A., & Shiels, B. (2011). Epidemiological analysis of tick-borne diseases in Zambia. *Veterinary Parasitology, 175*(3–4), 331–342.

Slunge, D., Sterner, T., & Adamowicz, W. (2019). Valuation when baselines are changing: Tick-borne disease risk and recreational choice. *Resource and Energy Economics, 58*, Article 101119.

Smith, G., Wileyto, E. P., Hopkins, R. B., Cherry, B. R., & Maher, J. P. (2001). *Risk factors for lyme disease in Chester County*. Public Health Reports.

Springer, Y. P., Eisen, L., Beati, L., James, A. M., & Eisen, R. J. (2014). Spatial distribution of counties in the continental United States with records of occurrence of *Amblyomma americanum* (Ixodida: Ixodidae). *Journal of Medical Entomology, 51*(2), 342–351.

Stachurski, F., & Vial, L. (2018). Installation de la tique *Hyalomma marginatum*, vectrice du virus de la fièvre hémorragique de Crimée-Congo, en France continentale. *Bulletin Épidémiologique, Santé Animale – Alimentation*, (84), 5–9.

Standaert, S. M., Dawson, J. E., Schaffner, W., Childs, J. E., Biggie, K. L., Singleton, J., Jr., & Hutcheson, R. H. (1995). Ehrlichiosis in a golf-oriented retirement community. *New England Journal of Medicine, 333*(7), 420–425.

Šumilo, D., Bormane, A., Asokliene, L., Vasilenko, V., Golovljova, I., Avšič-Županc, T., Hubálek, Z., & Randolph, S. E. (2008). Socio-economic factors in the differential upsurge of tick-borne encephalitis in Central and Eastern Europe. *Reviews in Medical Virology, 18*(2), 81–95.

Tardy, O., Bouchard, C., Chamberland, E., Fortin, A., Lamirande, P., Ogden, N., & Leighton, P. (2021). Mechanistic movement models reveal ecological drivers of tick-borne pathogen spread. *Journal of the Royal Society Interface, 18*(179), 20210134.

Valente, S. L., Wemple, D., Ramos, S., Cashman, S. B., & Savageau, J. A. (2015). Preventive behaviors and knowledge of tick-borne illnesses: Results of a survey from an endemic area. *Journal of Public Health Management and Practice, 21*(3), E16–E23.

Volpato, G., Fontefrancesco, M. F., Gruppuso, P., Zocchi, D. M., & Pieroni, A. (2020). Baby pangolins on my plate: Possible lessons to learn from the COVID-19 pandemic. *Journal of Ethnobiology and Ethnomedicine, 16*, 1–12.

Voyiatzaki, C., Papailia, S. I., Venetikou, M. S., Pouris, J., Tsoumani, M. E., & Papageorgiou, E. G. (2022). Climate changes exacerbate the spread of *Ixodes ricinus* and the occurrence of Lyme borreliosis and tick-borne encephalitis in Europe—How climate models are used as a risk assessment approach for tick-borne diseases. *International Journal of Environmental Research and Public Health, 19*(11), 6516.

Wang, H. H., Grant, W. E., Teel, P. D., & Hamer, S. A. (2016). Tick-borne infectious agents in nature: Simulated effects of changes in host density on spatial-temporal prevalence of infected ticks. *Ecological Modelling, 323*, 77–86.

Wimberly, M. C., Yabsley, M. J., Baer, A. D., Dugan, V. G., & Davidson, W. R. (2008). Spatial heterogeneity of climate and land-cover constraints on distributions of tick-borne pathogens. *Global Ecology and Biogeography, 17*(2), 189–202.

Wood, C. L., & Lafferty, K. D. (2013). Biodiversity and disease: A synthesis of ecological perspectives on Lyme disease transmission. *Trends in Ecology and Evolution, 28*(4), 239–247.

Wu, T., Perrings, C., Kinzig, A., Collins, J. P., Minteer, B. A., & Daszak, P. (2017). Economic growth, urbanization, globalization, and the risks of emerging infectious diseases in China: A review. *Ambio, 46*, 18–29.

Yamaji, K., Aonuma, H., & Kanuka, H. (2018). Distribution of tick-borne diseases in Japan: Past patterns and implications for the future. *Journal of Infection and Chemotherapy, 24*(7), 499–504.

Zanet, S., Ferroglio, E., Battisti, E., & Tizzani, P. (2020). Ecological niche modelling of *Babesia* spp. infection in wildlife experimentally evaluated in northern Italy with reference to questing *Ixodes ricinus* ticks. *International Journal for Parasitology: Parasites and Wildlife, 12*, 1–7.

Zhang, X., Sun, B., & Lou, Y. (2021). Dynamics of a periodic tick-borne disease model with co-feeding and multiple patches. *Journal of Mathematical Biology, 82*(4), 41.

Zhang, L., Yang, L., Wang, L., Gu, L., Li, H., & Malkhazova, S. (2022). Natural and socioeconomic conditions influence tick-borne encephalitis cases in Russia. *Polish Journal of Environmental Studies, 32*(1), 427–437.

Zhao, G., Wang, Y., Fan, Z., Ji, Y., Liu, M., Zhang, W., Li, X., Zhou, S., Li, H., Liang, S., Liu, W., Yang, Y., & Fang, L. (2021). Mapping ticks and tick-borne pathogens in China. *Nature Communications, 12*(1), 1075.

Chapter 4
Emerging Zoonotic Threats: Climate, Ticks, and Disease Transmission

Zahid Fareed, Maxwell Nwachukwu Opara, Sahar Younis, Mudassir Ahmad, Nosheen Munir, and Iqra Munir

Abstract Most of the disease-causing agents in humans have originated from animals. These emerging zoonotic threats are directly related to the climate change. Deforestation, intensive agriculture, and excessive grazing are responsible for climate change. The dynamics of host and tick populations and the ensuing dynamics of disease transmission constitute an extremely complicated system influenced by both abiotic and biotic variables. Because tick lifecycles are climate-sensitive, tick-borne diseases are also climate-sensitive. Abiotic factors frequently have an immediate impact on tick numbers and distributions. Zoonotic diseases associated with ticks include babesiosis, anaplasmosis, Lyme disease, ehrlichiosis, Rocky Mountain spotted fever, tularemia, and Q fever. The majority of them are affected by climate change. Temperature change is a key factor involved in the distribution of ticks causing these diseases. As a result, some of the countries which were free of these diseases are now not free of these diseases. There is a strict need to adapt to mitigation strategies such as host-targeted interventions. Moreover, surveillance and monitoring strategies involve an interdisciplinary approach to control the zoonotic transmission of diseases. Multidisciplinary cooperation among ecologists, climate

---

Z. Fareed
Veterinary Research Institute, Lahore, Pakistan

M. N. Opara
Department of Parasitology and Entomology, Faculty of Veterinary Medicine, University of Abuja, Abuja, Nigeria

S. Younis (✉)
Department of Parasitology, Faculty of Veterinary Science, University of Agriculture, Faisalabad, Pakistan
e-mail: saharyounis14@gmail.com

M. Ahmad
School of Science and Technology, Nottingham Trent University, England, UK

N. Munir
Anesthesiology, CMH Lahore Medical College and Institute of Dentistry, Lahore, Pakistan

I. Munir
Faculty of Allied Health Sciences, University Institute of Medical Laboratory Technology, The University of Lahore, Lahore, Pakistan

H. M. Rizwan and M. S. Sajid (eds.), *Ticks in a Changing Climate*, Biometeorology 7,
https://doi.org/10.1007/978-3-032-12259-9_4

change experts, city planners, government representatives, parasitologists, and social scientists will help to control the zoonotic transmission of diseases.

Keywords Tick ecology · Environmental determinants · Zoonotic diseases · Disease spread · Surveillance · Mitigation

4.1 Introduction to Emerging Zoonotic Threats

The complex relationship of human, animal, and environmental health has been highlighted by recurring epidemics of emerging and re-emerging zoonosis. An integrated, multidisciplinary, cross-sectoral approach is necessary to react early to emerging zoonotic pathogens. Emerging diseases are a bigger hazard as our globe gets more interconnected, necessitating local, regional, and international cooperation. One Health is a multisectoral, and collaborative strategy pushed to address these complex health problems more effectively (Kelly et al., 2020). For global health, emerging zoonotic diseases are a great challenge (Bird & Mazet, 2018). The World Health Organization reports that 60% of disease-causing agents in humans originate from animals. Over 70% of infectious agents responsible for new and reemerging animal diseases can cross species barriers. This creates conditions that support genetic exchange and the emergence of new, highly pathogenic strains. Animals often serve as the primary host reservoirs for these agents (Rong & Liu, 2023). As a result, any public health preventive strategy must address the health issue holistically, considering the interplay between the environment, animals, and humans (El Amri et al., 2020).

There is competition for increased food and supply production as a result of population growth and industrialization. Consequently, there is an increase in the human occupation of forest regions, contact with animals, parasites and vectors that accompany it, wildlife trafficking and consumption, water source degradation, and garbage accumulation (Tajudeen et al., 2022). At the same time, the production of animals and crops for human use has increased frequently in an undisciplined manner, which has destroyed forests that are vital to the planet's ecological and climatic equilibrium (Aju et al., 2015).

The transmission and dissemination of zoonotic diseases are indirectly influenced by changes in environmental conditions brought about by human activity and climate change (Estrada-Peña et al., 2014). Arthropod vectors frequently play a major role in the spread of these disease. Since vectors are ectotherms, environmental temperature variations have a significant impact on them (Chikezie et al., 2024). These facts are related to the recent reports of an increase in the occurrence of vector-borne zoonosis (Yadav & Upadhyay, 2023). Additionally, the vector density is frequently supported by other environmental elements, particularly major weather events like storms and floods. Landscape changes are caused by deforestation, excessive grazing, drying out wetlands, and intensive agriculture and industry (Xiong et al., 2023). The ecology may change as a result of modifications to local precipitation and temperature patterns brought about by changes in plant diversity and water management. This impacts animal and plant ecosystems, which in turn impacts the spread of vectors and diseases (Jánová, 2019).

Ticks cause the majority of vector-borne diseases (VBDs) cases and spread the widest variety of arthropod-borne infections. New tick-borne pathogens have been identified, and the number of documented cases of identifiable TBDs has been continuously rising in recent decades (Eisen et al., 2017). The past few decades have been greatly impacted by emerging diseases. The epidemiological landscapes of Africa, Asia, and South America have changed as a result of the introduction and reemergence of zoonotic and TBDs (Tazerji et al., 2022). Due to migration and travel, the effect of these diseases and the emergence of local transmission in places that were previously non-endemic have been made evident in recent years (Rodriguez-Morales et al., 2021). The difficulty in managing tick populations and identifying and treating infections brought on by the pathogens they spread hinders measures to control these new diseases (Madison-Antenucci et al., 2020).

4.2 Impact of Climate Change on Tick Ecology

Every organism interacting with the outside world is part of its climate envelope, establishing the range of temperatures in which its populations can thrive. Because some ticks have a narrow climatic tolerance and require specific environmental conditions, their geographic distribution is relatively limited. As a result, climate change is more likely to impact their population size and range compared to ticks with broader ecological adaptability. On the other hand, some tick species are widely distributed geographically (Gilbert, 2021). Since vectors are ectothermic, changes in the climate and weather (such as temperature, precipitation, and humidity) can have an impact on their ability to reproduce, survive, and spread diseases (de Souza & Weaver, 2024). Abiotic variables like humidity and temperature also affect the dynamics of tick populations and the spread of pathogens. These elements influence tick survival and questing behavior in addition to determining the type of vegetation and the population of related hosts. For example, this could impact when the transmission season begins or how closely the tick life stages synchronize their activities, which in turn determines when nymphs and larvae can co-feed (Sirotkin & Korenberg, 2019).

Together, the dynamics of tick and host populations and the ensuing dynamics of disease transmission constitute an extremely complicated system that is influenced by both biotic and abiotic variables (Hartemink & Takken, 2016). Determining whether climate change is having an impact on ticks and TBDs is even more uncertain. Environmental factors have a significant effect on tick physiology and behavior since they are poikilothermic, somewhat long-lived (some species take more than five years to go through one generation), and typically spend the majority of their life cycle off-host (Apanaskevich et al., 2014). The majority of tick species will consume any vertebrate that is available to them, although they frequently exhibit preferences related to whether they are exophilic or endophilic. Through altered climatic circumstances and host availability brought on by climate change, ticks, and TBDs are predicted to be impacted (Nuttall, 2022).

Because tick lifecycles are climate-sensitive, ticks and TBDs are climate-sensitive by nature. Individual tick survival, tick development time and tick host-seeking activity are important climatic and weather sensitivities. Because of these sensitivities, a warming climate may boost tick survivability, shorten tick life cycles, and extend tick activity seasons in some areas (Burtis et al., 2019). Changes in tick abundance brought on by indirect impacts of climate change on host communities may make it easier for tick-borne infections to spread. Climate change is expected to bring with it high temperatures and extreme weather events (hot, cold, and flooding), which in some places may make it harder for ticks to survive and spread diseases (Ogden et al., 2021).

Ticks must react to both biotic (such as humid leaves, dense vegetation, dense shade, and host interaction) and abiotic (such as temperature, humidity, and soil moisture) local variables since they spend a large portion of their life cycle exposed to the environment (Fig. 4.1). Compared to biotic factors, abiotic factors frequently have an immediate impact on tick numbers and distributions (Conte et al., 2007). Given that recent decades have seen noticeably greater winter temperatures in the US, Canada, and Mexico due to climate change. North America is thought to be the area that will be most impacted by global warming (Frelich & Reich, 2010). Many tick species' potential distributions will likely expand and move as a result of rising temperatures brought on by climate change, typically moving towards greater latitudes and elevations (Alkishe et al., 2021). Cooler temperatures also lengthen the time it takes for ticks to develop and generate, which increases the fraction of tick deaths before reproduction (Deshpande et al., 2023). Additionally, host-seeking activity is significantly influenced by rainfall and moisture availability. Exposure to low humidity significantly increases tick mortality and suppresses host-seeking behavior (Ginsberg et al., 2020).

Climate change also facilitates the co-feeding transmission (Nah et al., 2020). Rising temperatures will have an impact on ticks' hosts in addition to their direct targets, which is crucial when those hosts act as reservoir hosts for diseases spread by ticks (Mysterud et al., 2017). The distribution of the pathogen, *Babesia (B.) divergens*, is closely linked to that of cattle. There is some evidence, although indirect, of bovine babesiosis in more northern areas than in past decades, which is not surprising as infections seem to be caused by locally established tick populations, and that tick populations are moving northward (Gray & Ogden, 2021). In February 2019, temperatures in the south of England exceeded the average for the time of year by over 10 °C, resulting in exceptionally early tick activity and an outbreak of bovine babesiosis involving 20 cattle. This was a more localized but still distinct climate effect on the disease (Johnson et al., 2020). Ecologists, epidemiologists, and public health experts are challenged by the spread of TBDs to figure out the mechanisms behind their emergence (Wimms et al., 2023). Impact of climate change on tick ecology has been summarized in Table 4.1.

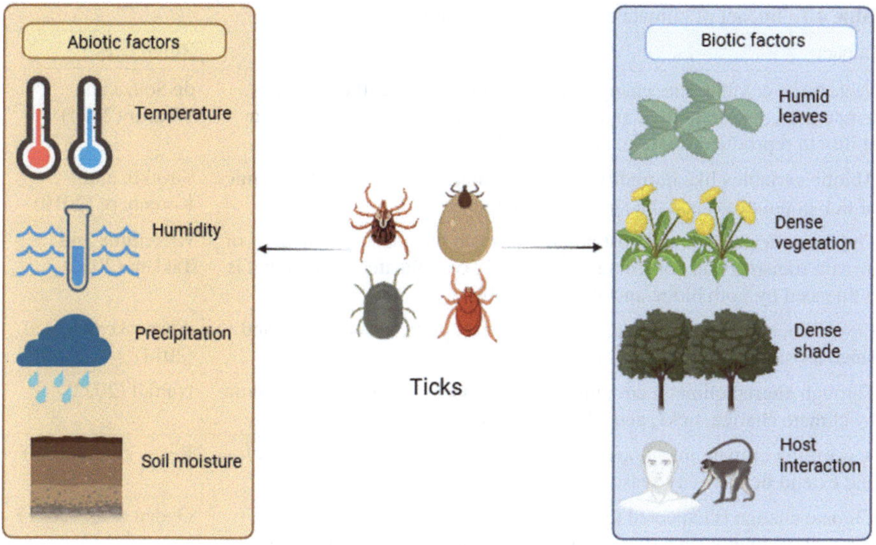

Fig. 4.1 Biotic and abiotic factors affecting distribution of ticks and tick-borne diseases

4.3 Environmental Determinants of Disease Spread

Tick vectors and the diseases they spread have a relationship that is especially vulnerable to changes in the world caused by human activity (Swei et al., 2020). The diverse, geographically, and temporally variable factors that contribute to the spectacular onset and recurrence of TBDs can have a compounding influence on disease biology and epidemiology (Ortiz et al., 2021). Socioeconomic changes, such as modern transportation and trade, unexpected and uncontrolled urbanization, and demographic changes, such as worldwide population movements and expansion are currently recognized primary drivers of TBD origin and recurrence (Palaniyandi, 2021). Illicit activities (such as illegal logging and cattle ranching, illegal drug trafficking), greater misuse of natural resources (such as changes in land use, forest degradation, biodiversity loss), farming methods, changes in host susceptibility, and pathogen adaptation (such as increased human and animal mobility, pathogen genetic variability), deterioration of public health infrastructure (such as ineffective vector control, disease surveillance, and prevention programs), and climate change (such as changes in regional temperature and rainfall patterns) alter vector dynamics (Dantas-Torres, 2015; Devine et al., 2021; Sheela et al., 2017). Climate change, forest fragmentation, and changes in the landscape can all raise tick populations and the risks associated with them. Humans, animal hosts, and vectors are all components of the complicated system that makes up the disease transmission process of TBDs (Elmieh, 2022).

The pace of forest destruction has significantly grown during the past few decades. Every year, millions of acres of forests disappear due to agriculture and timber

Table 4.1 Impact of climate change on tick ecology

Impact	References
Vectors are ectothermic, changes in the climate and weather (such as temperature, precipitation, and humidity) can have an impact on their ability to reproduce, survive, and spread diseases	de Souza and Weaver (2024)
Abiotic variables like humidity and temperature also affect the dynamics of tick populations and the spread of pathogens	Sirotkin and Korenberg (2019)
The dynamics of tick and host populations and the ensuing dynamics of disease transmission constitute an extremely complicated system that is influenced by both biotic and abiotic variables	Hartemink and Takken (2016)
Environmental factors have a significant effect on tick physiology and behavior since they are poikilothermic	Apanaskevich et al. (2014)
Through altered climatic circumstances and host availability brought on by climate change, ticks, and TBDs are predicted to be impacted	Nuttall (2022)
A warming climate may boost tick survivability, shorten tick life cycles, and extend tick activity seasons in some areas	Burtis et al. (2019)
Climate change is expected to bring with it high temperatures and extreme weather events (hot, cold, and flooding), which in some places may make it harder for ticks to survive and spread diseases	Ogden et al. (2021)
Compared to biotic factors, abiotic factors frequently have an immediate impact on tick numbers and distributions	Conte et al. (2007)
Given that recent decades have seen noticeably greater winter temperatures in the US, Canada, and Mexico due to climate change. North America is thought to be the area that will be most impacted by global warming	Frelich and Reich (2010)
Many tick species' potential distributions will likely expand and move as a result of rising temperatures brought on by climate change, typically moving towards greater latitudes and elevations	Alkishe et al. (2021)
Host-seeking activity is significantly influenced by rainfall and moisture availability	Ginsberg et al. (2020)
Climate change facilitates the co-feeding transmission	Nah et al. (2020)
Rising temperatures will have an impact on ticks' hosts in addition to their direct targets, which is crucial when those hosts act as reservoir hosts for diseases spread by ticks	Mysterud et al. (2017)
In February 2019, temperatures in the south of England exceeded by over 10 °C, resulting in exceptionally early tick activity and an outbreak of bovine babesiosis involving 20 cattle	Johnson et al. (2020)

harvesting, which has a major impact on climate change (McMahon et al., 2010). Tick vectors tend to benefit from changing microclimates with variations in temperature, moisture, and sunlight as well as the introduction of artificial breeding sites after deforestation (Loiseau & Sehgal, 2022). The bulk of TBD cases in temperate regions of Europe, North America, and Asia are caused by climate change, which have become more common in recent decades. (Diuk-Wasser et al., 2021).

Numerous anthropogenic factors, including land use modification, climate change, the expansion of important hosts, and changes in human behavior that result in increased human contact with ticks, have been linked to increases in the incidence and geographic distribution trends of ticks, hosts, and pathogens (Sonenshine, 2018; Wikel, 2018). Habitat-host-vector-pathogen relationships can be altered by shifts in the structure and function of ecosystems. Furthermore, the forest's resources and habitat have an impact on the number and makeup of host communities, which tick vectors feed on and may use to spread diseases (Blache et al., 2024). Complex and unpredictable dangers of TBDs for people, domestic animals, and wildlife might result from forest dynamics (Gottdenker et al., 2014).

Ecotone zones that have a combination of pasture, forest, and human-dwelling regions are inhabited or even beneficial to a variety of rodents, opossums, and bats (Bermúdez et al., 2017). These species may play a significant role in the ecology of pathogenic microbe transmission. Furthermore, blood-sucking arthropods like ticks that can spread infections to people are linked to synanthropic hosts near anthropogenic regions (Guégan et al., 2020). The prevalence of TBDs and tick abundance may be impacted by the changes that urbanization brings about in natural ecosystem components. The prevalence of tick-borne pathogens has grown in many urban regions of the United States, most likely as a result of these changes (Noden et al., 2022). Much of Southeast Asia is undergoing fast urbanization, which is changing the landscape's structure and function as well as the frequency and intensity of interactions between humans, wildlife, and the environment (Blasdell et al., 2022). Anthropogenic changes may increase vector population breeding grounds, pesticides or environmental pollutants may decrease parasite survival, urban noise or light may interfere with host-parasite interactions, increased impermeable surface cover may result in parasite mortality, and heat-island effects may affect parasite development or activity (deVore et al., 2020). The rate of urbanization that has increased globally in recent decades has had a substantial impact on tick-host communities, tick infection rates and biodiversity levels. Urban and peri-urban areas are portrayed as favorable environments for human-tick encounters in Europe due to the recent rise in peri-urbanization and the human activities associated with this change (Borşan et al., 2020). Environmental drivers of TBDs have been summarized in Fig. 4.2.

4.4 Zoonotic Diseases Associated with Ticks

4.4.1 Babesiosis

Babesiosis was recognized as a zoonotic pathogen some 60 years ago (Vannier & Krause, 2020). *Babesia bovis* and *B. bigemina* are causative agents of babesiosis in cattle. In sheep and goats, *B. ovis* is a causative agent, while in canines, *B. canis* and *B. gibsoni* are the causative agents (Muzaffar et al., 2024). Ticks are responsible for the transmission of the disease (Bajer et al., 2022). Human babesiosis is

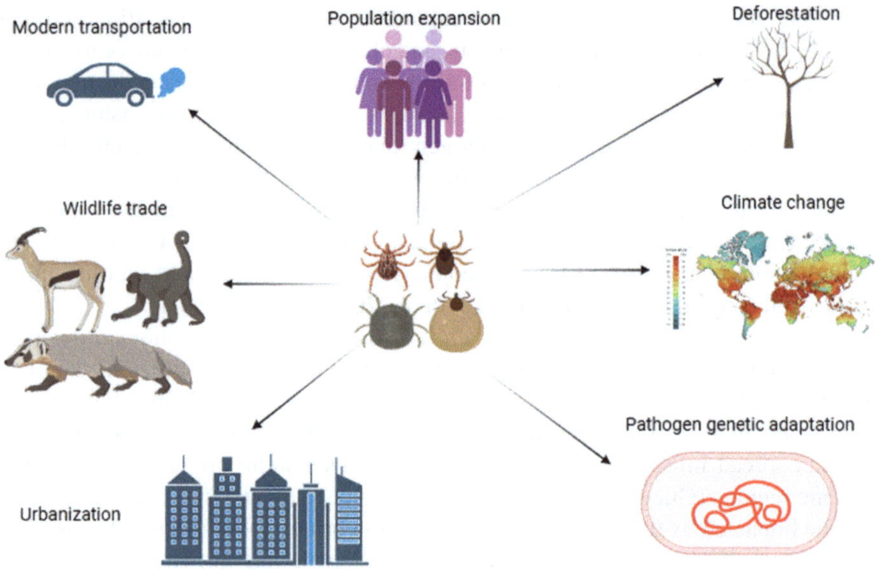

Fig. 4.2 Environmental drivers of ticks and tick-borne diseases

a newly discovered tick-borne zoonosis that is mostly spread by ticks, sporadically spreads through blood transfusions, and is rarely spread through organ donation or transplacental transmission (Waked & Krause, 2022). Despite being relatively uncommon in Europe, babesiosis is significant because most cases present as life-threatening infections, primarily in immunocompromised patients. The first case of human babesiosis was described in Europe in the late 1950s, and since then, more than 60 cases have been reported in Europe. Babesiosis is gaining attention as a global emerging zoonosis (Hildebrandt et al., 2021). Globally distributed and widespread in China and the United States, *B. microti* is the most common human infection. In China, endemic diseases are also caused by *B. venatorum* and *B. crassa*-like agents. Infections caused by *B. microti* have been on the rise worldwide in recent decades. Although the true number is believed to be substantially higher, over 2000 cases are reported annually in the United States (Kumar et al., 2021).

4.4.2 Lyme Disease

The most prevalent TBD in the US is Lyme disease, sometimes referred to as Lyme borreliosis (Bobe et al., 2021). Lyme disease *Borrelia* are invasive, persistent, tick-transmitted, obligatory parasitic bacterial pathogens that predominantly induce inflammation in humans and non-reservoir animals to cause disease. In humans, the organisms cause inflammatory reactions that lead to clinical signs and symptoms as they proliferate and disseminate locally (Coburn et al., 2021). Adults and children can contract Lyme disease from ticks carrying the *Borrelia (Bo.) burgdorferi* sensu lato complex genospecies. The disease, which is common in the Northern Hemisphere, is spreading around the world as a result of a number of environmental conditions and growing human encroachment into spirochete-hosting environments. *Ixodes ricinus* ticks are the vectors of *Bo. burgdorferi* sensu lato. *Bo. burgdorferi* is responsible for almost all infections in North America, while *Bo. afzelii* and *Bo. garinii* are most frequently linked to human disease in Europe (Radolf et al., 2021). The bacteria experience substantial alterations in gene expression during delivery from the infected tick, which leads to adaptability to the mammalian environment (Skotarczak, 2009).

4.4.3 Anaplasmosis

The agent that causes anaplasmosis belong to the anaplasmataceae family. *Anaplasma (A.) phagocytophilum* poses a huge threat to public health and causes significant financial losses, even if the majority of the species are not zoonotic (Peter, 2020). Granulocytic anaplasmosis in humans, pasture fever in cattle, and tick-borne fever in sheep are all caused by *A. phagocytophilum.* An estimated zoonotic danger is presented by the growing transboundary dissemination and prevalence of *A. phagocytophilum* in animals, livestock, and ticks (Bianchessi et al., 2023). Now *A. capra* has emerged as a tick-borne zoonotic pathogen (Altay et al., 2024).

4.4.4 Ehrlichiosis

Globally, several serious zoonotic TBDs, including ehrlichiosis, are becoming more prevalent (Adelabu et al., 2020). Over the past few decades, canine ehrlichiosis has drawn more attention from the veterinary and public health communities. Important zoonotic species have been found, including *Ehrlichia (E.) chaffeensis*, the etiologic agent of human monocytotropic ehrlichiosis (Marshet & Dessie, 2020), which is an emerging zoonotic TBD (Aziz et al., 2022). The majority of cases of the disease are probably asymptomatic, and symptomatic cases are easily missed or confused because they primarily present as a non-specific febrile illness. Infections with other *Ehrlichia* bacteria are uncommon (Gygax et al., 2024).

4.4.5 Rocky Mountain Spotted Fever

Rickettsia (R.) rickettsii is the causative agent of Rocky Mountain spotted fever, a serious zoonotic disease spread by ticks. Humans can contract the infection by being bitten by an infected hard tick belonging to the genera *Rhipicephalus, Dermacentor, Amblyomma,* and *Haemaphysalis* (Laukaitis & Macaluso, 2021). The disease continues to pose a hazard to public health, disproportionately impacting vulnerable groups like children, older people, migrants, and those living in poverty in several American nations (Álvarez-Hernández et al., 2024a). Since the 1920s, Rocky Mountain spotted fever has been a nationally reportable disease in the United States. As epidemiologic and diagnostic knowledge of the disease has changed throughout time, the definition of a case has changed as well. Reports of Rocky Mountain spotted fever are rising in the United States (Kjemtrup et al., 2022). In many Latin American countries, especially in northern Mexico, Rocky Mountain spotted fever, a severe and extremely deadly infectious disease has become a major public health concern for primarily vulnerable people. In contrast to many other tropical infectious diseases, Rocky Mountain spotted fever has historically received less attention, which has led to inadequate resource allocation and capability development for its prevention and control in endemic regions (Álvarez-Hernández et al., 2024b).

4.4.6 Tularemia

Humans, domestic animals, and wild animals are all susceptible to the recurrent bacterial disease tularemia. Although *Francisella tularensis*, the pathogen that causes the disease worldwide, however, North America and Asia are where cases are most frequently documented. Both sporadic and pandemic forms of the disease are possible (Pal et al., 2022). Ticks are among the vectors that carry the disease. Direct contact with infected animals can spread the sickness to humans (Yeni et al., 2021). The main vector of vector-borne tularemia is the tick. The clinical picture is determined by the mechanism of transmission and the bacterial subspecies involved. Flu-like symptoms are early indicators that can develop into various clinical manifestations of tularemia, which may or may not cause lymphadenopathy. Tick bites or contact with infected animals can result in the development of ulcero-glandular and glandular forms (Sharma et al., 2023).

4.4.7 Q Fever

In Pakistan and many other underdeveloped nations, Q fever is still a neglected zoonotic disease. Because the causative agent, *Coxiella burnetii*, is resistant to environmental conditions, both humans and animals are at long-term risk of infection. Since the infection is typically asymptomatic, it is rarely identified in animals until a herd experiences unfavorable pregnancy outcomes (Ullah et al., 2022). Since the pathogen was initially isolated from a *Dermacentor andersonii* tick, ticks have been considered potential vectors in addition to the primary route of transmission through inhalation of infected aerosols (Körner et al., 2021). *Haemaphysalis,*

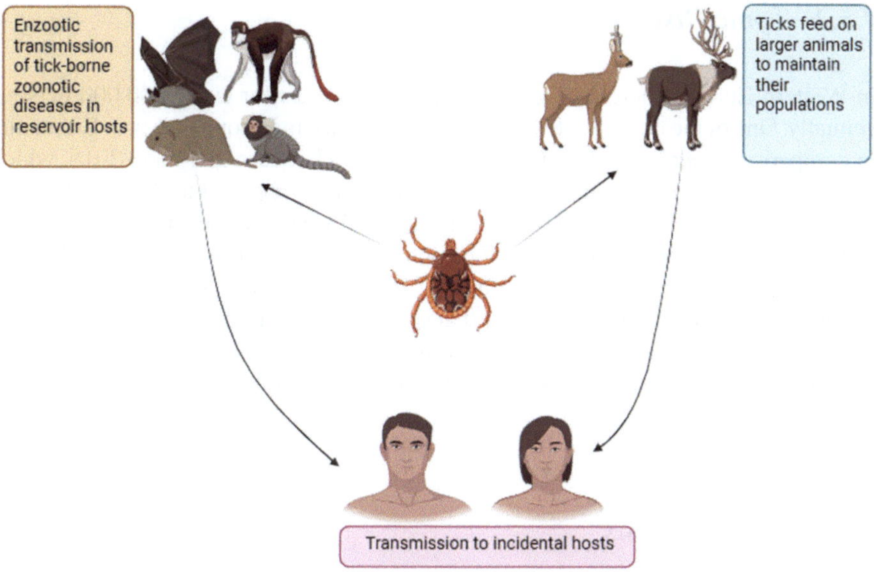

Fig. 4.3 Mode of transmission of zoonotic diseases associated with ticks

Ixodes and *Rhipicephalus* also transmit the disease (Behzadi et al., 2021). The mode of transmission of zoonotic diseases associated with ticks have been given in Fig. 4.3.

4.5 Impact of Climate Change on Disease Spread

4.5.1 Babesiosis

The climate is suitable for the ticks to spread babesiosis in North and South America including Brazil, Uruguay, Argentina, Mexico, and the USA. Climate is also ideal in sub-Saharan Africa, Australia, Western Europe, and Southeast Asia (Marques et al., 2020). Climate change has been proposed as a reason behind the emergence of babesiosis in the USA (Montero et al., 2022). In February 2019, temperatures in the south of England surpassed the average for the time of year by over 10 °C, resulting in exceptionally early tick movement and a spread of bovine babesiosis involving 20 cattle. This was a more localized but still distinct climate effect on the disease (Gray & Ogden, 2021). Babesiosis peaked during summer in England, with the highest prevalence in August and September (McFadzean et al., 2023).

4.5.2 Lyme Disease

In Western Europe, there are 22.05 Lyme disease cases for every 100,000 persons annually. One of the highest rates of disease prevalence in Europe is seen in Lithuania. In comparison, there were 7424 new cases reported between 2014 and 2016, which translates to an approximate incidence rate of 85.4, as opposed to 73.9–100.6 incidences per 100,000 people between 2011 and 2018. Due to their respective lowest and highest temperatures in Europe, the countries of Northern and Southwest Europe have lower incidence rates (Voyiatzaki et al., 2022). The climate in Canada is far more conducive to the feeding of the vector that spreads Lyme disease, which is why cases of the disease are being reported there (Ogden et al., 2024). *Ixodes ricinus* is also expanding into higher latitudes and elevations as a result of local climate change (Giesen et al., 2024).

4.5.3 Anaplasmosis

The results of the study show that between 2005 and 2013, the number of bovine anaplasmosis infections in Kansas increased steadily, and the disease also moved to newer geographic areas during that time. Greater space–time association for bovine anaplasmosis disease is seen in an area of central and south central counties (Spare, 2021). Three climate change indices including relative humidity, diurnal temperature range, and lowest land surface temperature determine the space–time pattern for bovine anaplasmosis in Kansas at the county level. (Hanzlicek et al., 2016). In other words, in regions with low minimum winter temperatures, more snowfall would likely result in higher anaplasmosis overwinter survival. In regions with higher minimum winter temperatures and humidity, such as the maritime climate of Long Island, the impact of snow on tick survivability may be negligible (O'Connor et al., 2021).

4.5.4 Ehrlichiosis

Ehrlichia canis is commonly observed between March and June (Ajith et al., 2024). As temperatures rise and conditions favor tick development and questing, it is anticipated that other tick species in temperate regions of the United States will continue to expand northward and to higher elevations. It is anticipated that tick-borne infections, such as *E chaffeensis*, will proliferate in parallel with their vectors (Nuttall, 2022). The risk of ehrlichiosis has grown globally as a result of changes in the temperature, land use, ecosystems, and wildlife reservoir populations, as well as increased human and canine contact (Gettings et al., 2020).

4.5.5 Rocky Mountain Spotted Fever

While *Amblyomma cajennense* is mostly found in South America, *Dermacentor variabilis* is found in North America, and all other continents, aside from Antarctica have an acceptable distribution (Minigan et al., 2018). *Amblyomma cajennense* is more likely to thrive in tropical areas, while *Dermacentor variabilis* is more likely to thrive in temperate ones. *Amblyomma cajennense* is sensitive to both temperature and precipitation, and it prefers tropical locations with hot and humid conditions, while *Dermacentor variabilis* is more sensitive to temperature (Ji et al., 2024).

4.6 Surveillance and Monitoring Strategies

A comprehensive understanding of the locations and timing of pathogen transmission is essential for the prevention of TBDs in humans and animals, including companions and livestock. Only monitoring that aims to determine the distribution of tick populations, the pathogens that are prevalent in those populations, and the times of year when tick activity occurs can provide this information (Johnson et al., 2022). Noted variations between the three surveillance techniques, with the Dutch project demonstrating a high degree of interdisciplinary cooperation, accurate vector identification, and public participation in research and instruction. There have been quantifiable results, like a decrease in tick bites and the identification of novel diseases and tick species (Onstwedder et al., 2022). Surveillance systems in Italy and Spain are based on mandatory notifications to health authorities; laws appear to be important, but law enforcement and financial resources are fairly dispersed and only apply to the most serious diseases. Collaborations are restricted to local efforts, and the non-scientific community is given only a passing consideration. The majority of research efforts in both nations have helped to advance our understanding of tick species distribution (Garcia-Vozmediano et al., 2022).

The seasonal, temporal, and geographic variation of ticks and tick-borne zoonotic diseases can be better understood through extensive and thorough surveillance of TBDs (Pak et al., 2019). Both active and passive tick monitoring techniques are frequently used while collecting tick samples. Active tick surveillance includes live animal capture, semiochemical-based trapping, dragging, and flagging. Typically less expensive, passive tick surveillance depends on farmers and volunteers submitting ticks for pathogen screening and identification (Nchu et al., 2020). Clinicians and public health officials can benefit from more detailed, county-level information about tick-borne zoonotic disease risk by using triangulation techniques to combine patient data on recall of TBDs with official human TBD data, tick presence, and established canine serological reports. In order to improve surveillance and provide a foundation for creating reliable models for human tick-borne disease risk, such data may be used in addition to public health sources (Tiffin et al., 2022).

4.7 Mitigation and Control Measures

Ticks are managed using a variety of techniques, including chemical and biological ones. Each approach has its own set of benefits and drawbacks. As demonstrated by past studies, chemical (acaricide) techniques are widely employed to control ticks, but they have some lasting impacts on the health of the animals. To eradicate ticks, a different and efficient control strategy should be used (Ramzan et al., 2021). To better understand how tick-borne pathogens are spread in the environment and how they may cause disease in various host species, researchers, policymakers, and medical and veterinary professionals must work together. This interdisciplinary approach will be essential for effective tick control and TBD mitigation strategies (Rowan et al., 2023). According to analysis, host-targeted interventions utilizing chemical pesticides may increase disease persistence and potentially transform a region free of disease into an endemic area in locations where the tick population is expanding quickly. Therefore, while developing tick control techniques, it is important to carefully consider the complex nature of pathogen transfer between ticks and hosts (Tosato et al., 2021).

4.8 Future Directions and Challenges

Knowledge of tick biology, tick-host–pathogen interactions, tick-borne pathogen circulation in natural foci, and disease epidemiology is still lacking (Sparagano et al., 2022). The proliferation of tick species is influenced by anthropogenic causes, climate, and animal movement. Numerous TBDs are predicted to spread from sub-Saharan Africa and other nations in the Mediterranean basin. The widespread prevalence of ticks is made worse by the emergence of acaricide resistance, which also presents a serious economic challenge (Abdelbaset et al., 2023). Predicting the effects of anthropogenic land use and climate change on the population and distribution of ticks and tick-associated diseases is one of the expected future difficulties facing the study of ticks and TBDs. Multidisciplinary cooperation among ecologists, climate change experts, city planners, government representatives, parasitologists, and social scientists will help overcome these obstacles. Additionally, better communication between veterinary and medical professionals is crucial (Steele et al., 2021). To maintain biosecurity, surveillance is also important (Irwin & Beadle, 2022).

The development of novel interventions to control tick infestations in both domestic and wild hosts is essential. Effective vaccines should be developed for use in both human and animal populations (de la Fuente et al., 2016). Preventing the global spread of ticks should be a priority (Tiffin et al., 2022). Additionally, modeling the impact of climate change and land use on the expansion of tick populations and wild hosts should be pursued (de la Fuente et al., 2023).

References

Abdelbaset, A. E., Kwak, M. L., Nonaka, N., & Nakao, R. (2023). Human-biting ticks and zoonotic tick-borne pathogens in North Africa: Diversity, distribution, and trans-Mediterranean public health challenges. *One Health, 16*, Article 100547.

Adelabu, O. A., Iweriebor, B. C., Okoh, A. I., & Obi, L. C. (2020). Phylogenetic profiling for zoonotic *Ehrlichia* spp. from ixodid ticks in the Eastern Cape, South Africa. *Transboundary and Emerging Diseases, 67*(3), 1247–1256.

Ajith, Y., Adithya, S., Panicker, V. P., Athira, N., Beena, V., Safeer, M. S., Preena, P., Nisha, A. R., Divya, C., Sangeetha, S. G., & Umesh, C. G. (2024). Biometeorological analysis on the molecular incidence of babesiosis and ehrlichiosis in dogs. *Theoretical and Applied Climatology, 155*(3), 1573–1582.

Aju, P. C., Iwuchukwu, J. J., & Ibe, C. C. (2015). Our forests, our environment, our sustainable livelihoods. *European Journal of Academic Essays, 2*(4), 6–19.

Alkishe, A., Raghavan, R. K., & Peterson, A. T. (2021). Likely geographic distributional shifts among medically important tick species and tick-associated diseases under climate change in North America: A review. *Insects, 12*(3), 225.

Altay, K., Erol, U., & Sahin, O. F. (2024). *Anaplasma capra*: A new emerging tick-borne zoonotic pathogen. *Veterinary Research Communications*, 1–12.

Álvarez-Hernández, G., López-Ridaura, R., Cortés-Alcalá, R., Rodríguez, G. G., Calleja-López, J. T., Rivera-Rosas, C. N., Alomía-Zegarra, J. L., Brophy, M., Brito-Lorán, C. B., del Carmen Candia-Plata, M., & Correa-Morales, F. (2024a). Rocky Mountain spotted fever in Mexico: A call to action. *The American Journal of Tropical Medicine and Hygiene, 111*(5), 1070.

Álvarez-Hernández, G., Paddock, C. D., Walker, D. H., Valenzuela, J. G., Calleja-López, J. T., Rivera-Rosas, C. N., & Sotelo-Mundo, R. R. (2024b). Rocky Mountain spotted fever is a neglected tropical disease in Latin America. *PLOS Neglected Tropical Diseases, 18*(7), Article e0012276.

Apanaskevich, D. A., Oliver, J. H., Sonenshine, D. E., & Roe, R. M. (2014). Life cycles and natural history of ticks. *Biology of Ticks, 1*, 59–73.

Aziz, M. U., Hussain, S., Song, B., Ghauri, H. N., Zeb, J., & Sparagano, O. A. (2022). Ehrlichiosis in dogs: A comprehensive review about the pathogen and its vectors with emphasis on South and East Asian countries. *Veterinary Sciences, 10*(1), 21.

Bajer, A., Beck, A., Beck, R., Behnke, J. M., Dwużnik-Szarek, D., Eichenberger, R. M., Farkas, R., Fuehrer, H. P., Heddergott, M., Jokelainen, P., & Leschnik, M. (2022). Babesiosis in Southeastern, Central and Northeastern Europe: An emerging and re-emerging tick-borne disease of humans and animals. *Microorganisms, 10*(5), 945.

Behzadi, M. Y., Mostafavi, E., Rohani, M., Mohamadi, A., Ahmadinezhad, M., Moazzezy, N., Shams-Ghahfarokhi, M., & Razzaghi-Abyaneh, M. (2021). A review on important zoonotic bacterial tick-borne diseases in the Eastern Mediterranean region. *Journal of Arthropod-Borne Diseases, 15*(3), 265.

Bermúdez, S. E., Gottdenker, N., Krishnvajhala, A., Fox, A., Wilder, H. K., González, K., Smith, D., López, M., Perea, M., Rigg, C., & Montilla, S. (2017). Synanthropic mammals as potential hosts of tick-borne pathogens in Panama. *PLoS ONE, 12*(1), Article e0169047.

Bianchessi, L., Rocchi, M. S., Maley, M., Allen, K., Ballingall, K., & Turin, L. (2023). Presence of *Anaplasma phagocytophilum* Ecotype I in UK ruminants and associated zoonotic risk. *Pathogens, 12*(2), 216.

Bird, B. H., & Mazet, J. A. (2018). Detection of emerging zoonotic pathogens: An integrated one health approach. *Annual Review of Animal Biosciences, 6*(1), 121–139.

Blache, N., Chalvet-Monfray, K., Déprés, C., & Morand, S. (2024). A scoping review of the impacts of forest dynamics on acari-borne diseases: Beyond forest fragmentation.

Blasdell, K. R., Morand, S., Laurance, S. G., Doggett, S. L., Hahs, A., Trinh, K., Perera, D., & Firth, C. (2022). Rats and the city: Implications of urbanization on zoonotic disease risk in Southeast Asia. *Proceedings of the National Academy of Sciences, 119*(39), Article e2112341119.

Bobe, J. R., Jutras, B. L., Horn, E. J., Embers, M. E., Bailey, A., Moritz, R. L., Zhang, Y., Soloski, M. J., Ostfeld, R. S., Marconi, R. T., & Aucott, J. (2021). Recent progress in Lyme disease and remaining challenges. *Frontiers in Medicine, 8*, Article 666554.

Borşan, S. D., Toma-Naic, A., Péter, Á., Sándor, A. D., Peştean, C., & Mihalca, A. D. (2020). Impact of abiotic factors, habitat type and urban wildlife on the ecology of hard ticks (Acari: Ixodidae) in urban and peri-urban habitats. *Parasites & Vectors, 13*, 476.

Burtis, J. C., Fahey, T. J., & Yavitt, J. B. (2019). Survival and energy use of Ixodes scapularis nymphs throughout their overwintering period. *Parasitology, 146*(6), 781–790.

Chikezie, F. M., Opara, K. N., & Ubulom, P. M. E. (2024). Impacts of changing climate on arthropod vectors and diseases transmission. *Nigerian Journal of Entomology, 40*, 179–192.

Coburn, J., Garcia, B., Hu, L. T., Jewett, M. W., Kraiczy, P., Norris, S. J., & Skare, J. (2021). Lyme disease pathogenesis. *Current Issues in Molecular Biology, 42*(1), 473–518.

Conte, A., Goffredo, M., Ippoliti, C., & Meiswinkel, R. (2007). Influence of biotic and abiotic factors on the distribution and abundance of Culicoides imicola and the Obsoletus Complex in Italy. *Veterinary Parasitology, 150*(4), 333–344.

Dantas-Torres, F. (2015). Climate change, biodiversity, ticks and tick-borne diseases: The butterfly effect. *International Journal for Parasitology: Parasites and Wildlife, 4*(3), 452–461.

de la Fuente, J., Estrada-Peña, A., Rafael, M., Almazán, C., Bermúdez, S., Abdelbaset, A. E., Kasaija, P. D., Kabi, F., Akande, F. A., Ajagbe, D. O., & Bamgbose, T. (2023). Perception of ticks and tick-borne diseases worldwide. *Pathogens, 12*(10), 1258.

de la Fuente, J., Kopáček, P., Lew-Tabor, A., & Maritz-Olivier, C. (2016). Strategies for new and improved vaccines against ticks and tick-borne diseases. *Parasite Immunology, 38*(12), 754–769.

de Souza, W. M., & Weaver, S. C. (2024). Effects of climate change and human activities on vector-borne diseases. *Nature Reviews Microbiology*, 1–16.

Deshpande, G., Beetch, J. E., Heller, J. G., Naqvi, O. H., & Kuhn, K. G. (2023). Assessing the influence of climate change and environmental factors on the top tick-borne diseases in the United States: A systematic review. *Microorganisms, 12*(1), 50.

Devine, J. A., Wrathall, D., Aguilar-González, B., Benessaiah, K., Tellman, B., Ghaffari, Z., & Ponstingel, D. (2021). Narco-degradation: Cocaine trafficking's environmental impacts in Central America's protected areas. *World Development, 144*, Article 105474.

DeVore, J. L., Shine, R., & Ducatez, S. (2020). Urbanization and translocation disrupt the relationship between host density and parasite abundance. *Journal of Animal Ecology, 89*(4), 1122–1133.

Diuk-Wasser, M. A., VanAcker, M. C., & Fernandez, M. P. (2021). Impact of land use changes and habitat fragmentation on the eco-epidemiology of tick-borne diseases. *Journal of Medical Entomology, 58*(4), 1546–1564.

Eisen, R. J., Kugeler, K. J., Eisen, L., Beard, C. B., & Paddock, C. D. (2017). Tick-borne zoonoses in the United States: Persistent and emerging threats to human health. *ILAR Journal, 58*(3), 319–335.

El Amri, H., Boukharta, M., Zakham, F., & Ennaji, M. M. (2020). *Emergence and reemergence of viral zoonotic diseases: Concepts and factors of emerging and reemerging globalization of health threats.* Academic Press.

Elmieh, N. (2022). *The impacts of climate and land use change on tick-related risks.*

Estrada-Peña, A., Ostfeld, R. S., Peterson, A. T., Poulin, R., & de la Fuente, J. (2014). Effects of environmental change on zoonotic disease risk: An ecological primer. *Trends in Parasitology, 30*(4), 205–214.

Frelich, L. E., & Reich, P. B. (2010). Will environmental changes reinforce the impact of global warming on the prairie–forest border of central North America? *Frontiers in Ecology and the Environment, 8*(7), 371–378.

Garcia-Vozmediano, A., De Meneghi, D., Sprong, H., Portillo, A., Oteo, J. A., & Tomassone, L. (2022). A One Health evaluation of the surveillance systems on tick-borne diseases in the Netherlands, Spain and Italy. *Veterinary Sciences, 9*(9), 504.

Gettings, J. R., Self, S. C., McMahan, C. S., Brown, D. A., Nordone, S. K., & Yabsley, M. J. (2020). Local and regional temporal trends (2013–2019) of canine *Ehrlichia* spp. seroprevalence in the USA. *Parasites & Vectors, 13*, 153.

Giesen, C., Cifo, D., Gomez-Barroso, D., Estévez-Reboredo, R. M., Figuerola, J., & Herrador, Z. (2024). The role of environmental factors in Lyme disease transmission in the European Union: A systematic review. *Tropical Medicine and Infectious Disease, 9*(5), 113.

Gilbert, L. (2021). The impacts of climate change on ticks and tick-borne disease risk. *Annual Review of Entomology, 66*, 373–388.

Ginsberg, H. S., Rulison, E. L., Miller, J. L., Pang, G., Arsnoe, I. M., Hickling, G. J., Ogden, N. H., LeBrun, R. A., & Tsao, J. I. (2020). Local abundance of *Ixodes scapularis* in forests: Effects of environmental moisture, vegetation characteristics, and host abundance. *Ticks and Tick-Borne Diseases, 11*(1), Article 101271.

Gottdenker, N. L., Streicker, D. G., Faust, C. L., & Carroll, C. R. (2014). Anthropogenic land use change and infectious diseases: A review of the evidence. *EcoHealth, 11*, 619–632.

Gray, J. S., & Ogden, N. H. (2021). Ticks, human babesiosis and climate change. *Pathogens, 10*(11), 1430.

Guégan, J. F., Ayouba, A., Cappelle, J., & De Thoisy, B. (2020). Forests and emerging infectious diseases: Unleashing the beast within. *Environmental Research Letters, 15*(8), Article 083007.

Gygax, L., Schudel, S., Kositz, C., Kuenzli, E., & Neumayr, A. (2024). Human monocytotropic ehrlichiosis—A systematic review and analysis of the literature. *PLOS Neglected Tropical Diseases, 18*(8), Article e0012377.

Hanzlicek, G. A., Raghavan, R. K., Ganta, R. R., & Anderson, G. A. (2016). Bayesian space-time patterns and climatic determinants of bovine anaplasmosis. *PLoS ONE, 11*(3), Article e0151924.

Hartemink, N., & Takken, W. (2016). Trends in tick population dynamics and pathogen transmission in emerging tick-borne pathogens in Europe: An introduction. *Experimental and Applied Acarology, 68*, 269–278.

Hildebrandt, A., Zintl, A., Montero, E., Hunfeld, K. P., & Gray, J. (2021). Human Babesiosis in Europe. *Pathogens, 10*(9), 1165.

Irwin, P., & Beadle, J. (2022). The 'other' epidemic: Canine ehrlichiosis in Australia. *Microbiology Australia, 43*(4), 156–159.

Jánová, E. (2019). Emerging and threatening vector-borne zoonoses in the world and in Europe: A brief update. *Pathogens and Global Health, 113*(2), 49–57.

Ji, H., Wei, X., Ma, D., Wang, X., & Liu, Q. (2024). Predicting the global potential distribution of two major vectors of Rocky Mountain Spotted Fever under conditions of global climate change. *PLOS Neglected Tropical Diseases, 18*(1), Article e0011883.

Johnson, N., Paul Phipps, L., McFadzean, H., & Barlow, A. M. (2020). An outbreak of bovine babesiosis in February, 2019, triggered by above average winter temperatures in southern England and co-infection with *Babesia divergens* and *Anaplasma phagocytophilum*. *Parasites & Vectors, 13*, 1–5.

Johnson, N., Phipps, L. P., Hansford, K. M., Folly, A. J., Fooks, A. R., Medlock, J. M., & Mansfield, K. L. (2022). One Health approach to tick and tick-borne disease surveillance in the United Kingdom. *International Journal of Environmental Research and Public Health, 19*(10), 5833.

Kelly, T. R., Machalaba, C., Karesh, W. B., Crook, P. Z., Gilardi, K., Nziza, J., Uhart, M. M., Robles, E. A., Saylors, K., Joly, D. O., & Monagin, C. (2020). Implementing One Health approaches to confront emerging and re-emerging zoonotic disease threats: Lessons from PREDICT. *One Health Outlook, 2*, 1–7.

Kjemtrup, A. M., Padgett, K., Paddock, C. D., Messenger, S., Hacker, J. K., Feiszli, T., Melgar, M., Metzger, M. E., Hu, R., & Kramer, V. L. (2022). A forty-year review of Rocky Mountain spotted fever cases in California shows clinical and epidemiologic changes. *PLOS Neglected Tropical Diseases, 16*(9), Article e0010738.

Körner, S., Makert, G. R., Ulbert, S., Pfeffer, M., & Mertens-Scholz, K. (2021). The prevalence of *Coxiella burnetii* in hard ticks in Europe and their role in Q fever transmission revisited—A systematic review. *Frontiers in Veterinary Science, 8*, Article 655715.

Kumar, A., O'Bryan, J., & Krause, P. J. (2021). The global emergence of human babesiosis. *Pathogens, 10*(11), 1447. https://doi.org/10.3390/pathogens10111447
Laukaitis, H. J., & Macaluso, K. R. (2021). Unpacking the intricacies of *Rickettsia*–vector interactions. *Trends in Parasitology, 37*(8), 734–746.
Loiseau, C., & Sehgal, R. N. (2022). *Consequences of deforestation and habitat degradation on wildlife mosquito-borne diseases.* Wageningen Academic Publishers.
Madison-Antenucci, S., Kramer, L. D., Gebhardt, L. L., & Kauffman, E. (2020). Emerging tick-borne diseases. *Clinical Microbiology Reviews, 33*(2), e00083-e118.
Marques, R., Krüger, R. F., Peterson, A. T., de Melo, L. F., Vicenzi, N., & Jiménez-García, D. (2020). Climate change implications for the distribution of the babesiosis and anaplasmosis tick vector, *Rhipicephalus (Boophilus) microplus. Veterinary Research, 51*, 94.
Marshet, B., & Dessie, D. (2020). A review on canine ehrlichiosis and its zoonotic implications.
McFadzean, H., Johnson, N., Phipps, L. P., Swinson, V., & Boden, L. A. (2023). Surveillance and risk analysis for bovine babesiosis in England and Wales to inform disease distribution. *Animals, 13*(13), 2118.
McMahon, S. M., Parker, G. G., & Miller, D. R. (2010). Evidence for a recent increase in forest growth. *Proceedings of the National Academy of Sciences, 107*(8), 3611–3615.
Minigan, J. N., Hager, H. A., Peregrine, A. S., & Newman, J. A. (2018). Current and potential future distribution of the American dog tick (Dermacentor variabilis, Say) in North America. *Ticks and Tick-Borne Diseases, 9*(2), 354–362.
Montero, E., Gray, J., Lobo, C. A., & González, L. M. (2022). *Babesia* and human babesiosis. *Pathogens, 11*(4), 399.
Muzaffar, H. A., Aqeel, M., Taimoor, M., Fawad, M., Ali, S., Rahim, H. M. T., Irshad, M. U., & Arshad, M. S. (2024). Overview of vaccination against babesiosis and theileriosis. In M. A. Alvi, M. Rashid, M. A. Zafar, M. A. S. Mughal, & S. I. Toor (Eds.), *Complementary and alternative medicine: Immunization/vaccinology.* Unique Scientific Publishers.
Mysterud, A., Jore, S., Østerås, O., & Viljugrein, H. (2017). Emergence of tick-borne diseases at northern latitudes in Europe: A comparative approach. *Scientific Reports, 7*(1), 16316.
Nah, K., Bede-Fazekas, Á., Trájer, A. J., & Wu, J. (2020). The potential impact of climate change on the transmission risk of tick-borne encephalitis in Hungary. *BMC Infectious Diseases, 20*, 1–10.
Nchu, F., Nyangiwe, N., Muhanguzi, D., Nzalawahe, J., Nagagi, Y. P., Msalya, G., Joseph, N. A., Kimaro, E. G., Mollel, M., Temba, V., & Harouna, D. V. (2020). Development of a practical framework for sustainable surveillance and control of ticks and tick-borne diseases in Africa. *Veterinary World, 13*(9), 1910.
Noden, B. H., Roselli, M. A., & Loss, S. R. (2022). Effect of urbanization on presence, abundance, and coinfection of bacteria and protozoa in ticks in the US Great Plains. *Journal of Medical Entomology, 59*(3), 957–968.
Nuttall, P. A. (2022). Climate change impacts on ticks and tick-borne infections. *Biologia, 77*(6), 1503–1512.
O'Connor, C., Prusinski, M. A., Jiang, S., Russell, A., White, J., Falco, R., Kokas, J., Vinci, V., Gall, W., Tober, K., & Haight, J. (2021). A comparative spatial and climate analysis of human granulocytic anaplasmosis and human babesiosis in New York State (2013–2018). *Journal of Medical Entomology, 58*(6), 2453–2466.
Ogden, N. H., Ben Beard, C., Ginsberg, H. S., & Tsao, J. I. (2021). Possible effects of climate change on ixodid ticks and the pathogens they transmit: Predictions and observations. *Journal of Medical Entomology, 58*(4), 1536–1545.
Ogden, N. H., Dumas, A., Gachon, P., & Rafferty, E. (2024). Estimating the incidence and economic cost of Lyme disease cases in Canada in the 21st century with projected climate change. *Environmental Health Perspectives, 132*(2), Article 027005.
Onstwedder, C., Lock-Wah-Hoon, J., Van Dorp, S., Braks, M., Van Asten, L., Zheng, Y., Krafft, T., Tong, Y., Van der Hoek, W., Liu, Q. Y., & Pilot, E. (2022). Comparing vector-borne disease surveillance and response in Beijing and The Netherlands. *Annals of Global Health, 88*(1), 59.

Ortiz, D. I., Piche-Ovares, M., Romero-Vega, L. M., Wagman, J., & Troyo, A. (2021). The impact of deforestation, urbanization, and changing land use patterns on the ecology of mosquito and tick-borne diseases in Central America. *Insects, 13*(1), 20.

Pak, D., Jacobs, S. B., & Sakamoto, J. M. (2019). A 117-year retrospective analysis of Pennsylvania tick community dynamics. *Parasites & Vectors, 12*, 189.

Pal, M., Shuramo, M. Y., & Gutama, K. P. (2022). Tularaemia: A re-emerging infectious zoonotic disease of public health significance. *International Journal of Clinical and Experimental Medicine Research, 6*, 48–51.

Palaniyandi, M. (2021). New, emerging, re-emerging tropical infectious and non-communicable diseases persistent to the climate, landscape, and environmental changes on the grounds of the urbanizations, industrializations, and globalisation. *International Journal of Environment and Climate Change, 11*(11), 32–46.

Peter, S. G. (2020). Zoonotic *Anaplasma* and *Ehrlichia* infections and their potential reservoirs: A review.

Radolf, J. D., Strle, K., Lemieux, J. E., & Strle, F. (2021). Lyme disease in humans. *Current Issues in Molecular Biology, 42*(1), 333–384.

Ramzan, M., Murtaza, G., Sattar, A., Munawar, N., Ullah, A., Ejaz, A., Ayaz, F., Anwar, S., Jameel, K., & Kamran, F. (2021). Techniques for managing ticks and tick-borne diseases under changing climate: A review. *Egyptian Academic Journal of Biological Sciences, B. Zoology, 13*(1), 117–128.

Rodriguez-Morales, A. J., Paniz-Mondolfi, A. E., Faccini-Martínez, Á. A., Henao-Martínez, A. F., Ruiz-Saenz, J., Martinez-Gutierrez, M., Alvarado-Arnez, L. E., Gomez-Marin, J. E., Bueno-Marí, R., Carrero, Y., & Villamil-Gomez, W. E. (2021). The constant threat of zoonotic and vector-borne emerging tropical diseases: Living on the edge. *Frontiers in Tropical Diseases, 2*, Article 676905.

Rong, N., & Liu, J. (2023). Development of animal models for emerging infectious diseases by breaking the barrier of species susceptibility to human pathogens. *Emerging Microbes & Infections, 12*(1), 2178242.

Rowan, S., Mohseni, N., Chang, M., Burger, H., Peters, M., & Mir, S. (2023). From tick to test: A comprehensive review of tick-borne disease diagnostics and surveillance methods in the United States. *Life, 13*(10), 2048.

Sharma, R., Patil, R. D., Singh, B., Chakraborty, S., Chandran, D., Dhama, K., Gopinath, D., Jairath, G., Rialch, A., Mal, G., & Singh, P. (2023). Tularemia–A re-emerging disease with growing concern. *Veterinary Quarterly, 43*(1), 1–16.

Sheela, A. M., Ghermandi, A., Vineetha, P., Sheeja, R. V., Justus, J., & Ajayakrishna, K. (2017). Assessment of relation of land use characteristics with vector-borne diseases in tropical areas. *Land Use Policy, 63*, 369–380.

Sirotkin, M. B., & Korenberg, E. I. (2019). Influence of abiotic factors on infectious agents environmentally associated with Ixodidae ticks on the example of borrelia and tick-borne encephalitis virus. *Biology Bulletin Reviews, 9*, 543–561.

Skotarczak, B. (2009). Adaptation factors of Borrelia for host and vectors. *Annals of Agricultural and Environmental Medicine, 16*(1), 1–8.

Sonenshine, D. E. (2018). Range expansion of tick disease vectors in North America: Implications for spread of tick-borne disease. *International Journal of Environmental Research and Public Health, 15*(3), 478.

Sparagano, O., Földvári, G., Derdáková, M., & Kazimírová, M. (2022). New challenges posed by ticks and tick-borne diseases. *Biologia, 77*(6), 1497–1501.

Spare, M. R. (2021). *Host, vector, environment and management: epidemiology of bovine anaplasmosis in the State of Kansas*. Kansas State University.

Steele, S. G., Booy, R., Manocha, R., Mor, S. M., & Toribio, J. A. L. (2021). Towards One Health clinical management of zoonoses: A parallel survey of Australian general medical practitioners and veterinarians. *Zoonoses and Public Health, 68*(2), 88–102.

Swei, A., Couper, L. I., Coffey, L. L., Kapan, D., & Bennett, S. (2020). Patterns, drivers, and challenges of vector-borne disease emergence. *Vector-Borne and Zoonotic Diseases, 20*(3), 159–170.

Tajudeen, Y. A., Oladunjoye, I. O., Bajinka, O., & Oladipo, H. J. (2022). Zoonotic spillover in an era of rapid deforestation of tropical areas and unprecedented wildlife trafficking: Into the wild. *Challenges, 13*(2), 41.

Tazerji, S. S., Nardini, R., Safdar, M., Shehata, A. A., & Duarte, P. M. (2022). An overview of anthropogenic actions as drivers for emerging and re-emerging zoonotic diseases. *Pathogens, 11*(11), 1376.

Tiffin, H. S., Rajotte, E. G., Sakamoto, J. M., & Machtinger, E. T. (2022). Tick control in a connected world: Challenges, solutions, and public policy from a United States border perspective. *Tropical Medicine and Infectious Disease, 7*(11), 388.

Tosato, M., Nah, K., & Wu, J. (2021). Are host control strategies effective to eradicate tick-borne diseases (TBD)? *Journal of Theoretical Biology, 508*, Article 110483.

Ullah, Q., Jamil, T., Saqib, M., Iqbal, M., & Neubauer, H. (2022). Q fever—A neglected zoonosis. *Microorganisms, 10*(8), 1530.

Vannier, E., & Krause, P. J. (2020). Babesiosis. In J. E. Bennett, R. Dolin, & M. J. Blaser (Eds.), *Hunter's tropical medicine and emerging infectious diseases.*

Voyiatzaki, C., Papailia, S. I., Venetikou, M. S., Pouris, J., Tsoumani, M. E., & Papageorgiou, E. G. (2022). Climate changes exacerbate the spread of *Ixodes ricinus* and the occurrence of Lyme borreliosis and tick-borne encephalitis in Europe—How climate models are used as a risk assessment approach for tick-borne diseases. *International Journal of Environmental Research and Public Health, 19*(11), 6516.

Waked, R., & Krause, P. J. (2022). Human babesiosis. *Infectious Disease Clinics of North America, 36*(3), 655–670.

Wikel, S. K. (2018). Ticks and tick-borne infections: Complex ecology, agents, and host interactions. *Veterinary Sciences, 5*(2), 60.

Wimms, C., Aljundi, E., & Halsey, S. J. (2023). Regional dynamics of tick vectors of human disease. *Current Opinion in Insect Science, 55*, Article 101006.

Xiong, Y., Mo, S., Wu, H., Qu, X., Liu, Y., & Zhou, L. (2023). Influence of human activities and climate change on wetland landscape pattern—A review. *Science of the Total Environment, 879*, Article 163112.

Yadav, N., & Upadhyay, R. K. (2023). Global effect of climate change on seasonal cycles, vector population and rising challenges of communicable diseases: a review. *Journal of Atmospheric Science Research, 6*(1).

Yeni, D. K., Büyük, F., Ashraf, A., & Shah, M. S. U. D. (2021). Tularemia: A re-emerging tick-borne infectious disease. *Folia Microbiologica, 66*(1), 1–14.

Chapter 5
Impact of Climate Change on Tick-Borne Viral Diseases

Mahvish Maqbool, Aysha Rana, Hira khan, Maara Raheel Spaul, and Faisal Rasheed Anjum

Abstract Climate change is affecting the prevalence of ticks and tick-borne diseases (TBDs) around the globe through changes in temperature patterns, precipitation values, and ecosystem variables. This is helping to expand the habitats of vectors, especially ticks. These changing scenarios need robust adaptive strategies that involve biologists, public health specialists, environmental specialists, and economists to design a framework for research initiatives. This chapter describes the impact of climate change on tick-borne viral diseases (TBVDs). It emphasizes the importance of enhanced surveillance systems, enhancing public education regarding ticks and tick-borne viruses, and taking interdisciplinary initiatives to control these infections. It reviews the TBVDs, their impact, how the geographical pattern of tick prevalence changes with climate change, surveillance systems available, what can be improved, and current ways to control ticks. It also points out some research areas which would help design a framework to overcome this issue, like enhancement needed in ecological modelling, climate resilient vector control methods, vaccine development, and enhancing the diagnostic field to diagnose the TBVDs at their initial stage. A central point of this review is to focus on interdisciplinary collaboration among the biologists, ecologists, economists, policy makers, veterinarians, and the public health sector for effective prevention and control of TBVDs.

M. Maqbool (✉)
Department of Entomology, Texas A & M, College Station, TX, USA
e-mail: mahvish21@tamu.edu

A. Rana
Foot and Mouth Disease Research Centre, Lahore, Pakistan

College of Veterinary Medicine, University of Minnesota, Twin-Cities, USA

H. khan · M. R. Spaul
Institute of Microbiology, University of Veterinary and Animal Sciences, Lahore, Pakistan

H. khan
College of Sciences, Department of Biology, United Arab Emirates University, Abu Dhabi, UAE

F. R. Anjum
Bristol Veterinary School, University of Bristol, Bristol, UK

H. M. Rizwan and M. S. Sajid (eds.), *Ticks in a Changing Climate*, Biometeorology 7,
https://doi.org/10.1007/978-3-032-12259-9_5

Keywords Tick activity · Climate factors · Surveillance systems · Tick-Borne viral diseases · Public health

5.1 Introduction

5.1.1 Ticks and Their Significance

Ticks are hematophagous ectoparasites, also having a key role in ecological systems and public health. For blood, ticks need vertebrate hosts to infest many types of hosts. Mammals, birds, reptiles, and amphibians are included in these hosts (Khan et al., 2021).Ticks may be found on every single continent, even Antarctica. They are, for the most part, abundant throughout both temperate and tropical regions. They gain ecological importance by maintaining pathogen life cycles and influencing host population dynamics inside enzootic reservoirs. Their presence can sometimes help regulate overpopulated wildlife species using parasitism with disease transmission (Eisen & Eisen, 2018; Randolph et al., 2014). Also, at this same time, ticks can be of significant concern within both medical and veterinary contexts since they can directly cause harm through loss of blood, injection of toxins, hypersensitivity reactions, and, most critically, transmission of infectious agents (de la Fuente et al., 2023a, b). Many pathogens are transmitted via ticks, giving them disproportionate importance even though they are but a small fraction within global arthropods, making them the second most important vector group after mosquitoes (de la Fuente et al., 2008).

Ticks have a complex type of life cycle that includes egg, larva, nymph, and fully adult developmental stages. Ticks often need three blood meals to finish their entire development process. Each mobile stage must include larva, nymph, and adult stages. This process can span a few months or up to years since environmental conditions and host availability often vary (Pfäffle et al., 2013). Ticks carry pathogens while being ecologically important to nature. They serve as a food source for species, including several birds, reptiles, and specific insects like parasitic wasps. Their parasitic interactions remain of great importance. Each one of them is what the ecosystem relies on for regulation. Disease agents can indirectly affect predator–prey dynamics and host biodiversity (Randolph et al., 2014). Ticks may also be indicators of the environment. Tick population dynamics will sometimes change because of broader ecological shifts that include climate change in conjunction with habitat fragmentation and alterations in host populations. The northward expansion of *Ixodes (I.) scapularis* within North America is related to warming temperatures as well as forest fragmentation, so ticks are valuable biological sentinels when monitoring environmental health (Estrada-Peña et al., 2013a; Ogden et al., 2006).

Ticks spreading diseases represent a growing worldwide health concern. These may spread bacteria such as *Borrelia burgdorferi*, causing Lyme disease; viruses like tick-borne encephalitis virus or Crimean-Congo hemorrhagic fever virus; and protozoa like *Babesia* spp., and rickettsiae like *Rickettsia rickettsii* (Estrada-Peña & de la Fuente, 2014; Parola & Raoult, 2001). Some TBDs can be zoonotic since they involve animal reservoirs. Important risks that arise from these diseases exist for livestock as well as for humans (Steere et al., 2004). Ticks can also transmit pathogens to animals and cause direct and indirect losses. Costly treatments and control measures are typically needed to address these infections (de la Fuente et al., 2008; Ghosh et al., 2006).

Tick-borne viral diseases (TBVDs) are a growing global health concern because some ecological and environmental changes have greatly expanded the geographic range of the population density of tick vectors (Eisen & Eisen, 2018; Süss, 2011), mainly due to a diverse group of viruses. Tick-borne viral diseases can be severe or even more so, though frequently less known than tick-borne protozoal or bacterial diseases, and many present high mortality and morbidity risks (Randolph, 2009). Ticks serve as both vectors and reservoirs for a wide array of viruses. Because wildlife hosts are involved, like rodents, birds, and ungulates, these pathogens are sustained in enzootic cycles. Human infection often occurs incidentally when people encounter infected ticks outdoors in endemic regions (Estrada-Peña et al., 2013a). Flavivirus, Nairovirus, Reovirus, Arenavirus, and others are included in these families (Labuda & Nuttall, 2004).

Linked factors contribute to the increasing incidence of TBVDs. Climate change lets tick habitats spread into formerly non-endemic areas by altering temperature and humidity patterns, since those patterns affect tick survival and reproduction (Ogden & Lindsay, 2016). Human exposure to ticks has also increased due to deforestation, urbanization, and changes in land use, which disrupt ecosystems and bring wildlife and humans into closer contact. Furthermore, international travel and trade contribute to the transboundary movement of infected ticks and hosts. Because ticks and hosts are infected, this eases the emergence of TBVDs in new regions (Hermance & Thangamani, 2017; Kilpatrick & Randolph, 2012). TBVDs can present important challenges for those public health systems. Many infections remain underdiagnosed and underreported because their symptoms are nonspecific, and endemic regions are often diagnosed with few symptoms. The clinical presentation of TBVDs often mimics other febrile illnesses, so differential diagnosis proves difficult in the absence of specific laboratory testing (El-Radhi, 2019). Because specific antiviral treatments or vaccines are not widely available for many TBVDs, people rely on supportive care and vector control to manage the disease (Kosoy et al., 2015).

5.2 Climate Change and Tick Ecology

5.2.1 Temperature Changes and Tick Activity

Temperature plays a central role in tick development, survival, and behavior. Rising global temperatures are expanding the geographic range for many tick species. Rising temperatures also strengthen the population dynamics of these species in areas where they already exist (Eisen & Eisen, 2018; Ogden et al., 2006). The blacklegged tick, *I. scapularis*, has been spreading northward into Canada and the higher United States elevations. Winters that are milder and periods of more extended favorable temperatures essentially allow ticks to complete their life cycles in some new regions (Estrada-Peña et al., 2013b). Likewise, the lone star tick *Amblyomma (A.) americanum*, significantly expanded its range throughout the Midwest and Northeast U.S. (Jones et al., 2008; Ogden & Lindsay, 2016).

Temperature influences the range; move to find a host, and questing behavior. Warmer temperatures increase questing frequency and duration, heightening the probability of ticks finding and attaching to a host (Eisen & Eisen, 2018; Kilpatrick & Randolph, 2012). This heightened probability results from increased questing frequency and duration. Furthermore, higher temperatures accelerate tick development to shorten how long the life cycle lasts, potentially increasing the number of generations (Randolph, 2009; Ogden & Lindsay, 2016). Tick survival has a thermal threshold, however. Extremely high temperatures can result in tick desiccation and mortality. This is quite true when adequate humidity is absent (Estrada-Peña et al., 2013b; Kilpatrick & Randolph, 2012).

5.2.2 Impact of Seasonal Shifts on Tick Life Cycle

Seasonality is undoubtedly a key factor that influences tick ecology. Usual seasonal patterns regulate molting, host-seeking, and reproduction time. Climate change causes milder winters, earlier springs, and extended autumns. The seasonal timing regarding these processes is being disrupted (Jones et al., 2008; Randolph, 2009). Higher overwinter survival rates occur during mild winters, especially for adult and nymph tick stages. Tick populations are more likely to survive by resuming activity earlier in the spring, as winter temperatures rise and snowfall becomes less frequent (Eisen & Eisen, 2018). With an earlier spring onset, nymphal ticks become active sooner. This earlier emergence especially concerns people because nymphs transmit human disease most dangerously. They are small, hard to detect, and efficiently transmit pathogens (Estrada-Peña et al., 2013a).

Climate-induced changes can cause an overlap in activity periods among adults, nymphs, and larvae (Fig. 5.1). This overlap eases co-feeding transmission since ticks feed on a single host and share pathogens, so disease spreads more as no systemic infection exists in the host (Ogden & Lindsay, 2016; Eisen & Eisen, 2018).

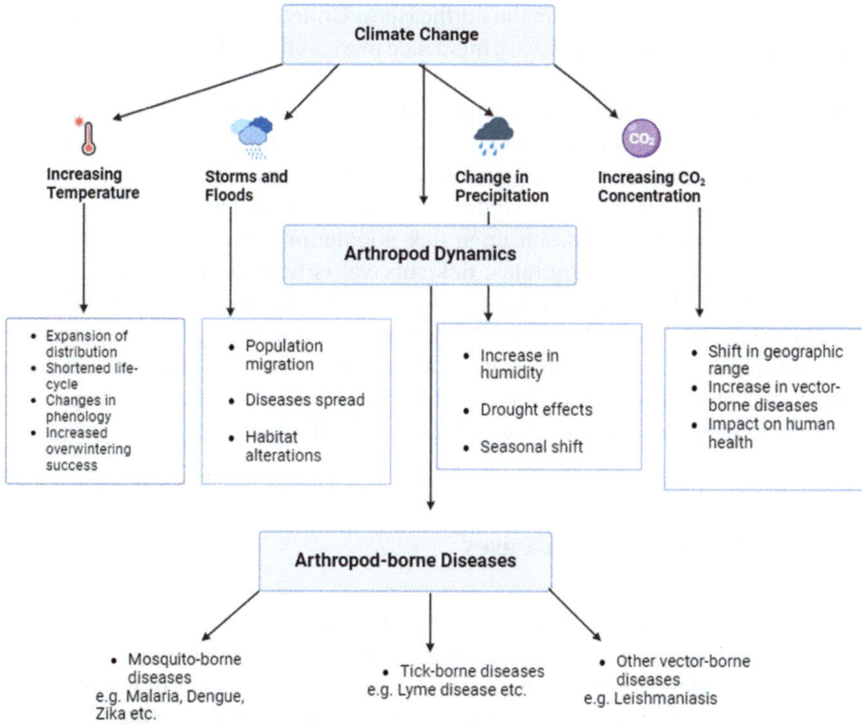

Fig. 5.1 Climate factors affecting arthropod dynamics

Seasonal shifts can affect host behavior. Deer, rodents, and birds are important hosts for immature and adult ticks. They could shift breeding or migratory patterns due to temperature and vegetation changes. These shifts have the potential to desynchronize the availability of hosts with tick activity, and this may disrupt life cycle timing and reduce or increase tick reproductive success depending on the local context (Jones et al., 2008; Kilpatrick & Randolph, 2012).

5.2.3 Effects of Humidity and Precipitation on Tick Habitats

Ticks remain relatively sensitive to moisture and humidity. Relative humidity is vital since it helps ticks maintain water balance when off-host, primarily when they pursue and are exposed to open air. Precipitation is what indirectly controls both tick distribution and density. Precipitation maintains or disrupts moisture in microhabitats like leaf litter and shaded vegetation (Eisen & Eisen, 2018; Patz et al., 2005). Tick survival conditions in regions experiencing increased rainfall from climate change have been more favorable. Environments that are moist, lower the risk of desiccation. In these environments, ticks actively seek hosts through extended periods (Ogden & Lindsay,

2016). Researchers noticed this in the northeastern United States and central Europe, where tick densities and Lyme disease incidence increased with higher humidity (Patz et al., 2005).

On the other hand, excessive precipitation and flooding can reduce tick populations temporarily by drowning immature stages or washing away leaf litter. Tick populations may rebound quickly after the habitat recovers, especially if host availability is high (Jones et al., 2008; Ogden & Lindsay, 2016). Droughts, on the other hand, pose a significant constraint upon tick populations. Due to lower soil moisture plus increased evaporation rates, tick survival is limited in arid areas. This is especially relevant for *I. ricinus* and *A. americanum*, species dependent on humid microenvironments (Eisen & Eisen, 2018; Estrada-Peña et al., 2013b). Nevertheless, some species adapt when they alter questing behaviors or when they seek hosts during cooler and more humid times of the day, or when they burrow deeper into vegetation so that they retain moisture (Jones et al., 2008).

5.3 Tick-Borne Viral Diseases

There is a global increase in TBVDs owing to the expansion of tick populations into new geographic regions and enhanced human exposure brought on by environmental and climatic changes (Ergunay et al., 2023). Numerous factors, some of which may be local, can lead to an increase in cases, the discovery of new diseases, or the emergence of new epidemics in previously pathogen-free areas. For instance, climate change has been connected to the rise of some tick species, and metagenomics has facilitated the identification of previously unidentified tick-borne viruses. Some ideas suggest that human-caused landscape changes, such as altered cultural patterns, the abandonment of cultural sites, and deforestation in particular South American regions, are the primary source of TBVDs outbreaks in both humans and veterinary populations (de la Fuente et al., 2023a, b).

5.3.1 Tick-Borne Encephalitis (TBE)

Tick-borne encephalitis virus (TBEV) is a small, enveloped *flavivirus* with a diameter of about 50 nm. TBEV contains an unsegmented single-stranded positive RNA genome of approximately 11 kb. Three structural proteins—capsid (C), membrane (M), and envelope protein (E)—as well as seven non-structural proteins, are produced by co- and post-translational processing of the single polyprotein it encodes (Stibraniova et al., 2025). Non-structural proteins are vital in viral propagation, structural protein processing, and maintenance of host cell physiology. The M glycoprotein, formed explicitly as a precursor (prM), binds to the envelope protein to prevent the fusion loop from activating prematurely. The genome and capsid protein belong to the nucleocapsid in the viral envelope of lipids produced by the host cells, as do the

M and E glycoproteins (Pulkkinen et al., 2018). Glycoprotein E is the main antigen of TBEV and it has a role in membrane fusion and receptor binding. The glycoprotein E coding gene is frequently sequenced and analysed. However, a paired distance analysis revealed that it follows unique evolutionary trends compared to other TBEV genomic domains (Kwasnik et al., 2023). A phylogenetic analysis showed the presence of three TBEV subtypes. The TBEV subtypes have a molecular divergence of 5% to 6%, indicating they are closely linked. Despite considerable genetic similarities, each subtype's infections differ from the others (Bogovic et al., 2015).

The primary way that TBEV is transmitted to humans and animals is through tick bites from *I. ricinus* and *I. persulcatus*. TBEV persists in small geographically defined areas termed "natural foci". TBEV is primarily spread by small rodents, insectivores, and domestic or wild animals (Böhm et al., 2017). Ticks and warm-blooded hosts can engage in horizontal sexual transmission. Along with nonviremic, sexual, and other modes, TBEV can spread by the transovarial and transstadial routes (Bakhvalova et al., 2009). The virus can be seen from one health perspective because it transmits between ticks, animals, and humans. Humans are accidental and dead-end hosts, infected primarily through the bites of hard ticks. Consumption of contaminated milk or dairy products from infected animals may also lead to TBEV transmission. (Ilic et al., 2020; Kwasnik et al., 2023). Furthermore, new virus infection routes have been documented, including organ transplants and blood transfusions, while dissemination through air remains a theoretical possibility (Yu et al., 2024). The overview of tick-borne encephalitis virus (TBEV) is given in Fig. 5.2.

After a tick bite, the TBEV undergoes primary local replication. Dendritic skin cells (Langerhans cells) are proposed as the first line of defense, capturing the virus

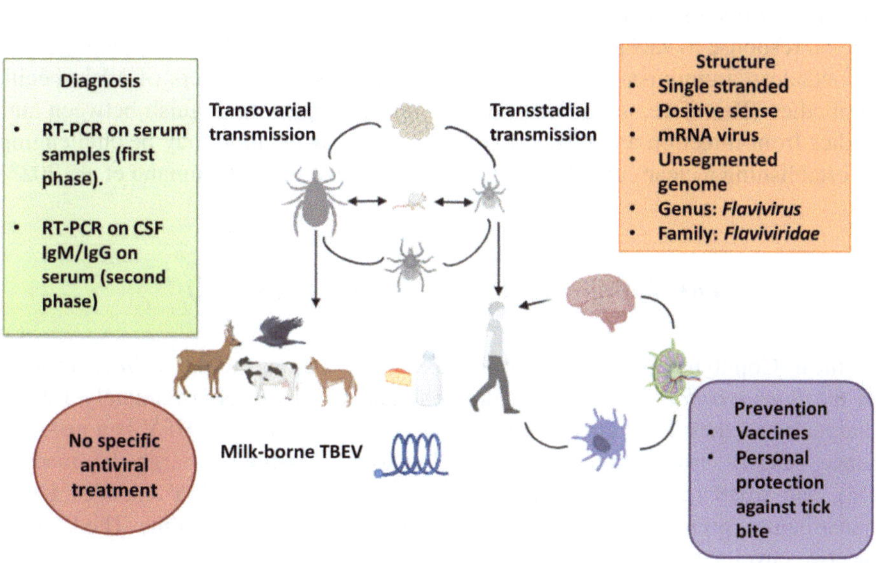

Fig. 5.2 Overview of tick-borne encephalitis virus (TBEV)

and transporting it to regional lymph nodes. TBEV is disseminated from this primary site to extra-neural organs, where sustained replication of the TBEV maintains viremia for several days. The viremia phase is most likely when the virus enters the brain (Mandl, 2005). The precise way TBEV crosses the blood–brain barrier remains speculative; four routes have been suggested: through the peripheral nerves, susceptible olfactory neurons, transcytosis, or interstitial diffusion. The primary targets of TBEV infection are neurons in the CNS (Bogovic et al., 2015).

TBE is a clinical condition induced by TBEV. It has an incubation period of 7–14 days and can last up to 28 days (Callaby et al., 2025). The disease produced by TBEV infection consists of two stages: a first viremic phase, which might progress to a second neurological phase. The abortive clinical pattern indicates that disease progression may cease after the early stages. This kind of TBE might be asymptomatic or cause a mild febrile illness that does not progress to a neurological form. About one-third of TBE patient's progress to the second stage of the disease. This phase is divided into meningeal and focal forms (Brêchet et al., 2025).

Diagnosing TBEV is relatively simple. The first serum samples collected when CNS symptoms arise in the 2nd phase of the disease include TBEV-IgM and more broadly, TBEV-IgG antibodies (Holzmann, 2003). During the initial stage of disease, isolation and detection of the virus can be done using RT-PCR. However, TBEV is seldom found in CSF at the start of the second phase and only very rarely in people with progressing disease (Lindquist & Vapalahti, 2008).

There is presently no treatment for TBE. The most effective strategies for preventing TBEV infections are active vaccination and tick bite avoidance through personal protection (Ackermann-Gäumann et al., 2024). TBE immunizations include traces of viral nonstructural protein 1 (NS1). As a result, the identification of NS1-specific antibodies has been proposed as a method for distinguishing the immunological response to vaccination from spontaneous infection (Albinsson et al., 2024). Nonetheless, immunisation results in dose-dependent, mild titers of NS1-specific antibodies. Therefore, using NS1-specific IgG testing to distinguish between antibodies from infection and those obtained through vaccination may be challenging, so establishing a clear cut-off point is essential (Ackermann-Gäumann et al., 2023).

5.3.2 Crimean-Congo Hemorrhagic Fever (CCHF)

Crimean-Congo hemorrhagic fever virus (CCHFV) belongs to the *Nairovirus* genus of the *Bunyaviridae* family. These are spherical virions of approximately 100 nm diameter and their lipid envelope is obtained from the host. CCHFV is a negative-sense, ssRNA virus with a tri-segmented genome classified as large (L), medium (M), and small (S) based on size (Aslam et al., 2016). CCHFV encodes 1 or 2 nonstructural proteins (NSs) in addition to the four structural proteins. The CCHFV has two envelope glycoproteins (Shayan et al., 2015).

Even though animals do not appear to exhibit any symptoms of the disease, wild and domesticated animals are crucial to the cycle of disease transmission because they serve as a reservoir for ongoing tick infection. Many tick species have been linked to the CCHFV. However, members of the genus *Hyalomma* are thought to be the principal vectors and most prevalent ticks capable of transmitting the virus to humans. They flourish in hot, dry, or semi-arid environments and may be found in Africa, Asia, and Europe (Messina et al., 2015). The tick *H. marginatum* has two hosts. After hatching, larvae look for their first blood meal among tiny animal hosts such as rodents, birds, and lagomorphs. Following engorgement, the larvae remain on the same host and grow into nymphs. The nymphs continue to eat the same animal until they get engorged and molt into adults. Ticks look for larger vertebrate hosts, such as livestock to feed and reproduce. The engorged females then separate and ovulate within their surroundings (Celina et al., 2025).

People who live near animals or tick vector habitats are more susceptible to infected tick bites, whereas those working in slaughter houses are at risk of blood-borne infection. Human-to-human transmission is possible (Fig. 5.3), particularly among healthcare personnel or close relatives who come into contact with contaminated blood or body fluids from CCHFV patients (Messina et al., 2015).

The four primary phases of a Crimean-Congo Hemorrhagic Fever (CCHF) infection are incubation, pre-hemorrhagic, hemorrhagic, and convalescence (Karanam et al., 2025). The disease's incubation period can range from 3 to 14 days, contingent on the host's immune condition, the viral load, and the virus's mode of transmission (Gürbüz et al., 2021). When the disease is not treated, individuals may die because of multiple organ failure. Survivors of CCHF may require a whole year to recover completely (Ergönül, 2006). Although the pathophysiology and immunobiology of CCHF are not fully known, it seems that many pathways contribute to viral entry,

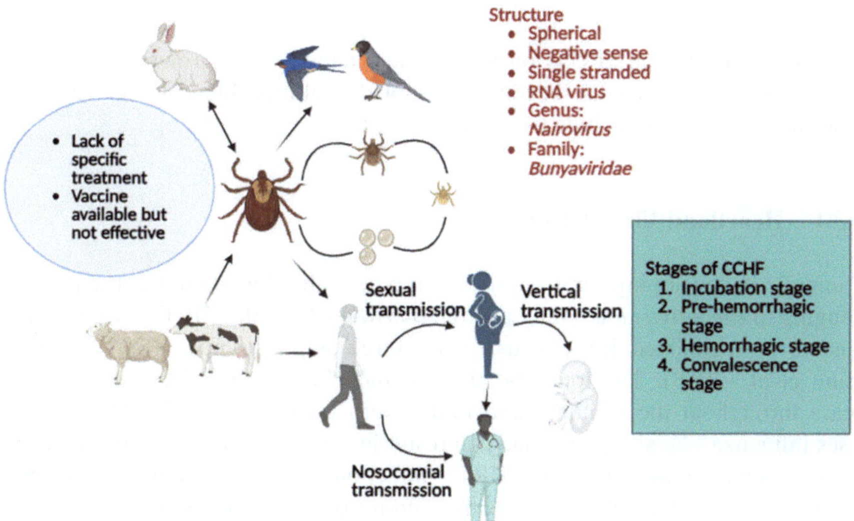

Fig. 5.3 Overview of Crimean-Congo Hemorrhagic Fever (CCHF)

replication, and immune response (Appannanavar et al., 2011). CCHFV invades the epithelium after transmission by infected tick and reaches the basolateral compartment of the epidermis, infecting endothelial cells in local capillaries and tiny blood vessels, dendritic cells, and macrophages (Frank et al., 2024).

Two main techniques for detecting CCHF during the first week following the onset of symptoms are viral antigen detection assays and nucleic acid amplification testing (Raabe, 2020). The antigen-capture enzyme immunoassay (EIA) is less sensitive than the nested RT-PCR for viral antigen detection. Furthermore, immunological testing is more effective at detecting CCHF after the first week (Pirincal & Doymaz, 2024).

CCHF is a severe viral disease that significantly threatens public health. Understanding the pathophysiology and immunology of CCHF is crucial for finding effective countermeasures due to its high death rate and the lack of a specific treatment or vaccination. The genomic variability of CCHFV makes it challenging to develop a vaccine that protects against all virus strains. Furthermore, most CCHF vaccination experiments used the prototype IbAr10200 CCHFV strain, which was obtained from ticks and has unknown human toxicity (Spik et al., 2006). To overcome these obstacles, heterologous challenge trials are necessary to create more potent vaccines capable of defending against a wide range of CCHFV strains. Despite encouraging preclinical research, no human trials of these vaccines have been conducted, and it is unclear if the findings can be effectively used in human clinical trials (Ozdarendeli, 2023).

5.3.3 Other Emerging Viral Threats

Due to their substantial effects on humans and animals and growing geographic range, tick-borne viruses are becoming a global health problem. Other emerging viruses in this category include the following viruses.

5.3.3.1 Heartland Virus (HRTV)

An unnoticed developing viral disease with potentially devastating consequences, belonging to *Phenuiviridae* family and the *Bunyavirales* order. HRTV is a negative-stranded, tri-segmented RNA virus. It has three single-stranded RNA segments (Brault et al., 2018). The lone star tick (*A. americanum*) reliably transmits this disease throughout the southern and midwestern United States. HRTV infection causes influenza-like symptoms and can result in severe disease, even death. There is no recognized treatment for the HRTV infection, nor a vaccine as a therapeutic strategy. Because of a lack of prospective and retrospective research, the disease is most likely grossly underreported (Dembek et al., 2024).

5.3.3.2 Powassan Virus (POWV)

North America's sole tick-borne encephalitis serotype. It is a single-stranded, neuroinvasive, positive-sense RNA flavivirus (*Flaviviridae).* In 1958, POWV was discovered in the brain of a five-year-old kid who died of encephalitis in Powassan, Ontario. POWV is an uncommon and severe virus that produces a rapidly spreading disease. Two essential tick species, *I. scapularis* and *I. cookei,* provide natural support for the virus (Hermance & Thangamani, 2017). The interactions between pathogen, host, vector, and environment are intricate and interwoven (Hassett et al., 2021).

5.3.3.3 Bourbon Virus (BRBV)

Single-stranded negative-sense RNA genome of 10–11 kb. Electron imaging of BRBV virions reveals various genomic regions that are certainly internally coated with NPs, as well as GP molecules on the virion surface (Bai et al., 2019). Like HRTV, BRBV appears to be vectored by *A. americanum.* Infected individuals have displayed symptoms such as thrombocytopenia, leukopenia, maculopapular rashes, fever, headache, nausea, and vomiting. Currently, no vaccines or particular antiviral medications are available to treat BRBV-related diseases (Shah et al., 2023).

5.3.3.4 Kyasanur Forest Disease (KFD)

Flavivirus-borne disease caused by the KFD virus (KFDV). KFDV is a virus carried by hard ticks known as *Haemaphysalis spinigera* (Varma et al., 1960). KFD often develops after a 3–8-day incubation period, with symptoms comparable to viremia. While the sickness is self-limiting and requires a 7–10 day recovery period, around 10–20% of those infected with this virus showed hemorrhagic, encephalitic, and multi-organ complications (John et al., 2014).

5.3.3.5 Omsk Haemorrhagic Fever

Single-stranded, a positive-sense RNA genome that is wrapped in a capsid protein (Gritsun et al., 2003). There is no specific vaccination for Omsk haemorrhagic fever. However, the TBEV vaccine may provide some protection against it. The virus is mainly transferred by *Dermacentor reticulatus* ticks, while people can become infected through contact with infected muskrats (*Ondatra zibethicus*). Muskrats are particularly vulnerable to the Omsk hemorrhagic fever virus (Růžek et al., 2010).

5.3.3.6 Louping ill

Louping ill virus (LIV) belongs to the family *Flaviviridae*. The virus measures 40–50 nm in length and has one strand of RNA. The virus induces encephalitis and is spread by the sheep tick *I. ricinus*. Human infection was first recorded in I934 (Davidson et al., 1991). Apart from the likelihood of continual virus spread, few pathogenic studies have been carried out. LIV infection in mice induces significant neuropathological changes and ultimately death (Yu et al., 2024).

5.3.3.7 Dabie Bandavirus

Previously referred to as severe fever with thrombocytopenia syndrome (SFTS). Virus belongs to *bandavirus* genus of *phenuiviridae* family (Ji et al., 2024). The genome of SFTSV is a segmented ssRNA. Even though the transmission cycle and point of emergence of SFTSV are unclear, *Haemaphysalis longicornis* is likely to spread the virus, and the increasing prevalence of these ticks increases the likelihood of SFTSV emergence. Tick bites are not the only way to prevent disease. The virus can spread horizontally from an infected tick to a host, vertically from a female parent tick to its offspring, or from an infected animal to another tick (Kim & Park, 2023).

5.3.3.8 Thogoto Virus (THOV)

Tick-borne orthomyxovirus whose genome consists of 6 negative-polarity ssRNA segments (Kochs et al., 2000). Initially discovered in the tick species *Rhipicephalus decoloratus*, the virus has now been reported in *H. longicornis*, which might serve as transmission vectors. However, no additional human illnesses have been detected since the initial findings. Abortions in sheep have now been linked to THOV infection, and both domestic and wild animals have shown positive seroprevalence (Yu et al., 2024).

5.3.3.9 Colorado Tick Fever Virus (CTFV)

This virus belongs to the *Coltivirus* genus and the *Reoviridae* family, with a genome consisting of 12 double-stranded RNA segments. CTFV is predominantly disseminated by *Dermacentor andersoni*, which use chipmunks, squirrels, and other rodents as natural hosts (Williamson et al., 2019). The agent is widely distributed across the western United States and southern Canada, particularly in hilly regions. Currently, the virus does not appear to be divided into multiple clades. Nevertheless, reassortment is made possible by its segmented genome (Alkishe & Peterson, 2022).

5.4 Recent Outbreaks and Case Studies

The rising frequency of ticks and tick-borne diseases poses significant public health challenges. Difficulties in controlling tick populations, along with the complexities of diagnosing and treating infections caused by the pathogens they transmit, hinder effective management of these emerging diseases (Madison-Antenucci et al., 2020).

5.4.1 Analysis of Recent TBE Outbreaks

Among Europe's, *I. ricinus* ticks is most significant vectors and carriers of TBEV. This dangerous arboviral zoonotic infection affects the CNS and is commonly seen in Asia and Europe (Pustijanac et al., 2023). In addition to Russia, Mongolia, and portions of China and Japan, this virus is endemic in central, eastern, and northern Europe. According to research, about 170,000 human cases were documented in Russia and Europe between 1990 and 2009 (Süss, 2011). According to the survey, 51,519 confirmed cases were reported in Europe between 2000 and 2019. The cases decreased in 2014 but increased again after 2015. The host community, host migration, environmental factors, and human mobility in the foci area are the primary causes of TBEV incidence. Over the whole period from 2000 to 2009, the average incidence rate was 3.27 (Ijaz et al., 2023).

A study on the prevalence of TBEV has illustrated that the period (2015–2019) showed that the Baltic Sea region and Central Europe are vulnerable to TBEV infection. Even with COVID-19 and the subsequent restrictions, the incidence of TBE increased in over 50% of the countries under study. The notable rise in the incidence of TBE during the pandemic (2020) occurred among countries with different climate zones. Considerable differences in pre- and post-pandemic status were evident in a 2022 research study that investigated the occurrence of TBE among European countries in the aftermath of the COVID-19 pandemic. At 25.45 per 100,000, the highest rate occurred in Lithuania with TBE. An exceptionally high incidence of TBE was seen in Central European nations. 12 of 23 countries under analysis depicted that the rate of TBE spiked during the COVID-19 pandemic in 2020, in contrast to 2015–2019 (Zając et al., 2022).

A study was conducted to analyze the epidemiology of TBE in animals in the endemic areas of Croatia's mainland during the period 2017–2020. The overall seroprevalence rates were 14.3% in 2017, 10.0% in 2018, 7.3% in 2019, and 17.1% in 2020. These differences were notable from year to year. The north-western and eastern counties had the highest seropositivity. According to a county-by-county analysis, the seroprevalence rates in 2017 ranged from 0 to 28.5%, in 2018, from 4.3 to 20.5%, in 2019, from 2.8 to 14.4%, and in 2020, from 5.5 to 35.7% (Vilibic-Cavlek et al., 2024).

A study was conducted to determine the epidemiological situation of tick-borne diseases in Latvia between 1993 and 2002. The yearly rate of field-collected adult tick infection with TBEV varied from 1.7% to 26.6% for *I. ricinus* adults, and 0% to 37.3% for *I. persulcatus*. Many ticks were collected from humans, exceeding 30% during this period. TBE virus isolates from ticks and patient blood samples were analyzed, and all TBEV subtypes were discovered to be co-circulating in Latvia (Bormane et al., 2004).

A retrospective study of the epidemiology patterns of TBE in Lithuania and clinical manifestations in adults indicated that incidence rates spiked by 8.5% each year between 1970 and 2014. TBE was linked to meningoencephalitis in 81.3% of cases. A total of 18% of patients experienced severe illness. The usual neurological symptoms found were ataxia (81.3%), meningeal signs (66.5%), and tremor (47.56.3%) of patients had limb paresis. 76.7% of patients were discharged with problems, whilst 0.7% of patients died (Radzišauskienė et al., 2018).

5.4.2 CCHF Outbreaks and Public Health Response

Among viral hemorrhagic fevers, CCHF is a newly identified infectious disease. Case fatality rate could be anywhere between 10 and 40%. Public health is seriously threatened by such a disease, especially in low- and middle-income nations where poverty is widespread and human-animal contact is frequent (Sravani, 2025). Avoiding close contact with sick people and animals and restricting or avoiding contact with infected ticks and biomaterials are ways to prevent and control the CCHF virus. Healthcare personnel are especially susceptible to infection. One of the most important endemic zoonotic illnesses in Afghanistan is CCHF. Of Afghanistan's 34 provinces, 26 reported cases of CCHF in 2022. The endemicity of CCHF makes sense, as approximately 21 million cattle in 34 districts of a country of 40 million people can harbor the vector (Gaina et al., 2023).

In Pakistan's Balochistan province, 2542 cases of CCHF were recorded between 2000 and August 10, 2021; 1418 cases had laboratory confirmation. Males made up 89% of the cases (n = 1262). Over the past ten years, the case fatality rate has risen from 5 to 13%. With 172 cases, 2017 has the most CCHF cases, followed by 2005 with 108 and 2004 with 107. 61% of cases included direct animal interaction, with farmers and butchers accounting for 22% of these cases (Naseer et al., 2024). With the help of farmers, hunters, cattle breeders, and bushmeat traffickers in 15 districts spread across the nation's three Agro-ecological zones, a cross-sectional study was conducted in Benin between 2022 and 2024. Serum samples from 366 animals, 254 from domestic animals, and 112 from wild species were examined. Antibodies against the CCHF virus were found in 1.18% of the wild animals examined. The frequencies of seroprevalence were 1.19% in gigantic rats, 5.88% in hares, and 3.7% in squirrels. 38 of 112 domestic animal samples had anti-CCHFV antibodies, with a seroprevalence of 33.93%. Specifically, 34 of 81 cattle and 4 of 24 horses showed antibodies. No positive cases were found in pigeons (Yessinou et al., 2025).

Variations in the frequency of CCHFV antibodies in animals by breed and age suggest different levels of exposure and susceptibility. Younger animals, such as calves, are particularly vulnerable to catching the virus because they are more likely to contact infected ticks when grazing. On the other hand, because they are exposed more often, older animals typically have high serum antibody levels. Crossbred cattle can occasionally be more seropositive than local breeds for genetic or environmental reasons. According to longitudinal research, animals with tick infestations and antibodies may be susceptible to reinfection. Since CCHFV antibodies can remain in sheep that have contracted the virus on their own for up to two months, effective treatment and surveillance strategies are crucial (Celina et al., 2025). A survey was carried out on the seroprevalence of CCHF in domesticated animals in Northwestern Senegal. CCHF-specific antibodies were found in 32.5% of 283 animal sera. The infection rates for CCHFV in horses, cattle, sheep, dogs, donkeys, and goats were 70.3%, 57.1%, 22.1%, 18.2%, 17.2%, and 6.9%, respectively. Age and gender were shown to have a substantial impact on antibody titers. This study found a considerable seroprevalence of CCHFV among animals in Senegal's northern area, showing that the infection is endemic in the region, and there is a need to extend CCHF monitoring measures in Senegal (Mangombi et al., 2020).

5.5 Impact of Climate Change on Surveillance Systems

Climate change is reshaping the landscape of global public health, infrastructure, and disease surveillance systems (Liao et al., 2024). As global temperatures rise, precipitation patterns shift, and extreme weather events occur more frequently, because the earth's climate is increasingly volatile, the climate's influence on technological and public health surveillance infrastructure deepens (Paz et al., 2024). Surveillance systems are critical in monitoring disease outbreaks, security threats, and environmental changes. To an increasing extent, surveillance systems are vulnerable to both the direct and the indirect effects of climate change (De La Rocque et al., 2008). TBDs present prominent complexity, especially TBVDs, emerging more frequently in novel areas because of environmental changes, host availability, and vector ecology shifts (Swei et al., 2020). On surveillance systems, physical damage inflicted on the infrastructure is one of climate change's most immediate and observable impacts. Extreme weather events like hurricanes, floods, heatwaves, and damage surveillance equipment such as CCTV cameras, weather sensors, communication towers, and radar systems; global warming increases these events in frequency and severity (Valavanidis, 2022). Rising sea levels, with storm surges, particularly threaten surveillance systems as systems are in coastal and low-lying regions, often totally failing or needing costly relocation and repair (Azevedo de Almeida & Mostafavi, 2016). For instance, flooding can potentially submerge the camera systems and data centers. Also, it corrodes electrical components, disrupting operations (Thatikonda, 2023). Likewise, excessive heat from heat waves can impair the function of sensitive electronics. This heat can cause sensors to drift, reduce image quality, and shut systems down (Khanna, 2023). Due to climate change, wildfires, which are strengthening now, destroy critical infrastructure, such as poles, fiber cables, and solar panels, which support remote surveillance systems (Chatterjee et al., 2024). Since studies have revealed that the physical resilience of technological infrastructure is compromised under climate-induced stress, strong adaptation and maintenance strategies are necessary (Liu et al., 2018).

Climate change and surveillance intersect most urgently within public health, particularly while monitoring and responding to TBVDs. Climate change dramatically alters the ecology of the main vectors of TBVDs like CCHF (Hewson, 2024). Because of warmer temperatures and milder winters that have extended the geographical range for tick species into higher latitudes and altitudes, TBVDs are now in new and previously unaffected regions (Nuttall, 2022). TBE cases are now reported throughout parts of Europe where ticks were historically absent in those areas. The environment changes because of humidity and vegetation, so ticks reproduce and seek hosts more easily (Lindgren et al., 2000; Medlock & Leach, 2015). Because existing disease surveillance systems are often based upon geographic trends and historical ecological data, they are challenged by these shifts.

Additionally, ticks are seasonally active for extended periods, resulting in longer windows for transmission and more frequent exposure to humans (Wilhelmsson et al., 2013). Surveillance systems must adapt to monitor year-round risks rather than just seasonal outbreaks. This requires improved field data collection, including real-time vector mapping with predictive modeling based on climate variables (Anh Tuan & Dang, 2025). However, limitations regarding diagnostics obstruct knowing how effective current surveillance is, particularly for viral infections that require biosafety level 3 or 4 laboratories. Many rural and forested regions, in which TBVDs are most likely to emerge, lack adequate infrastructure for surveillance and diagnostic testing, and this leads to outbreaks that are delayed or underreported (Braks et al., 2022; Gubler, 2002).

Another problem is that TBVDs often remain underreported or misdiagnosed, especially in resource-poor regions. Misclassification plus poor case reporting happen because TBVD symptoms resemble bacterial or parasitic infections (Eshetu, 2015). Climate-induced stress affects healthcare systems, so vector surveillance is overlooked (Arias et al., 2024). Outbreaks of diseases such as COVID-19 or dengue also divert attention and resources. This delay is a hindrance to the outbreak response and containment (Hasan et al., 2022). Surveillance frameworks integrated are needed more. They combine entomological data, climate modeling, real-time diagnostics, and public health reporting.

A foundation for controlling TBDs is establishing strong surveillance systems for monitoring vector populations, animal hosts, and human cases (Fournet et al., 2018). Surveillance enables early outbreak detection, disease trend tracking, and targeted intervention guidance. Effective surveillance must passively rely on clinicians' and laboratories' reports, actively sample ticks, survey sera, and then monitor sentinel animals (Rowan et al., 2023).Surveillance has been further strengthened by the integration of geographic information systems (GIS) and remote sensing technologies so health authorities can correlate environmental conditions with tick density and infection prevalence (Eisen et al., 2017; Estrada-Peña et al., 2013b).

5.6 Assessing the Risk to Human Populations Under Climate Change

An important factor influencing human risk is the prolonged duration of human exposure, which increases seasonal activity and propagation of ticks. Ambient temperature and moisture levels significantly affect tick life cycles (Ogden et al., 2021). Ticks commence questing sooner on account of warmer summers. Extended summers also entail ticks remaining active for a longer duration. This prolonged duration elevates the aggregated hazard of tick bites, so groups involved in outside labor or leisure are notably susceptible to possible contagion (Eisen & Eisen, 2018; Ogden et al., 2013). TBVD outbreak hazard likewise escalates outside ecological dynamics via land-use changes and human incursion into tick habitats (Diuk-Wasser et al., 2021). Deforestation, agricultural expansion, and suburban development have increased interaction between humans, wildlife hosts, and tick vectors. Ecological niches are fractured; therefore, tick-host interactions are strengthened proximately to human residences. Consequently, zoonotic viruses show elevated transmission (Estrada-Peña et al., 2014; Ostfeld et al., 2006). Livestock translocation alongside shifts within grazing habits additionally introduce infected ticks into novel territories, especially applicable for CCHF propagation through *Hyalomma* ticks throughout agrarian sections of Africa, the Middle East, and Central Asia (Spengler et al., 2019).

TBVD risk is also indirectly affected by climate change through socioeconomic vulnerability and public health infrastructure. Healthcare infrastructures are presently overburdened in resource-scarce environments. Identifying, ascertaining, and reacting toward TBVD proliferations in that location remains challenging. CCHF and analogous viral ailments require clinical attention and high-security laboratories, notably. Within regions where health surveillance is weak, they underreport and misdiagnose commonly, so recognition of and response to outbreaks are delayed (Leblebicioglu et al., 2016; Spengler et al., 2019).

As a result of climate and environmental changes driving the emergence and re-emergence of TBVDs. These diseases pose an increasing challenge in forecasting and risk assessment. Powassan virus throughout North America, a nascent TBVD vectored via *I. scapularis*, has engendered lethal encephalomyelitis in recent years. Restricted clinical cognizance alongside testing accessibility implies it stays underdiagnosed (Ebel, 2010; Hermance & Thangamani, 2017). Alkhurma hemorrhagic fever virus, an emerging issue within the Arabian Peninsula, represents a tick-borne flavivirus throughout Saudi Arabia, also without strong epidemiological data (Charrel et al., 2005). These instances stress that integrated monitoring is vital and merges clinical, entomological, and ecological information. Global trade and travel have rapidly facilitated the spread of TBDs across continents, significantly contributing to their widespread transmission. Due to these factors, in human population, a One Health approach is essential for risk assessment, as it recognizes the interconnectedness of animal, human, and environmental health. Environmental determinants such as temperature and rainfall must be integrated within hazard models alongside tick and host ecology, land use, human behavior, and pandemic health system capacity. Elevated-risk zones can be discerned via GIS, ecological niche modeling, and climate forecasting. Targeted vaccination, public education, coupled with vector control, represent proactive interventions that these systems can inform (Ogden & Lindsay, 2016; Eisen et al., 2017; Semenza & Suk, 2018).

References

Ackermann-Gäumann, R., Brêchet, A., Smetana, J., Salát, J., Lienhard, R., Croxatto, A., & Růžek, D. (2023). Vaccination against tick-borne encephalitis elicits a detectable NS1 IgG antibody response. *Journal of Virological Methods, 322*, Article 114831.

Ackermann-Gäumann, R., Lang, P., & Zens, K. D. (2024). Defining the "Correlate(s) of Protection" to tick-borne encephalitis vaccination and infection–key points and outstanding questions. *Frontiers in Immunology, 15*, 1352720.

Albinsson, B., Hoffman, T., Kolstad, L., Bergström, T., Bogdanovic, G., Heydecke, A., & Petersson, A. (2024). Seroprevalence of tick-borne encephalitis virus and vaccination coverage of tick-borne encephalitis, Sweden, 2018 to 2019. *Eurosurveillance, 29*(2), 2300221.

Alkishe, A., & Peterson, A. T. (2022). Climate change influences on the geographic distributional potential of the spotted fever vectors *Amblyomma maculatum* and *Dermacentor andersoni. PeerJ, 10*, Article e13279.

Anh Tuan, D., & Dang, T. N. (2025). Climate-driven dynamics of infectious diseases: machine learning-based predictive modeling in Ba Ria-Vung Tau, Vietnam. *Tran Ngoc, Climate-Driven Dynamics of Infectious Diseases: Machine Learning-Based Predictive Modeling in Ba Ria-Vung Tau, Vietnam.*

Appannanavar, S. B., & Mishra, B. (2011). An update on Crimean Congo hemorrhagic fever. *Journal of Global Infectious Diseases, 3*(3), 285–292.

Arias, M. J., Kulkarni, N., Le, A., Holder, C. L., Unlu, I., & Fu, E. S. (2024). Climate change, emerging vector-borne illnesses, and anesthetic considerations. *Cureus, 16*(4).

Aslam, S., Latif, M. S., Daud, M., Rahman, Z. U., Tabassum, B., Riaz, M. S., & Husnain, T. (2016). Crimean-Congo hemorrhagic fever: Risk factors and control measures for the infection abatement. *Biomedical Reports, 4*(1), 15–20.

Azevedo de Almeida, B., & Mostafavi, A. (2016). Resilience of infrastructure systems to sea-level rise in coastal areas: Impacts, adaptation measures, and implementation challenges. *Sustainability, 8*(11), 1115.

Bai, C., Qi, J., Wu, Y., Wang, X., Gao, G. F., Peng, R., & Gao, F. (2019). Postfusion structure of human-infecting Bourbon virus envelope glycoprotein. *Journal of Structural Biology, 208*(2), 99–106.

Bakhvalova, V., Potapova, O., Panov, V., & Morozova, O. (2009). Vertical transmission of tick-borne encephalitis virus between generations of adapted reservoir small rodents. *Virus Research, 140*(1–2), 172–178.

Bogovic, P., & Strle, F. (2015). Tick-borne encephalitis: A review of epidemiology, clinical characteristics, and management. *World Journal of Clinical Cases, 3*(5), 430–441.

Böhm, B., Schade, B., Bauer, B., Hoffmann, B., Hoffmann, D., Ziegler, U., & Böttcher, J. (2017). Tick-borne encephalitis in a naturally infected sheep. *BMC Veterinary Research, 13*, 267.

Bormane, A., Lucenko, I., Duks, A., Mavtchoutko, V., Ranka, R., Salmina, K., & Baumanis, V. (2004). Vectors of tick-borne diseases and epidemiological situation in Latvia in 1993–2002. *International Journal of Medical Microbiology Supplements, 293*, 36–47.

Braks, M., Schaffner, F., Medlock, J. M., Berriatua, E., Balenghien, T., Mihalca, A. D., & Sprong, H. (2022). VectorNet: Putting vectors on the map. *Frontiers in Public Health, 10*, Article 809763.

Brault, A. C., Savage, H. M., Duggal, N. K., Eisen, R. J., & Staples, J. E. (2018). Heartland virus epidemiology, vector association, and disease potential. *Viruses, 10*(9), 498.

Brêchet, A., Kohler, P., Dörr, T., Grässli, F., Vock, M., Salát, J., & Croxatto, A. (2025). Tick-borne encephalitis virus seroprevalence and infection incidence in Switzerland, 2020–2021. *Scientific Reports, 15*(1), 8346.

Callaby, H., Beard, K. R., Wakerley, D., Lake, M. A., Osborne, J., Brown, K., & Davies, N. W. (2025). Tick-borne encephalitis: From tick surveillance to the first confirmed human cases, the United Kingdom, 2015 to 2023. *Eurosurveillance, 30*(5), 2400404.

Celina, S. S., Italiya, J., Tekkara, A. O., & Černý, J. (2025). Crimean-Congo haemorrhagic fever virus in ticks, domestic, and wild animals. *Frontiers in Veterinary Science, 11*, 1513123.

Charrel, R. N., Zaki, A. M., Fakeeh, M., Yousef, A. I., De Chesse, R., Attoui, H., & De Lamballerie, X. (2005). Low diversity of Alkhurma hemorrhagic fever virus, Saudi Arabia, 1994–1999. *Emerging Infectious Diseases, 11*(5), 683–688.

Chatterjee, P., Kadir, S. U., Srivastava, A., & Laszka, A. (2024). Grid resilience against wildfire with machine learning. *Big Data Application in Power Systems*, 393.

Davidson, M. M., Williams, H., & Macleod, J. A. (1991). Louping ill in man: A forgotten disease. *Journal of Infection, 23*(3), 241–249.

de la Fuente, J., Estrada-Peña, A., Rafael, M., Almazán, C., Bermúdez, S., Abdelbaset, A. E., & Ajagbe, D. O. (2023a). Perception of ticks and tick-borne diseases worldwide. *Pathogens, 12*(10), 1258.

de la Fuente, J., Estrada-Peña, A., Rafael, M., Almazán, C., Bermúdez, S., Abdelbaset, A. E., & Dahal, A. (2023b). Perception of ticks and tick-borne diseases worldwide. *Pathogens, 12*(10), 1258.

de la Fuente, J., Estrada-Peña, A., Venzal, J. M., Kocan, K. M., & Sonenshine, D. E. (2008). Overview: Ticks as vectors of pathogens that cause disease in humans and animals. *Frontiers in Bioscience, 13*(13), 6938–6946.

De La Rocque, S., Rioux, J. A., & Slingenbergh, J. (2008). Climate change: Effects on animal disease systems and implications for surveillance and control. *Revue Scientifique Et Technique, 27*(2), 339–354.

Dembek, Z. F., Mothershead, J. L., Cirimotich, C. M., & Wu, A. (2024). Heartland virus disease—An underreported emerging infection. *Microorganisms, 12*(2), 286.

Diuk-Wasser, M. A., VanAcker, M. C., & Fernandez, M. P. (2021). Impact of land use changes and habitat fragmentation on the eco-epidemiology of tick-borne diseases. *Journal of Medical Entomology, 58*(4), 1546–1564.

Ebel, G. D. (2010). Update on Powassan virus: Emergence of a North American tick-borne flavivirus. *Annual Review of Entomology, 55*, 95–110.

Eisen, R. J., & Eisen, L. (2018). The blacklegged tick, *Ixodes scapularis*: An increasing public health concern. *Trends in Parasitology, 34*(4), 295–309.

Eisen, R. J., Kugeler, K. J., Eisen, L., Beard, C. B., & Paddock, C. D. (2017). Tick-borne zoonoses in the United States: Persistent and emerging threats to human health. *ILAR Journal, 58*(3), 319–335.

El-Radhi, A. S. (2019). Differential diagnosis (DD) of febrile diseases. *Clinical manual of fever in children* (pp. 263–285). Springer International Publishing.

Ergönül, Ö. (2006). Crimean-Congo haemorrhagic fever. *The Lancet Infectious Diseases, 6*(4), 203–214.

Ergunay, K., Bourke, B. P., Reinbold-Wasson, D. D., Nikolich, M. P., Nelson, S. P., Caicedo-Quiroga, L., & Long, L. S. (2023). The expanding range of emerging tick-borne viruses in Eastern Europe and the Black Sea Region. *Scientific Reports, 13*(1), 19824.

Eshetu, E. (2015). A review on the diagnostic and control challenges of major tick-borne haemoparasite diseases of cattle. *Journal of Biology Agriculture Health, 5*(11), 160–172.

Estrada-Peña, A., & de la Fuente, J. (2014). The ecology of ticks and epidemiology of tick-borne viral diseases. *Antiviral Research, 108*, 104–128.

Estrada-Peña, A., et al. (2013a). Effects of environmental change on zoonotic disease risk: An ecological primer. *Trends in Parasitology, 29*(4), 205–214.

Estrada-Peña, A., Gray, J., Kahl, O., Lane, R. S., & Nijhof, A. M. (2013b). Research on the ecology of ticks and tick-borne pathogens—Methodological principles and caveats. *Frontiers in Cellular and Infection Microbiology, 3*, 29.

Estrada-Peña, A., Ostfeld, R. S., Peterson, A. T., Poulin, R., & de la Fuente, J. (2014). Effects of environmental change on zoonotic disease risk: An ecological primer. *Trends in Parasitology, 30*(4), 205–214.

Fournet, F., Jourdain, F., Bonnet, E., Degroote, S., & Ridde, V. (2018). Effective surveillance systems for vector-borne diseases in urban settings and translation of the data into action: A scoping review. *Infectious Diseases of Poverty, 7*, 1–14.

Frank, M. G., Weaver, G., & Raabe, V. (2024). Crimean-Congo hemorrhagic fever virus for clinicians—Virology, pathogenesis, and pathology. *Emerging Infectious Diseases, 30*(5), 847.

Gaina, A., Tahoun, M., Mashal, O., Safi, H., Alizai, F., Jalil, H., & Abouzeid, A. (2023). The largest reported outbreak of CCHF in hospital settings: Lessons from Kandahar, Afghanistan. *The Lancet Infectious Diseases, 23*(9), e330–e331.

Ghosh, S., Azhahianambi, P., & de la Fuente, J. (2006). Control of ticks of ruminants, with special emphasis on livestock farming systems in India: Present and future possibilities for integrated control—a review. *Experimental & Applied Acarology, 40*, 49–66.

Gritsun, T., Nuttall, P., & Gould, E. A. (2003). Tick-borne flaviviruses. *Advances in Virus Research, 61*, 317–371.

Gubler, D. J. (2002). The global emergence/resurgence of arboviral diseases. *Nature Medicine, 8*(12), S73–S75.

Gürbüz, E., Ekici, A., Ünlü, A. H., & Yilmaz, H. (2021). Evaluation of seroprevalence and clinical and laboratory findings of patients admitted to health institutions in Gümüşhane with suspicion of Crimean-Congo hemorrhagic fever. *Turkish Journal of Medical Sciences, 51*(4), 1825–1832.

Hasan, M. M., Sahito, A. M., Muzzamil, M., Mohanan, P., Islam, Z., Billah, M. M., & Essar, M. Y. (2022). Devastating dengue outbreak amidst COVID-19 pandemic in Bangladesh: An alarming situation. *Tropical Medicine and Health, 50*(1), 11.

Hassett, E. M., & Thangamani, S. (2021). Ecology of Powassan virus in the United States. *Microorganisms, 9*(11), 2317.

Hermance, M. E., & Thangamani, S. (2017). Powassan virus: An emerging arbovirus of public health concern in North America. *Vector-Borne and Zoonotic Diseases, 17*(7), 453–462.

Hewson, R. (2024). Understanding viral haemorrhagic fevers: Virus diversity, vector ecology, and public health strategies. *Pathogens, 13*(10), 909.

Holzmann, H. (2003). Diagnosis of tick-borne encephalitis. *Vaccine, 21*(S1), S36–S40.

Ijaz, S., Bhatti, M., Shahid, S., Faiz, A., Asad, K., Arshad, M., & Mushtaq, A. (2023). Tick-borne encephalitis—A threat to life. *One Health Triad, 1*, 8–11.

Ilic, M., Barbic, L., Bogdanic, M., Tabain, I., Savic, V., Licina, M. L. K., & Angelov, V. (2020). Tick-borne encephalitis outbreak following raw goat milk consumption in a new micro-location, Croatia, June 2019. *Ticks and Tick-Borne Diseases, 11*(6), Article 101513.

Ji, S.-R., Byun, H.-R., Rieu, M.-S., Han, S.-W., Nam, H.-Y., Seo, S., & Cho, S.-Y. (2024). First detection of *Bandavirus dabieense* in ticks collected from migratory birds in the Republic of Korea. *Acta Tropica, 257*, Article 107279.

John, J. K., Kattoor, J. J., Nair, A. R., Bharathan, A. P., Rekha Valsala, R. V., & Sadanandan, G. V. (2014). Kyasanur forest disease: A status update. *Advances in Animal and Veterinary Sciences, 2*(1), 1–9.

Jones, K. E., Patel, N. G., Levy, M. A., Storeygard, A., Balk, D., Gittleman, J. L., & Daszak, P. (2008). Global trends in emerging infectious diseases. *Nature, 451*(7181), 990–993.

Karanam, S. K., Nagvishnu, K., Uppala, P. K., Edhi, S., & Varri, S. R. (2025). Crimean-Congo hemorrhagic fever: Pathogenesis, transmission and public health challenges. *World Journal of Virology, 14*(1), Article 100003.

Khan, B. A., Imran, M., & Yousaf, S. (2021). Biology and ecology of ticks of medical and veterinary importance. *Veterinary Pathobiology & Public health*, 36.

Khanna, V. K. (2023). *Extreme-temperature and harsh-environment electronics: Physics, technology and applications.* IOP Publishing.

Kilpatrick, A. M., & Randolph, S. E. (2012). Drivers, dynamics, and control of emerging vector-borne zoonotic diseases. *The Lancet, 380*(9857), 1946–1955.

Kim, E.-H., & Park, S.-J. (2023). Emerging tick-borne Dabie bandavirus: Virology, epidemiology, and prevention. *Microorganisms, 11*(9), 2309.

Kochs, G., Weber, F., Gruber, S., Delvendahl, A., Leitz, C., & Haller, O. (2000). Thogoto virus matrix protein is encoded by a spliced mRNA. *Journal of Virology, 74*(22), 10785–10789.

Kosoy, O. I., et al. (2015). Bourbon virus in the United States: A new Thogotovirus associated with human illness. *Emerging Infectious Diseases, 21*(5), 762–765.

Kwasnik, M., Rola, J., & Rozek, W. (2023). Tick-borne encephalitis—Review of the current status. *Journal of Clinical Medicine, 12*(20), 6603.

Labuda, M., & Nuttall, P. A. (2004). Tick-borne viruses. *Parasitology, 129*(S1), S221–S245.

Leblebicioglu, H., Ozaras, R., Irmak, H., & Sencan, I. (2016). Crimean-Congo hemorrhagic fever in Turkey: Current status and future challenges. *Antiviral Research, 126*, 21–34.

Liao, H., Lyon, C. J., Ying, B., & Hu, T. (2024). Climate change, its impact on emerging infectious diseases and new technologies to combat the challenge. *Emerging Microbes & Infections, 13*(1), 2356143.

Lindgren, E., Tälleklint, L., & Polfeldt, T. (2000). Impact of climatic change on the northern latitude limit and population density of the disease-transmitting European tick *Ixodes ricinus. Environmental Health Perspectives, 108*(2), 119–123.

Lindquist, L., & Vapalahti, O. (2008). Tick-borne encephalitis. *The Lancet, 371*(9627), 1861–1871.

Liu, Y., Huang, L., & Chen, Y. (2018). Environmental impacts on IoT sensor networks: Resilience and sustainability perspectives. *Sensors, 18*(9), 2901.

Madison-Antenucci, S., Kramer, L. D., Gebhardt, L. L., & Kauffman, E. (2020). Emerging tick-borne diseases. *Clinical Microbiology Reviews, 33*(2), e00083-e118.

Mandl, C. W. (2005). Steps of the tick-borne encephalitis virus replication cycle that affect neuropathogenesis. *Virus Research, 111*(2), 161–174.

Mangombi, J. B., Roqueplo, C., Sambou, M., Dahmani, M., Mediannikov, O., Comtet, L., & Davoust, B. (2020). Seroprevalence of Crimean-Congo hemorrhagic fever in domesticated animals in Northwestern Senegal. *Vector-Borne and Zoonotic Diseases, 20*(10), 797–799.

Medlock, J. M., & Leach, S. A. (2015). Effect of climate change on vector-borne disease risk in the UK. *The Lancet Infectious Diseases, 15*(6), 721–730.

Messina, J. P., Pigott, D. M., Golding, N., Duda, K. A., Brownstein, J. S., Weiss, D. J., & Wint, G. W. (2015). The global distribution of Crimean-Congo hemorrhagic fever. *Transactions of the Royal Society of Tropical Medicine and Hygiene, 109*(8), 503–513.

Naseer, U., Larik, E. A., Ahmad, W., Ahmed, Z., Riaz, D., Arif, M., & Saeed, A. (2024). Epidemiological analysis of event-based surveillance data of Crimean-Congo hemorrhagic fever (CCHF) outbreaks across Balochistan Province, Pakistan, from 2000 to August 2021. *Qatar Journal of Public Health, 2024*(1), 3.

Nuttall, P. A. (2022). Climate change impacts on ticks and tick-borne infections. *Biologia, 77*(6), 1503–1512.

Ogden, N. H., Ben Beard, C., Ginsberg, H. S., & Tsao, J. I. (2021). Possible effects of climate change on ixodid ticks and the pathogens they transmit: Predictions and observations. *Journal of Medical Entomology, 58*(4), 1536–1545.

Ogden, N. H., & Lindsay, L. R. (2016). Effects of climate and climate change on vectors and vector-borne diseases: Ticks are different. *Trends in Parasitology, 32*(8), 646–656.

Ogden, N. H., Maarouf, A., Barker, I. K., Bigras-Poulin, M., Lindsay, L. R., Morshed, M. G., & Charron, D. F. (2006). Climate change and the potential for range expansion of the Lyme disease vector *Ixodes scapularis* in Canada. *International Journal for Parasitology, 36*(1), 63–70.

Ogden, N. H., Mechai, S., & Margos, G. (2013). Changing geographic ranges of ticks and tick-borne pathogens: Drivers, mechanisms and consequences for pathogen diversity. *Frontiers in Cellular and Infection Microbiology, 3*, 46.

Ostfeld, R. S., Canham, C. D., Oggenfuss, K., Winchcombe, R. J., & Keesing, F. (2006). Climate, deer, rodents, and acorns as determinants of variation in Lyme disease risk. *PLoS Biology, 4*(6), Article e145.

Ozdarendeli, A. (2023). Crimean-Congo hemorrhagic fever virus: Progress in vaccine development. *Diagnostics, 13*(16), 2708.

Parola, P., & Raoult, D. (2001). Ticks and tickborne bacterial diseases in humans: An emerging infectious threat. *Clinical Infectious Diseases, 32*(6), 897–928.

Patz, J. A., Campbell-Lendrum, D., Holloway, T., & Foley, J. A. (2005). Impact of regional climate change on human health. *Nature, 438*(7066), 310–317.

Paz, S., Díaz, J., Negev, M., & Linares, C. (2024). Climate change and global health. In *Handbook of epidemiology*. New York, NY: Springer New York.

Pfäffle, M., Littwin, N., Muders, S. V., & Petney, T. N. (2013). The ecology of tick-borne diseases. *International Journal for Parasitology, 43*(12–13), 1059–1077.

Pirincal, A., & Doymaz, M. Z. (2024). The role of nucleocapsid protein (NP) in the immunology of Crimean-Congo hemorrhagic fever virus (CCHFV). *Viruses, 16*(10), 1547.

Pulkkinen, L. I., Butcher, S. J., & Anastasina, M. (2018). Tick-borne encephalitis virus: A structural view. *Viruses, 10*(7), 350.

Pustijanac, E., Buršić, M., Talapko, J., Škrlec, I., Meštrović, T., & Lišnjić, D. (2023). Tick-borne encephalitis virus: A comprehensive review of transmission, pathogenesis, epidemiology, clinical manifestations, diagnosis, and prevention. *Microorganisms, 11*(7), 1634.

Raabe, V. N. (2020). Diagnostic testing for Crimean-Congo hemorrhagic fever. *Journal of Clinical Microbiology, 58*(4), e01580-e1619.

Radzišauskienė, D., Žagminas, K., Ašoklienė, L., Jasionis, A., Mameniškienė, R., Ambrozaitis, A., & Mockienė, E. (2018). Epidemiological patterns of tick-borne encephalitis in Lithuania and clinical features in adults in the light of the high incidence in recent years: A retrospective study. *European Journal of Neurology, 25*(2), 268–274.

Randolph, S. E. (2009). Tick-borne disease systems emerge from the shadows: The beauty lies in molecular detail, the message in epidemiology. *Parasitology, 136*(12), 1403–1413.

Randolph, S. E., Sonenshine, D. E., & Roe, R. M. (2014). Ecology of non-nidicolous ticks. In D. E. Sonenshine & R. M. Roe (Eds.), *Biology of ticks*. Oxford University Press.

Rowan, S., Mohseni, N., Chang, M., Burger, H., Peters, M., & Mir, S. (2023). From tick to test: A comprehensive review of tick-borne disease diagnostics and surveillance methods in the United States. *Life, 13*(10), 2048.

Růžek, D., Yakimenko, V. V., Karan, L. S., & Tkachev, S. E. (2010). Omsk haemorrhagic fever. *The Lancet, 376*(9758), 2104–2113.

Semenza, J. C., & Suk, J. E. (2018). Vector-borne diseases and climate change: A European perspective. *FEMS Microbiology Letters, 365*(2), fnx244.

Shah, T., Li, Q., Wang, B., Baloch, Z., & Xia, X. (2023). Geographical distribution and pathogenesis of ticks and tick-borne viral diseases. *Frontiers in Microbiology, 14*, 1185829.

Shayan, S., Bokaean, M., Shahrivar, M. R., & Chinikar, S. (2015). Crimean-Congo hemorrhagic fever. *Laboratory Medicine, 46*(3), 180–188.

Spengler, J. R., Bergeron, É., & Spiropoulou, C. F. (2019). Crimean-Congo hemorrhagic fever and expansion from endemic regions. *Current Opinion in Virology, 34*, 70–78.

Spik, K., Shurtleff, A., McElroy, A. K., Guttieri, M. C., Hooper, J. W., & Schmaljohn, C. (2006). Immunogenicity of combination DNA vaccines for Rift Valley fever virus, tick-borne encephalitis virus, Hantaan virus, and Crimean-Congo hemorrhagic fever virus. *Vaccine, 24*(21), 4657–4666.

Sravani, D. (2025). A review on infectious disease management and their challenges. *Sustainable Nanomaterials for Treatment and Diagnosis of Infectious Diseases*, 23–50.

Steere, A. C., Coburn, J., & Glickstein, L. (2004). The emergence of Lyme disease. *The Journal of Clinical Investigation, 113*(8), 1093–1101.

Stibraniova, I., Bartikova, P., & Dzubara, J. (2025). Response of host immune system to tick borne encephalitis virus. *Acta Virologica, 68*, 12936.

Süss, J. (2011). Tick-borne encephalitis 2010: Epidemiology, risk areas, and virus strains in Europe and Asia—An overview. *Ticks and Tick-Borne Diseases, 2*(1), 2–15.

Swei, A., Couper, L. I., Coffey, L. L., Kapan, D., & Bennett, S. (2020). Patterns, drivers, and challenges of vector-borne disease emergence. *Vector-Borne and Zoonotic Diseases, 20*(3), 159–170.

Thatikonda, K. (2023). *Integrating electrical systems with intelligent computing and applications.* Academic Guru Publishing House.

Valavanidis, A. (2022). Extreme weather events exacerbated by the global impact of climate change.

Varma, M., Webb, H., & Pavri, K. M. (1960). Studies on the transmission of Kyasanur Forest disease virus by *Haemaphysalis spinigera* Newman. *Transactions of the Royal Society of Tropical Medicine and Hygiene, 54*(6), 509–516.

Vilibic-Cavlek, T., Krcmar, S., Bogdanic, M., Tomljenovic, M., Barbic, L., Roncevic, D., & Hunjak, B. (2024). An overview of tick-borne encephalitis epidemiology in endemic regions of continental Croatia, 2017–2023. *Microorganisms, 12*(2), 386.

Wilhelmsson, P., Lindblom, P., Fryland, L., Nyman, D., Jaenson, T. G., Forsberg, P., & Lindgren, P. E. (2013). Ixodes ricinus ticks removed from humans in Northern Europe: Seasonal pattern of infestation, attachment sites and duration of feeding. *Parasites & Vectors, 6*, 1–11.

Williamson, B. N., Fischer, R. J., Lopez, J. E., Ebihara, H., & Schwan, T. G. (2019). Prevalence and strains of Colorado tick fever virus in Rocky Mountain wood ticks in the Bitterroot Valley, Montana. *Vector-Borne and Zoonotic Diseases, 19*(9), 694–702.

Yessinou, R. E., Farougou, S., Olopade, J. O., Oluwayelu, D. O., Happi, A., Happi, C., & Groschup, M. (2025). Seroprevalence and risk factors of Crimean-Congo hemorrhagic fever exposure in wild and domestic animals in Benin. *Viruses, 17*(3), 387.

Yu, K. -M., & Park, S. -J. (2024). Tick-borne viruses: Epidemiology, pathogenesis, and animal models. *One Health, 100903.*

Zając, Z., Bartosik, K., Kulisz, J., & Woźniak, A. (2022). Incidence of tick-borne encephalitis during the COVID-19 pandemic in selected European countries. *Journal of Clinical Medicine, 11*(3), 803.

Chapter 6
Impact of Climate Change on Tick-Borne Bacterial Diseases

Muhammad Sohail Sajid, Muhammad Hussnain Raza, Syed Soban Hassan, and Ian Daniel

Abstract Ticks are important vectors of bacterial diseases such as Lyme disease, anaplasmosis, Rocky Mountain spotted fever, ehrlichiosis, and Q fever. Animals serve as reservoir hosts for these diseases while humans become incidental hosts through contact with the infected ticks. Temperature, moisture, and seasonal changes affect the lifecycles of ticks. Ticks react to biotic factors such as humid leaves, dense shade, dense vegetation, and host interaction, and abiotic factors such as temperature, humidity, soil moisture, and precipitation patterns. *Ixodes, Rhipicephalus, Dermacentor,* and *Ambylomma* ticks are responsible for the transmission of these bacterial diseases. Recent outbreaks of Lyme disease have been reported in North America. Challenges in the management and prevention of these tick-borne bacterial diseases include the limitation of traditional control methods because of the emergence of antiparasitic drug resistance and climate change. Climate change such as deforestation, urbanization, increased mobility of humans and animals, modern transportation, illegal wildlife trade, and genetic variations within ticks which cause them to adapt to the environmental conditions are responsible for the emergence of these tick-borne bacterial diseases. People should be educated along with the adaptation of surveillance strategies to control the spread of these tick-borne bacterial diseases.

Keywords Tick behavior · Geographic shifts · Bacterial diseases · Management challenges · Adaptive measures

---

M. S. Sajid (✉)
Department of Parasitology, Faculty of Veterinary Science, University of Agriculture, Faisalabad, Pakistan
e-mail: drsohailuaf@hotmail.com; drsohailuaf@uaf.edu.pk

Climate Change Graduate Group, Institute of Soil and Environmental Science, University of Agriculture, Faisalabad, Pakistan

M. H. Raza · S. S. Hassan
Faculty of Veterinary Sciences, KBCMA College of Veterinary and Animal Sciences, Sub-Campus UVAS, NarowalLahore, Pakistan

I. Daniel
Department of Veterinary Pathobiology, College of Veterinary Medicine & Biomedical Sciences, Texas A&M University, College Station, TX, USA

H. M. Rizwan and M. S. Sajid (eds.), *Ticks in a Changing Climate*, Biometeorology 7,
https://doi.org/10.1007/978-3-032-12259-9_6

6.1 Introduction

Ticks are obligate ectoparasites that feed on blood, parasitize every vertebrate of the world including humans and belong to subclass Acari of class Arachnida (Parola & Raoult, 2001a). Hard ticks (Ixodidae), soft ticks (Argasidae) and Nuttallielidae (which is only limited to Tanzania and South Africa), are the three classifications of ticks (Nava et al., 2009). Ixodes comprises more than 700 species and Argasidae comprises about 200 species which are responsible for transmission of wide range of disease causing agents throughout the world (Kernif et al., 2016). Ticks can lead to a lot of diseases with multiple pathogens like bacteria, viruses, rickettsia and protozoa (Spach et al., 1993).

Ticks have been describes as vectors of bacterial diseases. After feeding on the bacteremic animals ticks may get infected with bacteria. Ticks may get bacteria by transovarial or transstadial transmission. As ticks feed on the host, they transmit the infection to other host (de la Fuente et al., 2008a). *Borrelia (B.) burgdorferi* is responsible to cause Lyme disease and *B. miyamotoi* causes relapsing fever. Ixodes ticks e.g., *Ixodes (I.) scapularis, I. ricinus* are responsible to transmit *Borrelia* spp. *Rickettsia (R.) rickettsii* causes Rocky Mountain spotted fever, *R. africae* causes African tick bite fever, and *R. conorii* causes Mediterranean spotted fever. Different species of ticks like *Dermacentor, Amblyomma*, and *Rhipicephalus* are responsible to transmit the *Rickettsia. Amblyomma americanum* commonly known as Lone star tick can transmit *Ehrlichia (E.) chaffeensis* (causes Human monocytic ehrlichiosis) and *E. ewingii* (causes ehrlichiosis in humans and dogs. *Ixodes* spp. also transmit *Anaplasma (A.) phagocytophilum* and responsible to cause human granulocytic anaplasmosis, however *Rhipicephalus (Boophilus) microplus* can transmit *A. marginale* in animals. *Rhipicephalus* spp., and *Dermacentor* spp. can also transmit *Coxiella (C.) burnetii,* the causing agent of Q fever (Parola & Raoult, 2001b).

Serious social unrest, population dislocation, economic suffering, and environmental degradation are all threats posed by global climate change. Increased variability and long-term changes in temperature, rainfall patterns, storm intensity, the frequency of droughts or flooding, and sea level rise may all have an impact on human and animal health. Tick-borne diseases (TBDs) are particularly likely to be impacted by climate change and increased climatic variability (Gage et al., 2008). Suitable climate is a necessity for emergence of TBDs. Any change in climate can affect the geographical spread, dynamics of transmission and re-emergence of TBDs (Rocklöv & Dubrow, 2020). The incubation period of a vector-borne infectious agent within its vector organism is notably sensitive to temperature variations, often exhibiting an exponential relationship (Patz et al., 2003).

6.2 Environmental Influences on Tick Ecology

6.2.1 Temperature Effects on Tick Behavior

Inanimate environmental factors like temperature, humidity, and light are believed to affect the ticks during their off-host period. Evidence indicates that alterations in the spatial distribution and abundance of ticks, along with the transmission of associated TB pathogens, are significantly affected by climate variability and fluctuations in tick host population density (Esteve-Gassent et al., 2016). The fundamental conditions for ticks to be "happy" are a humidity rate of over 85%, air temperatures of over 6 °C to 7 °C, and a substantial number of blood-delivering hosts (Süss et al., 2008). Temperature is one of the crucial factors necessary for the development, survival, and reproduction of ticks (Régnier et al., 2022). For example, *B. burgdorferi* is greatly affected by temperature change that serves as a cue resulting in turn on or off of such genes responsible for tick colonization, transmission from tick to host and its survival in host (Couret et al., 2022).

As a result of warmer weather, tick population can increase because this shortens the time between stages of development. However, because ticks are extremely sensitive to dehydration, conditions may be less optimal when temperatures rise, which could result in higher tick mortality. In addition, the extent to which tick abundance varies with the seasons and the degree to which the various lifecycle stages overlap determine whether elevated temperatures are linked to an increased risk of TBDs (Hancock et al., 2011). Generally, there is a nonlinear relationship between temperature and development length, with development taking longer at lower temperatures or ceasing entirely. Since tick populations can survive cold winters as long as their habitat offers a place to hide. It has been proposed that the key to understanding how climate change affects tick populations is how temperature affects lifecycle length through effects on interstadial development rates. According to the reasoning, if daily per-capita mortality rates are unaltered by temperature, then more larvae will survive to become reproducing adults because the lifecycle is shorter in warmer climates (Ogden et al., 2021).

6.2.2 Seasonal Changes and Tick Activity

Tick prevalence is often linked to seasonal weather patterns. Disease transmission and the development of tick and host life-history features are significantly influenced by the seasonal population dynamics of ticks. A large number of tick species spend the majority of their lifespan in low numbers, peak in abundance for a brief time, and then disappear for a few weeks (Oorebeek & Kleindorfer, 2008). The introduction of exotic tick species in new geographic areas may be facilitated by climate change. For example, *Hyalomma marginatum marginatum*, an Asian native that is now found in the United States and *Haemaphysalis longicornis* newly founded in France (Nuttall,

2022). It is reasonable to predict that seasonal intervals in central Europe without *I. ricinus* will shorten if winter temperatures rise in the future. The impact of winter activity on the remaining seasonal activity pattern remain ambiguous, as winter-active ticks use valuable energy. Furthermore, it is uncertain whether alterations in the seasonal activity of ticks e.g., *I. ricinus* nymphs and adults are advantageous or harmful to the persistence of TB infections. In many places, *I. ricinus* activity is expected to increase during the autumn and winter months due to increased global warming. Additionally, a larger percentage of ticks may be active during these seasons than they are now, which could result in a temporal shift in the risk of TBDs (Gray et al., 2009).

6.2.3 Impact of Moisture Levels on Tick Habitats

Researchers have proposed a link between tick species' natural habitats, physiological resistance to desiccation, and ability to maintain water balance (Hair et al., 1975). The unfed tick loses water through evaporation and gets it from moist air or water that comes into contact with its cuticle. While attached to the host, the tick gains water from the consumed blood and loses it in the excrement. On the other hand the engorged tick is unable to absorb water from humid air (Lees, 1946). One regional land surface condition that has been found to have an impact on tick habitat characteristics is the availability of moisture. The survival of nymphal blacklegged ticks has been demonstrated to be negatively impacted by prolonged periods of inadequate atmospheric moisture. Thus, land surface moisture estimations derived from remotely sensed hydrothermal observations should greatly enhance monitoring and forecasting of TBDs risks and offer a near-real-time characterization of tick habitat (Berger et al., 2013). Tick activity is significantly impacted by moisture level, which largely regulates how long a tick may spend questing. When a tick ascends off the mat and up the plants to quest, it becomes particularly prone to desiccation, leading the tick to retreat back within the mat to recover its water balance before questing again (Greenfield, 2011).

6.2.4 Geographic Shifts in Tick Distribution

Ticks must react to both biotic (such as humid leaves, dense vegetation, dense shade, and host contact) and abiotic (such as temperature, humidity, and soil moisture) variables (Fig. 6.1) since they spend a large portion of their life cycle exposed to the environment (Couret et al., 2022). Based on a geographical scale, the Centre for Disease Control and Prevention (CDC) has created risk maps for various tick species. The highest number of tick-associated disease cases since records have been kept was 48,610 in 2016 and 59,349 in 2017. These cases included Lyme disease, spotted fever group rickettsiosis (SFGR), babesiosis, Powassan virus, anaplasmosis/ehrlichiosis,

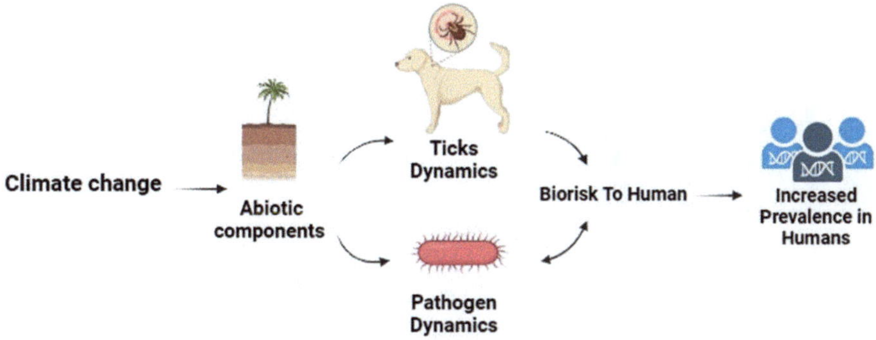

Fig. 6.1 Climate change influences abiotic components, creating favorable conditions for tick survival and expansion. This leads to changes in tick dynamics and pathogen transmission, increasing the risk of tick-borne diseases. As infected ticks interact with animals and humans, the likelihood of disease transmission rises, ultimately contributing to a higher prevalence of tick-borne illnesses in humans

and tularaemia. Over the past 20 years, seven novel pathogens have been identified in the United States: *E. ewingii*, *Heartland virus*, *R. parkeri*, *E. mayonii*, *E. miyamotoi,* and *E. muris eauclairensis* (Alkishe et al., 2021).

Research conducted in Switzerland has shown that, even at the same elevation, tick abundance is influenced by a mountainside's perspective. In the research sites in Switzerland, *I. ricinus* has been found up to 1450 m above sea level (m.a.s.l). The tick is much less common in Italy over 1300 m.a.s.l. In Switzerland, Austria, the Czech Republic, and Scotland, it can be found up to 1450 m.a.s.l, 1560 m.a.s.l., 1080–1270 m.a.s.l., and 700 m.a.s.l. respectively (Medlock et al., 2013). In many American cities, the number of ticks and TBDs has sharply increased (Noden et al., 2017). Figure 6.2. Explains the effect of deforestation induced urbanization on the transmission of ticks.

6.3 Key Bacterial Diseases Transmitted by Ticks

6.3.1 Lyme Disease

Lyme disease is one of the most common tick-transmitted disease caused by bacteria (*Borrelia*) belonging to spirochete family. The human pathogens *B. burgdorferi* sensu stricto, *B. afzelii, B. garinii*, and potentially *B. bissettii* are among the agents of Lyme borreliosis, which are grouped under the *B. burgdorferi* sensu lato species complex and further subdivided into at least 11 distinct genomic species (Stanek & Strle, 2003). According to Mead (2015), as the tick hatch from the egg it does not acquire the bacteria, rather it gains bacteria while feeding on infected hosts like mice, birds, shrews and other small mammals. So this infected tick is then

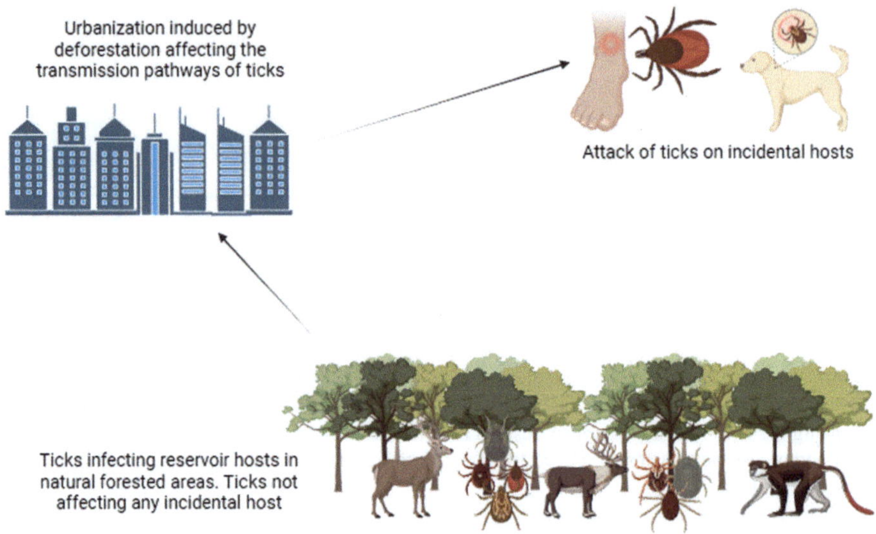

Fig. 6.2 Effects of deforestation induced urbanization on transmission of ticks and tick-borne diseases

responsible for the spread of the bacteria to other reservoir hosts and humans. The disease is basically transmitted by the symptomless bite of ticks belonging to species of *Ixodes*. The main *Borrelia* reservoirs are birds, various mouse strains, dormice, deer, voles, and lizards; some of these hosts may also contribute to the spread of ticks and *B. burgdorferi* to new areas (Trevisan et al., 2021). Erythema migrans, the most prevalent sign of early localised Lyme borreliosis, often develops three to thirty days after a tick bite. Along with erythema migrans, *B. burgdorferi* infection has also been linked to direct involvement of the eye, such as uveitis, keratitis, episcleritis, vitritis, and optic neuritis. Carditis, neuroborreliosis, and arthritis are the hallmarks of late extracutaneous Lyme borreliosis presentations (Hengge et al., 2003).

In Europe and Asia the incidence of Lyme disease ranges from low to negligible. On the other hand > 80 cases per 100,000 individuals are fallen prey to Lyme disease in Netherlands, Belgium, Slovenia, Austria, Estonia and Lithuania. While in the northeast and mid-Atlantic region and the north-central region of the United States account for more than 90% of cases recorded in North America (Steere et al., 2016). According to Nadelman and Wormser (1998) the clinical hallmark for the Lyme disease that occurs in 90% of the cases is erythema migrans (EM), while other symptoms associated with the disease are fatigue, myalgia, arthralgia, headache, fever, chills and stiff neck.

6.3.2 Anaplasmosis

Anaplasmosis is a tick-borne infectious disease characterized by progressive anemia. It is worldwide transmissible disease of cattle caused by rickettsia *A. marginale* (Ristic, 1981). Unlike other TBDs that affect cattle, *A. marginale* can spread both biologically by ticks and mechanically through biting flies or blood-contaminated fomites. Once entered the host (cattle) *A. marginale* grows within the erythrocytes and clinical disease is observed when 40 to 50% of erythrocytes are destroyed (Kocan et al., 2000). Reusing needles, dehorners, ear taggers, castrating knives or other surgical tools, and tattoo tools can all result in mechanical transfer. The bacterium is spread mechanically by biting flies' blood-contaminated mouthparts or by blood-contaminated tools (Abdisa, 2019). Ticks of the Ixodidae family serve as biological vectors and are crucial to the transmission and development of *Anaplasma* during different phases of its life cycle. However, *A. platys* infections in bitches during the first 25 to 35 days of pregnancy have also been known to spread vertically, and *A. phagocytophilum* infections have also been known to spread intrauterinally (Atif et al., 2021).

Over the previous 50 years, there hasn't been much change in anaplasmosis control methods. Current methods of control vary by region and include the use of acaricides to control arthropods, the administration of antibiotics, live vaccine premunization, dead vaccine immunisation, and the preservation of herds free of *Anaplasma* (Kocan et al., 2000). Numerous European nations have reported cases of anaplasmosis in a variety of animal species, with prevalence's ranging from 1.09 to 97.9%. The particular geographic location may be the cause of the prevalence variations. High prevalence could be connected to the area's ideal tick habitat, which includes sufficient flora and high populations of animal hosts (Karlsen et al., 2020). Despite animals, human infection with *A. phagocytophilum* has been recorded in patients across North America, Europe, and Asia, with a significant occurrence in the United States. Since 1999, anaplasmosis has become a nationally notifiable disease in the United States. Since then, the number of cases countrywide has increased more than 16 times, from 348 in 2000 to 5,762 in 2017 (Russell et al., 2021).

6.3.3 Rocky Mountain Spotted Fever

A life threatening TBDs caused by *R. rickettsia*, an obligate intracellular bacterium. The first clinical picture is presented by Edward E Maxey, who characterized it as a febrile disease with persistent moderately high fever, purpuric eruptions of skin starting from ankle and then spreading to the whole body (Dantas-Torres, 2007). By infecting the endothelial cells lining the tiny capillaries of all major tissues and organ systems, *R. rickettsii* causes a moderately severe to potentially fatal systemic disease in its host (Kim, 2022). The varied and frequently severe clinical manifestations linked to untreated illness are caused by fatal and irreversible damage to endothelium in the dermis, lungs, heart, kidneys, gastrointestinal tract, brain, skeletal muscle, and other locations. The rash, headache, myalgias, and gastrointestinal symptoms that are frequently linked to Rocky Mountain spotted fever (RMSF) are caused by these pathophysiologic events, which can also result in more severe symptoms like gangrene, pulmonary haemorrhage and oedema, acute respiratory distress syndrome, myocarditis, acute renal failure, meningoencephalitis, and cerebral oedema (Masters et al., 2003).

The ticks mainly responsible for the zoonoses of this bacterium are American dog tick (*Dermacentor variabilis*) or the wood tick (*Dermacentor andersoni*). In USA southeast and south-central states are the most common states to be affected with this illness (Treadwell et al., 2000). In Arizona State this tick is not commonly found and so RMSF is rarely reported in this state and according to reports only 3 cases were reported from 1981 to 2001. While from 2002 to 2004 16 patients have been identified with RMSF in rural eastern Arizona (Demma et al., 2005). Sonora (Mexico) recorded 1394 cases and 247 deaths between 2003 and 2016, with an overall case fatality rate of 18%, higher than that of other endemic diseases. Patients under the age of 19 made up for 60% of fatalities and 58% of all cases. A significant outbreak of RMSF in Mexicali, Baja California, between 2009 and 2010 brought more attention to the disease's resurgence in Mexico. From 2009 and 2016, there were 967 cases and 132 fatalities, and the overall case fatality rate during this time (14%) was higher than that of a number of other infectious diseases of regional significance (Álvarez-Hernández et al., 2017).

6.3.4 Other Important Tick-Borne Bacterial Infections

Ticks can harbor multiple agents at once and can spread bacterial, parasitic, and viral diseases. Tick-borne African Fever caused by *R. africae*, and Tick-borne lymphadenopathy (TIBOLA), associated with *R. slovaca*, *R. raoultii*, and *R. rioja*, are some important tick-borne bacterial diseases (Boulanger et al., 2019). *E. chaffeensis* causes a tick-borne bacterial disease known as human monocytic ehrlichiosis (Rikihisa, 2022). *Ixodidae* is the primary vector of this disease (Zhang et al., 2024). Human monocytic ehrlichiosis is an emerging tick-borne disease. White-tailed deer is a reservoir host while dogs and humans are incidental hosts (Madesh et al., 2024). Human ehrlichiosis is also caused by *E. ewingii* (Adams et al., 2025). *E. ewingii* also affects dogs and causes canine monocytic ehrlichiosis (Zu et al., 2025).

Q fever is also another important tick-borne bacterial disease caused by *Coxiella burnetti* (Ullah et al., 2022). Sheep and goats are reservoir hosts and humans become incidental hosts (Ullah et al., 2019a). Ticks play an important role in the transmission of this disease (Ullah et al., 2019b). Figure 6.3. Gives the important tick-borne bacterial diseases associated with specific ticks.

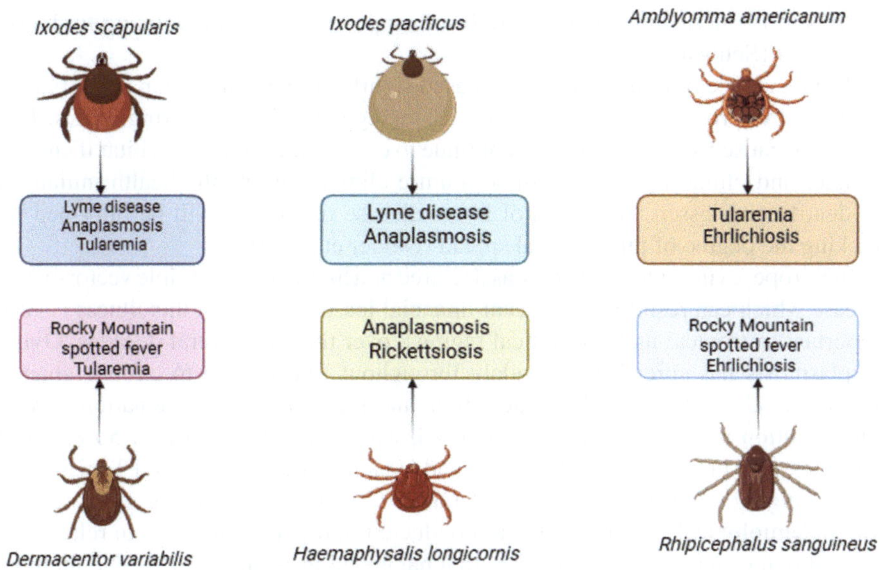

Fig. 6.3 Important tick-borne bacterial diseases associated with specific ticks

6.4 Case Studies and Recent Outbreaks

Lyme disease, a tick-borne bacterial disease caused by spirochetes belonging to the *B. burgdorferi* sensu lato complex, can impact various organ systems in humans (Steere et al., 2016). The majority of infections in the US are caused by *B. burgdorferi* sensu stricto, while *B. mayonii* has also been demonstrated to infect people in the upper Midwest (Stanek & Strle, 2018). It has also been newly discovered in temperate regions of North America and Europe. The latest canine borreliosis cases in Europe have been reported in practically every nation (Skotarczak, 2018).

The geographic distribution is spreading into neighbouring American states, despite being concentrated in historically high-incidence areas. Several factors may explain the stable or declining case counts of Lyme disease in many high-incidence states. These include the actual stabilization of disease incidence or changes in reporting procedures. Some states may have altered surveillance methods to reduce the resource burden, which could lead to underreporting or artificial trends (Sharareh et al., 2019). A total of 275,589 Lyme disease cases—208,834 confirmed and 66,755 probable—were reported to the CDC between 2008 and 2015. Between 2009 and 2010, the overall number of confirmed and probable cases varied between 38,468 and 30,158 per year. In 2009, the greatest number of confirmed cases (29,959) was reported, while 2012 saw the fewest (22,014). Forty-eight states and the District of Columbia reported confirmed cases. The age distribution of reported cases was bimodal, and they were more common in males and in the early summer months when North American nymphal stage *Ixodes* spp. vector ticks are looking for blood meal hosts (Schwartz et al., 2017).

The incidence of Lyme disease increased significantly in Ontario in 2017, particularly in the province's eastern region. Blacklegged ticks harbouring TBDs, like those that cause Lyme disease, will continue to expand into suitable habitat if current weather and climatic trends continue. Future clinical and public health initiatives to identify and lessen the effects of Lyme disease in Ontario will be informed by tracking the degree of this regional spread (Nelder et al., 2018).

In Europe, ovine anaplasmosis was declared as a highly transmissible vector-borne disease which can result in significant financial losses. Although this illness is very important in tropical and subtropical regions, over the past several decades. Ovine anaplasmosis has spread more widely throughout Europe due to climate change (Lacasta et al., 2021). Brazil, Mexico, Iran, and India are among the nations where this condition is most commonly diagnosed, with a prevalence of over 50%. Other countries with prevalence's between 20 and 50% include Morocco and Tunisia, and China ranging from 9.4% to 65.3% (Urán & Cardona, 2019; Wang et al., 2017). An epidemiological investigation was conducted in the impacted area in relation to the first ovine anaplasmosis epidemic that happened in Spain in 2014. The findings demonstrate that *A. ovis* is highly prevalent in the flocks under analysis, demonstrating the bacteria's strong transmissibility (de la Fuente et al., 2008b). Tick survival in the environment has been aided by climate change and rising global temperatures, which has led to the expansion of *A. ovis* in humans and other species. When ticks live in an environment that is conducive to their survival, rising temperatures extend their time of activity. This, along with the coexistence of domestic and wild ruminants, creates a variety of reservoirs and hosts that contribute to the disease's high prevalence (Bouchard et al., 2019).

6.5 Challenges in Management and Prevention

6.5.1 *Limitations of Traditional Control Methods*

Due to traditional tick control methods, the most emerging concern of the world is antparasitic resistance. Acaricide resistance has been found in veterinary-relevant tick populations in about 50% of nations, with many reports of cross-resistance between ticks and other parasite species, including helminths, mites, flies, and lice (FAO, 2004). Development of resistance builds in three steps. (1) Establishment of resistance, (2) development of resistance (3) emergence of resistance (Sutherst & Comins, 1979). The factors responsible mainly for the antiparasitic resistance are the parasite genetics, i.e., an increase in the frequency and variety of resistance alleles. Operational factors relate directly to the kind of drug to be used against a particular parasite to avoid cross-resistance, the exposure time to the drug, the frequency to be used, etc. The last one is the different biotic factors. The breeding seasons, the offspring intervals, and the life cycle stages of the parasite directly linking to resistance to the drugs (Maqbool et al., 2024; Rizwan et al., 2024). These factors are the reasons that directly take the traditional methods of tick control to a point of being ineffective.

The biology of ticks as well as the hosts and reservoirs involved in TBD cycles will be impacted by global climate change. Climate change not only affects vectors and diseases, but also affects them differently, resulting in nonlinear reactions and new interactions between hosts, infections, and vectors (Tiffin et al., 2022).

6.5.2 Developing Adaptive Strategies for Disease Control

Understanding the use, rotation, monitoring, and acaricide combinations may aid in maintaining the effectiveness of currently available drugs (Obaid et al., 2022). Target identification and vaccine development are made possible by recent developments in vector biology. The identification of novel tick-protective antigens will be done by the genome characterization efforts. It will be possible to find a tick vaccine quickly and methodically by combining the information with functional analysis using technologies such as bioinformatics, RNA interference, mutagenesis, immunomapping, transcriptomics, proteomics, and others (Ghosh et al., 2007).

Proteomics and tick genomics research will probably expand into initiatives that sequence, highlight, and functionally analyse whole tick genomes, yielding vital data for the creation of tick vaccines (Villar et al., 2017). What we have now must be preserved and used in conjunction with all of the integrated pest management concepts. Effective policies that guarantee sustainable services and market development should be developed by governments and various nodal agencies involved in livestock improvement, especially in developing nations (Mollong et al., 2025).

Plant-based products have demonstrated a variety of biological functions, such as growth regulation, toxicity, repellence, and antifeedant effects. The harmful potential of botanical acaricides is greatly influenced by the tick species, life stage, fed or unfed status, extraction technique, solvent type, harvest date, and concentration employed (Quadros et al., 2020). Neem tree derivatives have gained popularity as a natural defence against arthropod parasites and vectors and have shown efficacy against insect pests (Mulla & Su, 1999).

6.6 Health Implications

6.6.1 Assessing the Risks to Humans and Animals

Tick-borne pathogens are an obvious public health concern, even if the impact varies greatly among pathogens, both in terms of morbidity/severity of the disease and the number of cases (or individuals at risk) (Malkhazova et al., 2019). For instance, more than 24,000 cases of Lyme borreliosis are recorded each year in the Netherlands, while reports of other tick-borne infections are rare despite the presence of tick species (Hofhuis et al., 2015). In Europe, the risk to public and animal health is increasing due toTBDs in cities and their suburbs. Recently, bacterial TBDs are surfacing in endemic areas or emerging in new areas, raising concerns for biodiversity preservation, food security, and public health (Rizzoli et al., 2014). It is clear that climate change affects the spread of TBDs, but climate also affects tick species' geographic distribution, population densities and dynamics, potential for human and animal pathogen infection, and, lastly, the frequency of human and domestic animal contacts with infected ticks (Skotarczak, 2018).

Public parks, gardens, periurban regions, and graveyards are the main locations in European cities where people and their pets may come into contact with potentially infected ticks (Ginsberg & Faulde, 2008). In a research conducted in urban areas of Poland, ticks collected from cats and dogs were found to have high infestation rates and a high prevalence of *A. phagocytophilum*, *Neoerlichia micurensis*, and *Rickettsia* spp. (Król et al., 2016). The authors came to the conclusion that cats and dogs may have a major role in the spread of diseases and ticks in urban areas (Król et al., 2015). Molecular techniques are helpful in exact identification of spread of infections, precisely characterizing the frequency and geographic range of diseases spread by ticks, and determining the most effective treatment and management plans. Choosing molecular methodologies and markers for the identification of bacterial genomic DNA in tick-borne bacterial infections helps to make specific policies for their eradication (Skotarczak, 2018).

6.6.2 Strategies for Public Education and Awareness

Education is crucial for public health reasons as well as for identifying new ticks at surrounds. Since the general public is usually less aware of the risks, publicly accessible information is especially crucial in areas of high ticks density (Sakamoto, 2018). Despite widespread public awareness of Lyme disease, there may be a mismatch between disease information and preventative behaviour as the public may not fully comprehend or use tick preventative behaviours and control strategies (Hook et al., 2015).

The new concept of "One Health" takes into account the significance of small companion animals, particularly with respect to the tick-borne bacterial diseases that humans, dogs, and cats share, as well as zoonotic pathogens that arise from wildlife and farm animal species (Cantas & Suer, 2014). The most famous bacterial diseases are the borreliosis, ehrlichiosis, rickettsiosis, anaplasmosis, and bartonellosis (Mayne, 2014). The 'One Health' notion involves the impact of environmental changes (e.g., climate change, deforestation, and urbanization) plus lifestyle changes, such as increased worldwide human and animal movement (Li, 2017). Therefore, it appears that the only way to effectively eradicate tick-borne pathogen infection in urban and peri-urban areas is to implement a multidisciplinary "One Health" approach that combines the research outputs of experts from various disciplines, including veterinarians, molecular biologists, epidemiologists, physicians, sociologists, etc. (Mencke, 2013).

Educating pet owners on the dangers of diseases spread by ticks and the importance of prevention is crucial (Jones et al., 2018). Prevalence maps and research on bacterial tick-borne diseases are essential tools for educating people, veterinary professionals, and pet owners about this growing issue (Crist et al., 2022). Continued focus on vector population control, including development of innovative, workable methods that will restrict the spread of vector-borne disease pathogens, is necessary to safeguard human and animal health against vector-borne diseases. Targeting prevention and control efforts requires accurate mapping of the risk of Lyme borreliosis exposure in forests that are often visited by hosts (Skotarczak, 2018).

6.7 Future Directions and Adaptive Measures

6.7.1 Innovations in Tick Surveillance and Management

Understanding where and when pathogen transmission occurs is essential to prevent tick-borne infections in both humans and animals. The only way to obtain this information is through surveillance that aims to determine the distribution of tick populations, the pathogens that are prevalent in those populations, and the times of year when tick activity occurs. Developing a proactive risk management system requires both active and passive surveillance in addition to modelling predictions of tick and host species habitat appropriateness (Tiffin et al., 2022). A wide range of stakeholders, including public health experts, veterinarians, and citizen scientists, can be used to improve knowledge in order to do this (Den Broeder et al., 2018). Without this knowledge, it is more difficult to monitor how climate change or landscape modification affects the risk of TBDs and to apply mitigation methods to lessen pathogen transmission and impact. Monitoring disease trends in high-risk groups is the most often utilized surveillance technique. This mostly refers to the clinical manifestation of the disease and depends on samples from the patient or infected animal providing diagnostic proof of the infection. Data collection throughout time and space can reveal patterns in the occurrence of diseases and defining treatments. Dragging is a popular technique for gathering ticks on plants, especially for tick species that attach to their host while searching on the tops of vegetation. This involves covering the vegetation with a blanket fastened to a pole. Ticks cling to the fabric and will stay gathered. Ticks can also be collected directly from animals, including humans, as an alternative to gathering them from the environment (Johnson et al., 2022).

Using non-specialists to help collect data is becoming more and more popular in surveillance; this practice is known as "citizen science." Surveys are the conventional approach for enlisting non-specialists to produce data on disease prevalence. A number of helpful parameters related to a certain issue are found by analyzing the answers to a series of questions. Approaching TBDs in humans and animals can be accomplished in this way (Scharlemann et al., 2008). In the UK, the UK Health Security Agency (previously Public Health England) has implemented a nationwide tick surveillance program to supplement this (Cull et al., 2018, 2020). This programme invites the public to submit ticks and associated data for identification to start further study. The information produced offers details on the host affiliations, seasonal activity, and distribution of several tick species within the United Kingdom. This not only identifies the important species for both human and animal health, but it also shows when the risk may be at its peak. Furthermore, the programme gives information back to the public and encourages One Health to respond to concerning results and shares this information with other government agencies. This method is in use throughout Europe (Johnson et al., 2022).

6.7.2 Research Needs and Interdisciplinary Approaches

Interdisciplinary approach is becoming a popular phenomenon emerging in response to the imperative need to move beyond restricted professional perspectives towards a broader view of health and the environment (Kent & Thompson, 2012). Medical professionals, epidemiologists, environmental scientists, geographers, and experts in economics, business, law, social and political sciences, and religion must all have a common knowledge of the complex and interconnected crises facing the globe today (Bardosh et al., 2020). To guarantee continuous improvements in the health of people, domestic and wild animals, plants, soils, and ecosystems, partnerships in research, practice, policy, planning, and adaptive management are required (Aguirre et al., 2019). A common knowledge of the functional requirements of health is supported by the interrelated fields of ecosystem health, conservation medicine, ecohealth, one health, planetary health, and geohealth (Whitmee et al., 2015).

Developing new, more intricate models of transmission of tick-borne bacterial diseases and the impact of climate on the broader habitats of the ticks, finding additional behavioral proxies when needed. Effectively bridging the gap between laypeople and specialist advisors is frequently a difficult procedure that calls both socially conscious natural scientists and medically knowledgeable social researchers (Goodwin et al., 2012). As globalization continues to enhance the frequency and speed of trade and travel, collaborations using a One Health framework both within and across agencies and political borders will become more crucial. The World Health

Organization (WHO), the Food and Agriculture Organization of the United Nations (FAO), and the World Organization for Animal Health (OIE) have all accepted the "One Health" paradigm as a tripartite project.

6.8 Key Findings and Implications

The emergence and empowering of ticks in new habitats and regions is a silent alarm for the upcoming global threats to the health of humans and animals (de la Fuente et al., 2023). Tick-borne bacterial diseases like Lyme disease and rocky mountain fever are being found to be present and expanding in a more exponential way round the years in a pattern linking to the change of climatic conditions (Gilbert, 2021). Temperature, humidity and other physical parameters paradoxically lead to expansion of ticks in areas where they were not found before (Mertens, 2025). All these problems become more threatening with the emergence of antiparasitic resistance making the traditional control methods and medicines ineffective (Hu et al., 2024). Disease diagnostic surveillance, interdisciplinary collaborations and advanced tick control strategies are the need of the future (Machtinger et al., 2024). Developing a proactive risk management system requires both active and passive surveillance in addition to modelling predictions of tick and host species habitat appropriateness (Chakraborty et al., 2025). Tick and TBD detection, mitigation, and control will be strengthened by a greater focus on professional identification training, enhanced detection of exotic species invasions, and enhanced public education (Rowan et al., 2023). Finally, in order to better understand and control ticks and TBDs in a world that is becoming more interconnected, more study is needed on integrated tick management strategies as well as animal and environmental connections to disease cycles (Zortman et al., 2023).

6.9 Call for Global Collaboration and Action

An interdisciplinary and cooperative approach is necessary to manage such health emergencies since disease control involves biological, social, and political components (Choi & Pak, 2006). The WHO, FAO, and OIE have all accepted the "One Health" paradigm as a Trigonal project. This idea acknowledges the interdependencies and connections between people, animals, and the environment in the context of global health security. It also highlights the significance of breaking down academic and professional barriers. This strategy entails multisectoral, interdisciplinary system development, increasing efforts to identify, address, and recover from threats to the health of people, animals, and the environment (Minakshi et al., 2020).

Tick-borne diseases can have a major influence on public health and the livestock industry, affecting both human and animal health. For efforts to manage and prevent

TBDs to be successful and long-lasting, a one health approach is essential. Multisectoral coordination of TBD control at the national and local levels should incorporate this information since it is essential to community needs, priorities, and capacities (Johnson et al., 2022).

References

Abdisa, T. (2019). Epidemiology of bovine anaplasmosis. *SOJ Veterinary Sciences, 5*(1), 1–6.

Adams, S. N., Bestul, N. C., Calloway, K. N., Kersh, G. J., & Salzer, J. S. (2025). National surveillance of human ehrlichiosis caused by Ehrlichia ewingii, United States, 2013–2021. *Emerging Infectious Diseases, 31*(2), 222.

Aguirre, A. A., Basu, N., Kahn, L. H., Morin, X. K., Echaubard, P., Wilcox, B. A., & Beasley, V. R. (2019). Transdisciplinary and social-ecological health frameworks—Novel approaches to emerging parasitic and vector-borne diseases. *Parasite Epidemiology and Control, 4*, Article e00084.

Alkishe, A., Raghavan, R. K., & Peterson, A. T. (2021). Likely geographic distributional shifts among medically important tick species and tick-associated diseases under climate change in North America: A review. *Insects, 12*(3), 225.

Álvarez-Hernández, G., Roldán, J. F. G., Milan, N. S. H., Lash, R. R., Behravesh, C. B., & Paddock, C. D. (2017). Rocky Mountain spotted fever in Mexico: Past, present, and future. *The Lancet Infectious Diseases, 17*(6), e189–e196.

Atif, F. A., Mehnaz, S., Qamar, M. F., Roheen, T., Sajid, M. S., Ehtisham-ul-Haque, S., Kashif, M., & Ben Said, M. (2021). Epidemiology, diagnosis, and control of canine infectious cyclic thrombocytopenia and granulocytic anaplasmosis: Emerging diseases of veterinary and public health significance. *Veterinary Sciences, 8*(12), 312.

Bardosh, K. L., de Vries, D. H., Abramowitz, S., Thorlie, A., Cremers, L., Kinsman, J., & Stellmach, D. (2020). Integrating the social sciences in epidemic preparedness and response: A strategic framework to strengthen capacities and improve global health security. *Globalization and Health, 16*, 1–18.

Berger, K. A., Wang, Y., & Mather, T. N. (2013). MODIS-derived land surface moisture conditions for monitoring blacklegged tick habitat in southern New England. *International Journal of Remote Sensing, 34*(1), 73–85.

Bouchard, C., Dibernardo, A., Koffi, J., Wood, H., Leighton, P. A., & Lindsay, L. R. (2019). An increased risk of tick-borne diseases with climate and environmental changes. *Communicable Disease Reports, 45*, 83–89.

Boulanger, N., Boyer, P., Talagrand-Reboul, E., & Hansmann, Y. (2019). Ticks and tick-borne diseases. *Médecine Et Maladies Infectieuses, 49*(2), 87–97.

Cantas, L., & Suer, K. (2014). The important bacterial zoonoses in "one health" concept. *Frontiers in Public Health, 2*, 144.

Chakraborty, S., Lyons, L. A., Winata, F., Mateus-Pinilla, N., & Smith, R. L. (2025). Methods of active surveillance for hard ticks and associated tick-borne pathogens of public health importance in the contiguous United States: a comprehensive systematic review. *Journal of Medical Entomology*, tjaf031.

Choi, B. C. K., & Pak, A. W. P. (2006). Multidisciplinarity, interdisciplinarity and transdisciplinarity in health research, services, education and policy: 1. Definitions, objectives, and evidence of effectiveness. *Clinical and Investigative Medicine, 29*(6), 351–364.

Couret, J., Schofield, S., & Narasimhan, S. (2022). The environment, the tick, and the pathogen–It is an ensemble. *Frontiers in Cellular and Infection Microbiology, 12*, 1049646.

Crist, S. D., Kopsco, H., Miller, A., Gronemeyer, P., Mateus-Pinilla, N., & Smith, R. L. (2022). Knowledge, attitudes, and practices of veterinary professionals towards ticks and tick-borne diseases in Illinois. *One Health, 14*, Article 100391.

Cull, B., Pietzsch, M. E., Gillingham, E. L., McGinley, L., & Medlock, J. M. (2020). Seasonality and anatomical location of human tick bites in the United Kingdom. *Zoonoses and Public Health, 67*(1), 112–121.

Cull, B., Pietzsch, M. E., Hansford, K. M., Gillingham, E. L., & Medlock, J. M. (2018). Surveillance of British ticks: An overview of species records, host associations, and new records of *Ixodes ricinus* distribution. *Ticks and Tick-Borne Diseases, 9*, 605–614.

Dantas-Torres, F. (2007). Rocky Mountain spotted fever. *The Lancet Infectious Diseases, 7*(11), 724–732.

de la Fuente, J., Estrada-Peña, A., Rafael, M., Almazán, C., Bermúdez, S., Abdelbaset, A. E., Kasaija, P. D., Kabi, F., Akande, F. A., Ajagbe, D. O., & Dahal, A. (2023). Perception of ticks and tick-borne diseases worldwide. *Pathogens, 12*(10), 1258.

de la Fuente, J., Estrada-Peña, A., Venzal, J. M., Kocan, K. M., & Sonenshine, D. E. (2008a). Overview: Ticks as vectors of pathogens that cause disease in humans and animals. *Frontiers in Bioscience, 13*, 6938–6946.

de la Fuente, J., Ruiz-Fons, F., Naranjo, V., Torina, A., Rodríguez, O., & Gortázar, C. (2008b). Evidence of *Anaplasma* infections in European roe deer (*Capreolus capreolus*) from southern Spain. *Research in Veterinary Science, 84*(3), 382–386.

Demma, L. J., Traeger, M. S., Nicholson, W. L., Paddock, C. D., Blau, D. M., Eremeeva, M. E., Dasch, G. A., Levin, M. L., Singleton, J., Jr., Zaki, S. R., & Cheek, J. E. (2005). Rocky Mountain spotted fever from an unexpected tick vector in Arizona. *New England Journal of Medicine, 353*(6), 587–594.

Den Broeder, L., Devilee, J., Van Oers, H., Schuit, A. J., & Wagemakers, A. (2018). Citizen Science for public health. *Health Promotion International, 33*(3), 505–514.

Esteve-Gassent, M. D., Castro-Arellano, I., Feria-Arroyo, T. P., Patino, R., Li, A. Y., Medina, R. F., de León, A. A. P., & Rodríguez-Vivas, R. I. (2016). Translating ecology, physiology, biochemistry, and population genetics research to meet the challenge of tick and tick-borne diseases in North America. *Archives of Insect Biochemistry and Physiology, 92*(1), 38–64.

Food and Agriculture Organization (FAO). (2004). *Guidelines: Resistance management and integrated parasite control in ruminants*. FAO.

Gage, K. L., Burkot, T. R., Eisen, R. J., & Hayes, E. B. (2008). Climate and vectorborne diseases. *American Journal of Preventive Medicine, 35*(5), 436–450.

Ghosh, S., Azhahianambi, P., & Yadav, M. P. (2007). Upcoming and future strategies of tick control: A review. *Journal of Vector Borne Diseases, 44*(2), 79.

Gilbert, L. (2021). The impacts of climate change on ticks and tick-borne disease risk. *Annual Review of Entomology, 66*(1), 373–388.

Ginsberg, H. S., & Faulde, M. K. (2008). Ticks. In X. Bonnefoy, H. Kampen, & K. Sweeney (Eds.), *Public health significance of urban pests* (pp. 303–345). World Health Organization.

Goodwin, R., Schley, D., Lai, K. M., Ceddia, G. M., Barnett, J., & Cook, N. (2012). Interdisciplinary approaches to zoonotic disease. *Infectious Disease Reports, 4*(2), Article e37.

Gray, J. S., Dautel, H., Estrada-Peña, A., Kahl, O., & Lindgren, E. (2009). Effects of climate change on ticks and tick-borne diseases in Europe. *Interdisciplinary Perspectives on Infectious Diseases, 2009*(1), Article 593232.

Greenfield, B. P. J. (2011). Environmental parameters affecting tick (*Ixodes ricinus*) distribution during the summer season in Richmond Park, London. *Bioscience Horizons, 4*(2), 140–148.

Hair, J. A., Sauer, J. R., & Durham, K. A. (1975). Water balance and humidity preference in three species of ticks. *Journal of Medical Entomology, 12*(1), 37–47.

Hancock, P. A., Brackley, R., & Palmer, S. C. (2011). Modelling the effect of temperature variation on the seasonal dynamics of *Ixodes ricinus* tick populations. *International Journal for Parasitology, 41*(5), 513–522.

Hengge, U. R., Tannapfel, A., Tyring, S. K., Erbel, R., Arendt, G., & Ruzicka, T. (2003). Lyme borreliosis. *The Lancet Infectious Diseases, 3*(8), 489–500.

Hofhuis, A., Harms, M., van den Wijngaard, C., Sprong, H., & van Pelt, W. (2015). Continuing increase of tick bites and Lyme disease between 1994 and 2009. *Ticks and Tick-Borne Diseases, 6*(1), 69–74.

Hook, S. A., Nelson, C. A., & Mead, P. S. (2015). U.S. public's experience with ticks and tick-borne diseases: Results from National HealthStyles surveys. *Ticks and Tick-Borne Diseases, 6*(4), 483–488.

Hu, S., Batool, Z., Zheng, X., Yang, Y., Ullah, A., & Shen, B. (2024). Exploration of innovative drug repurposing strategies for combating human protozoan diseases: Advances, challenges, and opportunities. *Journal of Pharmaceutical Analysis*, 101084.

Johnson, N., Phipps, L. P., Hansford, K. M., Folly, A. J., Fooks, A. R., Medlock, J. M., & Mansfield, K. L. (2022). One Health approach to tick and tick-borne disease surveillance in the United Kingdom. *International Journal of Environmental Research and Public Health, 19*(10), 5833.

Jones, E. H., Hinckley, A. F., Hook, S. A., Meek, J. I., Backenson, B., Kugeler, K. J., & Feldman, K. A. (2018). Pet ownership increases human risk of encountering ticks. *Zoonoses and Public Health, 65*(1), 74–79.

Karlsen, A., Vojtek, B., Mojžišová, J., Prokeš, M., & Drážovská, M. (2020). Anaplasmosis in animals. *Folia Veterinaria, 64*, 17–26.

Kent, J., & Thompson, S. (2012). Health and the built environment: Exploring foundations for a new interdisciplinary profession. *Journal of Environmental and Public Health, 2012*(1), Article 958175.

Kernif, T., Leulmi, H., Raoult, D., & Parola, P. (2016). Emerging tick-borne bacterial pathogens. In *Emerging infections.*

Kim, H. K. (2022). *Rickettsia*-host-tick interactions: Knowledge advances and gaps. *Infection and Immunity, 90*(9), e00621-e721.

Kiss, T., Cadar, D., & Spînu, M. (2012). Tick prevention at a crossroad: New and renewed solutions. *Veterinary Parasitology, 187*(3–4), 357–366.

Kocan, K. M., Blouin, E. F., & Barbet, A. F. (2000). Anaplasmosis control: Past, present, and future. *Annals of the New York Academy of Sciences, 916*(1), 501–509.

Król, N., Kiewra, D., Szymanowski, M., & Lonc, E. (2015). The role of domestic dogs and cats in the zoonotic cycles of ticks and pathogens. Preliminary studies in the Wrocław Agglomeration (SW Poland). *Veterinary Parasitology, 214*(1–2), 208–212.

Król, N., Obiegała, A., Pfeffer, M., Lonc, E., & Kiewra, D. (2016). Detection of selected pathogens in ticks collected from cats and dogs in the Wrocław Agglomeration, South-West Poland. *Parasites & Vectors, 9*, 351.

Lacasta, D., Lorenzo, M., González, J. M., Ruiz de Arcaute, M., Benito, A. Á., Baselga, C., Milian, M. E., Lorenzo, N., Jiménez, C., Villanueva-Saz, S., & Ferrer, L. M. (2021). Epidemiological study related to the first outbreak of ovine anaplasmosis in Spain. *Animals, 11*(7), 2036.

Lees, A. D. (1946). The water balance in *Ixodes ricinus* L. and certain other species of ticks. *Parasitology, 37*(1–2), 1–20.

Li, A. M. (2017). Ecological determinants of health: Food and environment on human health. *Environmental Science and Pollution Research, 24*(10), 9002–9015.

Machtinger, E. T., Poh, K. C., Pesapane, R., & Tufts, D. M. (2024). An integrative framework for tick management: The need to connect wildlife science, One Health, and interdisciplinary perspectives. *Current Opinion in Insect Science, 61*, Article 101131.

Madesh, S., McGill, J., Jaworski, D. C., Ferm, J., Liu, H., Fitzwater, S., Hove, P., Ferm, D., Nair, A., Knox, C. A., & Ganta, R. R. (2024). Long-term protective immunity against Ehrlichia chaffeensis infection induced by a genetically modified live vaccine. *Vaccines, 12*(8), 903.

Malkhazova, S., Mironova, V., Shartova, N., Orlov, D., Malkhazova, S., Mironova, V., Shartova, N., & Orlov, D. (2019). Major natural focal disease distribution. *Mapping Russia's Natural Focal Diseases: History and Contemporary Approaches*, 69–197.

Maqbool, M., Sajid, M. S., Rizwan, H. M., Younus, M., Kamran, K., Zeeshan, M., & Usman, M. (2024). Navigating acaricidal resistance through implications in veterinary practice. In H. M. Rizwan, M. A. Naeem, M. Younus, M. S. Sajid, & X. Chen (Eds.), *Anti-parasitic drug resistance in veterinary practice*. CABI.

Masters, E. J., Olson, G. S., Weiner, S. J., & Paddock, C. D. (2003). Rocky Mountain spotted fever: A clinician's dilemma. *Archives of Internal Medicine, 163*(7), 769–774.

Mayne, P. J. (2014). Clinical determinants of Lyme borreliosis, babesiosis, bartonellosis, anaplasmosis, and ehrlichiosis in an Australian cohort. *International journal of general medicine*, 15–26.

Mead, P. S. (2015). Epidemiology of Lyme disease. *Infectious Disease Clinics, 29*(2), 187–210.

Medlock, J. M., Hansford, K. M., Bormane, A., Derdakova, M., Estrada-Peña, A., George, J. C., Golovljova, I., Jaenson, T. G., Jensen, J. K., Jensen, P. M., & Kazimirova, M. (2013). Driving forces for changes in geographical distribution of *Ixodes ricinus* ticks in Europe. *Parasites & Vectors, 6*, 1–11.

Mencke, N. (2013). Future challenges for parasitology: Vector control and 'One Health' in Europe: The veterinary medicinal view on CVBDs such as tick borreliosis, rickettsiosis and canine leishmaniosis. *Veterinary Parasitology, 195*, 256–271.

Mertens, J. E. (2025). The influence of climate change on vector-borne diseases in a wilderness medicine context. *Wilderness & Environmental Medicine, 36*(1), 44–60.

Minakshi, P., Ghosh, M., Kumar, R., Brar, B., Lambe, U. P., Banerjee, S., Ranjan, K., Kumar, B., Goel, P., Malik, Y. S., & Prasad, G. (2020). An insight into nanomedicinal approaches to combat viral zoonoses. *Current Topics in Medicinal Chemistry, 20*(11), 915–962.

Mollong, E., Lébri, M., Marie-Magdeleine, C., Lagou, S. M., Naves, M., & Bambou, J. C. (2025). Sustainable management of tick infestations in cattle: A tropical perspective. *Parasites & Vectors, 18*(1), Article 62.

Mulla, M. S., & Su, T. (1999). Activity and biological effects of neem products against arthropods of medical and veterinary importance. *Journal of the American Mosquito Control Association, 15*(2), 133–152.

Nadelman, R. B., & Wormser, G. P. (1998). Lyme borreliosis. *The Lancet, 352*(9127), 557–565.

Nava, S., Guglielmone, A. A., & Mangold, A. J. (2009). An overview of systematics and evolution of ticks. *Frontiers in Bioscience, 14*(8), 2857–2877.

Nelder, M. P., Wijayasri, S., Russell, C. B., Johnson, K. O., Marchand-Austin, A., Cronin, K., Johnson, S., Badiani, T., Patel, S. N., & Sider, D. (2018). Climate change and Lyme disease: The continued rise of Lyme disease in Ontario, Canada: 2017. *Canada Communicable Disease Report, 44*(10), 231.

Noden, B. H., Loss, S. R., Maichak, C., & Williams, F. (2017). Risk of encountering ticks and tick-borne pathogens in a rapidly growing metropolitan area in the US Great Plains. *Ticks and Tick-Borne Diseases, 8*(1), 119–124.

Nuttall, P. A. (2022). Climate change impacts on ticks and tick-borne infections. *Biologia, 77*(6), 1503–1512.

Obaid, M. K., Islam, N., Alouffi, A., Khan, A. Z., da Silva Vaz, I. Jr., Tanaka, T., & Ali, A. (2022). Acaricides resistance in ticks: Selection, diagnosis, mechanisms, and mitigation. *Frontiers in Cellular and Infection Microbiology, 12*, 941831.

Ogden, N. H., Beard, C. B., Ginsberg, H. S., & Tsao, J. I. (2021). Possible effects of climate change on ixodid ticks and the pathogens they transmit: Predictions and observations. *Journal of Medical Entomology, 58*(4), 1536–1545.

Oorebeek, M., & Kleindorfer, S. (2008). Climate or host availability: What determines the seasonal abundance of ticks? *Parasitology Research, 103*, 871–875.

Parola, P., & Raoult, D. (2001a). Tick-borne bacterial diseases emerging in Europe. *Clinical Microbiology and Infection, 7*(2), 80–83.

Parola, P., & Raoult, D. (2001b). Ticks and tickborne bacterial diseases in humans: An emerging infectious threat. *Clinical Infectious Diseases, 32*(6), 897–928.

Patz, J. A., Githeko, A. K., McCarty, J. P., Hussein, S., Confalonieri, U., & De Wet, N. (2003). Climate change and infectious diseases. In *Climate change and human health: Risks and responses.*

Quadros, D. G., Johnson, T. L., Whitney, T. R., Oliver, J. D., & Oliva Chávez, A. S. (2020). Plant-derived natural compounds for tick pest control in livestock and wildlife: Pragmatism or utopia? *Insects, 11*(8), 490.

Régnier, B., Legrand, J., & Rebaudo, F. (2022). Modeling temperature-dependent development rate in insects and implications of experimental design. *Environmental Entomology, 51*(1), 132–144.

Rikihisa, Y. (2022). The "biological weapons" of Ehrlichia chaffeensis: Novel molecules and mechanisms to subjugate host cells. *Frontiers in Cellular and Infection Microbiology, 11*, Article 830180.

Ristic, M. (1981). Anaplasmosis. In *Diseases of cattle in the tropics: Economic and zoonotic relevance.* Dordrecht: Springer Netherlands.

Rizwan, H. M., Naeem, M. A., Younus, M., Sajid, M. S., & Chen, X. (2024). *Anti-parasitic drug resistance in veterinary practice.* CABI.

Rizzoli, A., Silaghi, C., Obiegala, A., Rudolf, I., Hubálek, Z., Földvári, G., G., Plantard, O., Vayssier-Taussat, M., Bonnet, S., Špitalská, E., & Kazimírová, M. (2014). Ixodes ricinus and its transmitted pathogens in urban and peri-urban areas in Europe: new hazards and relevance for public health. *Frontiers in public health, 2*, 251.

Rocklöv, J., & Dubrow, R. (2020). Climate change: An enduring challenge for vector-borne disease prevention and control. *Nature Immunology, 21*, 479–483.

Rowan, S., Mohseni, N., Chang, M., Burger, H., Peters, M., & Mir, S. (2023). From tick to test: A comprehensive review of tick-borne disease diagnostics and surveillance methods in the United States. *Life, 13*(10), 2048.

Russell, A., Prusinski, M., Sommer, J., O'Connor, C., White, J., Falco, R., Kokas, J., Vinci, V., Gall, W., Tober, K., & Haight, J. (2021). Epidemiology and spatial emergence of anaplasmosis, New York, USA, 2010–2018. *Emerging Infectious Diseases, 27*(8), 2154.

Sakamoto, J. M. (2018). Progress, challenges, and the role of public engagement to improve tick-borne disease literacy. *Current Opinion in Insect Science, 28*, 81–89.

Scharlemann, J. P. W., Johnson, P. J., Smith, A. A., Macdonald, D. W., & Randolph, S. E. (2008). Trends in ixodid tick abundance and distribution in Great Britain. *Medical and Veterinary Entomology, 22*(3), 238–247.

Schwartz, A. M., Hinckley, A. F., Mead, P. S., Hook, S. A., & Kugeler, K. J. (2017). Surveillance for Lyme disease—United States, 2008–2015. *MMWR Surveillance Summaries, 66*(SS-22), 1–12.

Sharareh, N., Behler, R. P., Roome, A. B., Shepherd, J., Garruto, R. M., & Sabounchi, N. S. (2019). Risk factors of Lyme disease: An intersection of environmental ecology and systems science. *Healthcare (Basel), 7*(2), 66.

Skotarczak, B. (2018). The role of companion animals in the environmental circulation of tick-borne bacterial pathogens. *Annals of Agricultural and Environmental Medicine, 25*(3), 473–480.

Spach, D. H., Liles, W. C., Campbell, G. L., Quick, R. E., Anderson, D. E., Jr., & Fritsche, T. R. (1993). Tick-borne diseases in the United States. *New England Journal of Medicine, 329*(13), 936–947.

Stanek, G., & Strle, F. (2003). Lyme borreliosis. *The Lancet, 362*(9396), 1639–1647.

Stanek, G., & Strle, F. (2018). Lyme borreliosis—from tick bite to diagnosis and treatment. *FEMS Microbiology Reviews, 42*, 233–258.

Steere, A. C., Strle, F., Wormser, G. P., Hu, L. T., Branda, J. A., Hovius, J. W., Li, X., & Mead, P. S. (2016). Lyme borreliosis. *Nature Reviews Disease Primers, 2*(1), 1–19.

Süss, J., Klaus, C., Gerstengarbe, F. W., & Werner, P. C. (2008). What makes ticks tick? Climate change, ticks, and tick-borne diseases. *Journal of Travel Medicine, 15*(1), 39–45.

Sutherst, R. W., & Comins, H. N. (1979). The management of acaricide resistance in the cattle tick, *Boophilus microplus* (Canestrini) (Acari: Ixodidae), in Australia. *Bulletin of Entomological Research, 69*, 519–537.

Tiffin, H. S., Rajotte, E. G., Sakamoto, J. M., & Machtinger, E. T. (2022). Tick control in a connected world: Challenges, solutions, and public policy from a United States border perspective. *Tropical Medicine and Infectious Disease, 7*(11), 388.

Treadwell, T. A., Holman, R. C., Clarke, M. J., Krebs, J. W., Paddock, C. D., & Childs, J. E. (2000). Rocky Mountain spotted fever in the United States, 1993–1996. *The American Journal of Tropical Medicine and Hygiene, 63*(1), 21–26.

Trevisan, G., Cinco, M., Trevisini, S., Di Meo, N., Chersi, K., Ruscio, M., Forgione, P., & Bonin, S. (2021). *Borreliae* Part 1: *Borrelia* Lyme group and Echidna-reptile group. *Biology, 10*(10), 1036.

Ullah, Q., El-Adawy, H., Jamil, T., Jamil, H., Qureshi, Z. I., Saqib, M., Ullah, S., Shah, M. K., Khan, A. Z., Zubair, M., & Neubauer, H. (2019a). Serological and molecular investigation of Coxiella burnetii in small ruminants and ticks in Punjab, Pakistan. *International Journal of Environmental Research and Public Health, 16*(21), 4271.

Ullah, Q., Jamil, H., Qureshi, Z. I., Saqib, M., & Neubauer, H. (2019a). Sero-Epidemiology of Q Fever (Coxiellosis) in Small Ruminants Kept at Government Livestock Farms of Punjab, Pakistan. *Pakistan Journal of Zoology, 51*(1).

Ullah, Q., Jamil, T., Saqib, M., Iqbal, M., & Neubauer, H. (2022). Q fever—a neglected zoonosis. *Microorganisms, 10*(8), 1530.

Urán, J., & Cardona, J. A. (2019). Prevalencia de *Anaplasma* spp. en el ámbito mundial: Revisión sistemática 1978–2018. *Hechos Microbiológicos, 10*, 30–38.

Villar, M., Marina, A., & de la Fuente, J. (2017). Applying proteomics to tick vaccine development: Where are we? *Expert Review of Proteomics, 14*(3), 211–221.

Wang, Z., Yang, J., Niu, Q., Brayton, K. A., Luo, J., Liu, G., Yin, H., & Liu, Z. (2017). Identification of *Anaplasma ovis* appendage-associated protein (AAAP) for development of an indirect ELISA and its application. *Parasites & Vectors, 10*, Article 359.

Whitmee, S., Haines, A., Beyrer, C., Boltz, F., Capon, A. G., de Souza, F., Dias, B., Ezeh, A., et al. (2015). Safeguarding human health in the Anthropocene epoch: Report of The Rockefeller Foundation-Lancet Commission on planetary health. *The Lancet, 386*(10007), 1973–2028.

Zhang, X., Lv, W., Teng, Z., Zhao, N., Zhou, Y., Ma, D., & Qin, T. (2024). Molecular detection of Rickettsiales and a potential novel Ehrlichia species closely related to Ehrlichia chaffeensis in ticks (Acari: Ixodidae) from Shaanxi Province, China, in 2022 to 2023. *Frontiers in Microbiology, 14*, 1331434.

Zortman, I., de Garine-Wichatitsky, M., Arsevska, E., Dub, T., Van Bortel, W., Lefrançois, E., Vial, L., Pollet, T., & Binot, A. (2023). A social-ecological systems approach to tick bite and tick-borne disease risk management: Exploring collective action in the Occitanie region in southern France. *One Health, 17*, Article 100630.

Zu, H., Xiang, Z., Zhang, X., Cao, Q., Lin, Y., Ying, Z., Bhowmick, B., Xiang, H., Han, Q., & Wang, J. (2025). Prevalence of Ehrlichia spp. in dogs and ticks in Hainan Province, China. *BMC Veterinary Research, 21*(1), 239.

Chapter 7
Impact of Climate Change on Tick-Borne Protozoal Diseases

Haider Abbas, Muhammad Qasim, Hafiz Muhammad Rizwan, Malcolm K. Jones, and Ali Butt

Abstract Global environmental conditions are changing due to climate change, which has major consequences for vector-borne disease epidemiology. Among these, tick-borne protozoal diseases which are spread by tick vectors, are becoming a major public and veterinary health problem. The effects of climate change on tick prevalence, distribution, and transmission dynamics as well as tick-borne protozoal infection including theileriosis and babesiosis are explored in this chapter. Tick populations and their capacity to spread protozoal infections are impacted by changing habitats, changing precipitation patterns, and rising temperatures. Diseases that were previously limited to temperate regions now have a wider geographic range as a result of tick habitat expansion brought about by warmer climates. Longer transmission seasons may result from modifications to seasonal trends that impact tick life cycles and the protozoa they carry. This information sheds light on possible hazards to animals and people, emphasizing the necessity of proactive monitoring and flexible approaches in the fields of public health and veterinary field. To identify changes in ticks and tickborne disease occurrence, with reference to climate change attribution, and to establish predictive models of animal and public health concerns to prepare for tickborne disease emergence, more integrated One Health surveillance and modeling investigations are required.

Keywords Climate change · Protozoal pathogens · Vector ecology · Tick-borne diseases · Epidemiological shifts · Host-vector-environment interactions

H. Abbas (✉) · H. M. Rizwan
Department of Pathobiology (Parasitology Section), KBCMA College of Veterinary and Animal Sciences, Narowal, Sub-Campus UVAS, Lahore, Pakistan
e-mail: haider.abbas@uvas.edu.pk

M. Qasim
Grand Valley State University, MI, USA

M. K. Jones
School of Veterinary Science, The University of Queensland, Gatton Queensland, Australia

A. Butt
Veterinary Officer, Livestock and Dairy Development Department, Punjab, Pakistan

H. M. Rizwan and M. S. Sajid (eds.), *Ticks in a Changing Climate*, Biometeorology 7,
https://doi.org/10.1007/978-3-032-12259-9_7

7.1 Introduction to Tick-Borne Protozoal Diseases

Protozoa are microscopic, unicellular, eukaryotes, of which some live as parasites and others are free living. Protozoa are categorized into four groups as sporozoans (apicomplexans), flagellates (excavates), amebae (amebozoans) and ciliates (ciliophorans). These Protozoa are heteromorphic and adapted to different physiological processes. These perform complicated metabolic processes within the cell, which have complicated internal structure. Moreover, these protozoa typically have specialized locomotory organelles including cilia, flagella and pseudopodia (de Koning, 2020). A number of protozoan parasites have been found in the blood circulation of vertebrates and are transmitted from one individual host to another by invertebrate vectors. Based on morphological features and host specificity, almost 48 genera of haemaprotozoans have been identified. These include intracellular (apicomplexans variously termed the piroplasms, haemosporidia, haemogregarines and haemococcidia) and extracellular protozoans (the flagellated trypanosomatids) (O'Donoghue, 2017).

Ticks are obligate haematophagous parasites of mammals, birds and reptiles. Currently, about 970 tick species have been found worldwide. Of these, ~750 belong to the family Ixodidae and ~218 belong to the family Argasidae, while only one species falls in the Nuttalliellidae (Dantas-Torres and Otranto, 2022). Ticks and the diseases they transmit pose significant threats to the health and production of the livestock (Abubakar et al., 2018; Ghafar et al., 2020). A large proportion (80%) of cattle farmed worldwide are subject to tick-borne disease (Solomon and Tanga, 2020). Tick infestation and tick-borne pathogens lead to huge global economic loss (US$30 billion) annually (Nyangiwe et al., 2018). Tick-borne haemoprotozoal diseases have crucial impact on the health and production of ruminants, resulting in significant financial losses in meat and dairy industry globally (Juyal et al., 2005). Among these tick-borne protozoans, babesiosis and theileriosis are the most economically significant, challenging global animal production. Other tick-borne protozoal diseases include cytauxzoonosis in domestic cats and hepatozoonosis in dogs, respectively (Ceylan et al., 2021).

Babesiosis in livestock, wild and companion animals, and humans arises from intraerythrocytic infection with species of the piroplasm genus *Babesia*., Babesiosis often exists in an endemic state in ruminants in tropical and subtropical regions. *Babesia (B.) divergens, B. occultans, B. major, B. bigemina* and *B. bovis* are among the *Babesia* spp. which cause disease in bovines. The hard ticks, *Ixodes* and *Rhipicephalus* spp. act as vector for these *Babesia* spp., and *B. bigemina* has been found more prevalent in bovines (Ozubek et al., 2020). In acute phase of babesiosis, clinical signs in animals include high grade fever, severe anemia, red urine (hemoglobinuria), lack of appetite and fatigue.

Theileria (T.) parva and *T. annulata* are highly infective and economically significant species affecting cattle. Tropical theileriosis is a major tick-borne disease of cattle caused by *T. annulata* and spread by *Hyalomma* tick vectors (Ozubek and Aktas, 2019). In theileriosis, cattle suffer from fever, hemolytic anemia, enlarged lymph nodes, decrease in productivity and abortion leading to economic losses in cattle industry (Watts et al., 2016). This is a seasonal disease occurring May–September with peak prevalence in July in Northern Hemisphere (Ceylan et al., 2021).

The influence of future climate on vector-borne infections ranges has been extensively researched, possibly because of the growing connection between vectors and climate and worries about vector management challenges (Thomson and Stanberry, 2022). Research on vector dispersal alone, as a straightforward arthropod-climate interaction, may not be entirely useful unless the findings are interpreted in light of how the ecosystem as a whole will respond and change. Animals have historically changed their ranges and behaviors to accommodate for natural temperature variance, (Harley, 2011). Others might adjust by relocating, but they are constrained by walls, cities, or have attained their highest altitude (Keesing and Ostfeld, 2021).

7.2 Effects of Climate Change on Tick Ecology

According to the Intergovernmental Panel on Climate Change's fourth assessment report, global warming may cause rise of 1.5 to 2.5 °C temperature across northern temperate Europe during the coming decades (IPCC, 2022). In addition to improving living standards in the majority of countries, the economic growth has sparked concerns about climate change globally (Mardani et al., 2019). Humanity is facing a number of challenges as a result of global warming and resulting climate change, including serious health concerns caused by contagious infections (Lowe et al., 2018). The main reason of Earth's warming is the rise in atmospheric concentrations of greenhouse gases, including nitrous oxide, methane, and carbon dioxide which are brought on by anthropogenic activities (Lee, 2023). It is considered that droughts, heavy rainfall, enhanced frequency and rapid rise in heatwave magnitudes are the adverse outcomes of global warming due to anthropogenic activities (Allen et al., 2018).

The distribution and abundance of ticks may increase or decline as a result of climate change. The duration of activity of tick-questing for hosts may also vary. Additionally, alterations in rate of development of ticks as a result of changes in climate may result in modified seasonal activity patterns. This could occur through changes in the number of ticks in a population that are subjected to regulatory processes like diapause (Christensen et al., 2007).

The prevalence and abundance of tick populations can be influenced by climate in two ways: directly through changes in per-capita death and proliferation rates, or indirectly through changes in tick activity, habitat, and host populations. Ticks of some species subjected to temperatures below −5 °C in a lab setting quickly die (Ogden et al., 2004). Several tick species may withstand subzero temperatures, however, and it is thought that glycoproteins in the hemolymph act as antifreeze (Neelakanta et al., 2010). According to a previous investigation, the daily death rate of unfed ticks (*Ixodes scapularis*) in winter was equivalent to that in summer, despite prolonged periods of ambient temperatures below −20 °C. This suggests that the duff layer in woodland settings can serve as shelter for ticks from extremely cold temperatures (Burtis et al., 2019). Ticks will naturally die if they become desiccated, despite their physiological and physical resistance to dehydration (Rodgers et al., 2007). In humid conditions, ticks rehydrate by releasing saliva, which is hygroscopic in nature, onto the hypostome. Once saturated, the saliva is re-absorbed by the tick after absorbing the water from the humid atmosphere (Alarcon-Chaidez, 2014). Low atmospheric moisture is believed to have an indirect impact on the survival of questing exophilic ticks by making them descend to the duff layer more frequently to rehydrate. This accelerates the depletion of lipid reserves and lowers the probability of successfully finding a host. Tick populations may be impacted by low ambient humidity levels through changes in host-seeking activity and outcome (Burtis et al., 2016).

Temperature influences the speed at which engorged larvae and nymphs convert into seeking nymphs and adults, engorged females to ovipositing females, and egg development. Development length and temperature generally have a nonlinear relationship, with development taking longer, or ceasing entirely, at lower temperatures (Ogden et al., 2004). Since tick populations can survive cold winters as long as their habitat offers a place to hide, it has been proposed that the key to understanding how climate change affects tick populations lies in understanding how temperature affects lifecycle duration through impacts on interstadial growth rates (Ogden and Lindsay, 2016). Therefore, the lifecycle is shortened in hotter climates, and more larvae survive to become reproducing adults (Burtis et al., 2019).

Climate change can affect ticks directly, modifying their development, survival, reproduction, and activity patterns; indirectly, it alters habitats and has consequences for the distribution and abundance of host species. Climate change also alters pathogen dynamics and can affect human behavior, such as increased outdoor activities, which would then change human infection risk. The complexity and interaction of these pathways render it difficult to ascertain the relative importance of each in contributing to overall risk of disease (Fig. 7.1) (Gilbert, 2021). At the physiological level, environment fluctuations are recognized by nervous and sensory systems of the ticks. The vast ranges of endurance and flexibility of the tick neural/sensory system intervene its behavioural adaptability (Šimo et al., 2013). Similarly, it is probable that the tick microbiome will facilitate adaptation by favoring endosymbiont communities that improve tick survival (Gottlieb and Duron, 2021). Because molting between feeds lead to increase the external incubation period, it is therefore believed that changes in the surrounding temperature have no effect on tick-borne infections.

The development process of arthropod-borne infections in their vectors, on the other hand, is modulated by temperature and humidity, according to a study (Turell et al., 2020). There is evidence that rising temperatures may accelerate the growth and reproduction of infective agents carried by ticks and raise the danger of transmission. For instance, research on *Theileria parva*, the agent of the tick-borne East Coast disease in cattle, shows that transmission by nymphs decreased from around 72 to 24 h, when infected nymphal *Rhipicephalus (R.) appendiculatus* were incubated at 18 or 37 °C after feeding (Ochanda et al., 1988). Thus, the transmission

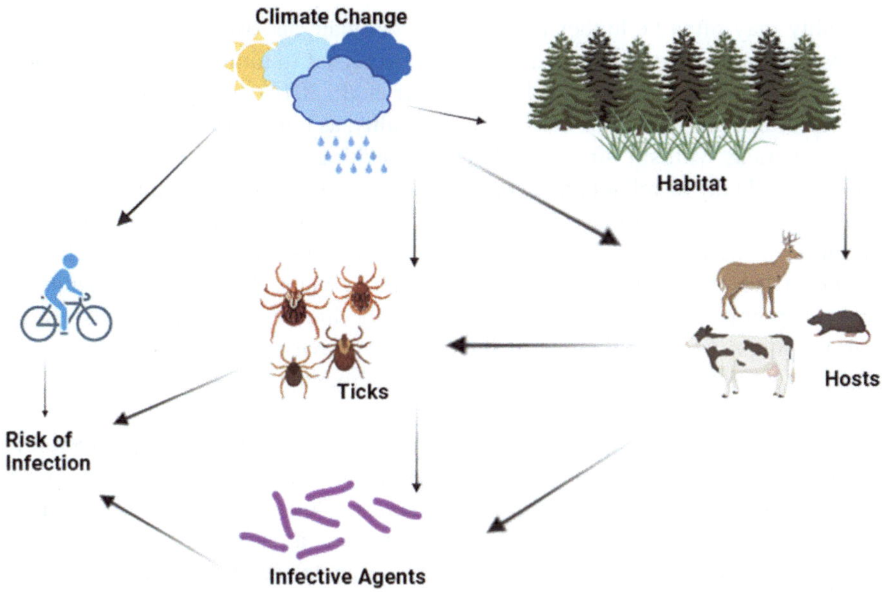

Fig. 7.1 Flowchart showing direct and indirect impact of climate change

process of infective agent in infected ticks was shortened by the higher environmental temperature.

Ticks exhibit species-specific reactions to temperature variations. *Haemaphysalis longicornis, Amblyomma americannum* and *Dermacentor* ticks, are more resilient to climatic stressors and can endure hotter and drier conditions (Gregory et al., 2022). According to these authors, who studied the tick in North America, although these tick species are not yet well-established in Canada, if climate change continues, they could establish permanent populations in this region (Elmieh, 2022). *Ixodes* ticks prefer off-host regions with higher humidity including beneath leaf litter or the forest canopy, and are more susceptible to atmospheric changes. By altering their activity levels, tick species can behaviorally adapt to unfavorable environmental situations (Ogden et al., 2021). For instance, dehydration has led to rise in the death rate among certain *Ixodes (I.) scapularis* ticks. In order to lessen the chance of dehydration, these ticks have consequently decreased their activity in hot and dry weather and sought refuge in the duff layer, which was located beneath the leaf litter. Because these ticks can seek out insulated habitats (such as snow and duff layer) to aid in overwintering, this trait also shields ticks from cold temperatures and lowers mortality. Depending on the situation, these behavioral shifts may lengthen or shorten the cycles of tick-host pathogen transmission (Ogden et al., 2021).

A rise in rainfall can result in decrease in nymph growth due to reduction in tick activity levels (Bouchard et al., 2019). Hence, it is assumed that rainfall may assist tick populations to establish in new regions. Tick survival may be impacted by extreme climate and weather fluctuations brought on by climate change. Tick mortality may often be higher in response to climate and weather variability than in absolute temperature. While droughts are expected to diminish tick populations due to dehydration of tick individuals (Berger et al., 2014), heavy floods caused by large rainfall events may also cause ticks mortality, as long as the flooding continues or burries ticks with silt hindering their host seeking activity (Weiler et al., 2017). Tick survival may be reliant on the prevailing weather patterns in certain ecosystems that offer few shelters from extreme fluctuations of temperature and humidity. For instance, in the northern and western regions of United Kingdom and Ireland, *I. ricinus*, a tick that typically inhabits woodland habitats, thrives on comparatively arid sheep-grazed uplands, partly because of the high rainfall in all seasons and warm winters. Tick populations in such exposed environments may be especially impacted by unpredictable weather that brings droughts and extremes of cold and heat (Ogden et al., 2021).

In its ecological niche, daylight hours, light intensity, the surrounding temperature, and ambient humidity are among the environmental stimuli that enable the majority of ixodid tick species to seek hosts for a bloodmeal (Belozerov et al., 2002). Climate change is expected to cause phenological changes in a particular area by delaying the time that overwintering or diapausing ticks start to seek host and/or when their eggs hatch. According to recorded data of juvenile *I. scapularis* feeding on small animals gathered over a 19-year period in New York revealed that their host-seeking behaviours extended by nearly 21 days in warmer years as opposed to colder years (Levi et al., 2016). Although the same phenological shift was not observed in *I. ricinus* nymphs over a 15-year study in a temperate Swiss forest, it was found that vapor pressure deficit in air had a significant impact on host-seeking of that species, with a higher threshold attained sooner in the spring that may have limited the rise in host-seeking nymph population (Hauser et al., 2018). There will be serious consequences for numerous tick-borne infective agents if climate change has a differential impact on the phenology of various tick stages, changing the extent of overlap in the activity patterns of larvae and nymphs (Gilbert, 2021). In terms of long vs. short-term diseases and local synchronized as well as asynchronized phenology of tick larvae and nymphs, there is presently insufficient evidence of a climate signal of change, despite the fact that these estimates align with the prevalence of various tick-borne diseases in North America (Diuk-Wasser, 2021).

7.3 Geographical Shifts in Tick Populations

Since the end of the 1980s, when the climate began to dramatically change, a study from Sweden has shown that the Northern distribution limit of *I. ricinus*, along with that of a number of other species of plants and animals, has moved northward (Tälleklint and Jaenson, 1998). Whereas in the 1980's, the northern range of *I. ricinus* was 61°N, the tick was found as far north as 66°N over the entire Baltic Sea coastline, as well as across the river valleys and the bigger lakes in the Northern areas (Jaenson et al., 1994). It has been demonstrated that increases in the number of days with temperatures essential for tick existence, activity, and growth throughout multiple seasons are linked to this shift in latitude distribution (Lindgren et al., 2000). New tick populations have been observed establishing even at lower than −12 °C cold winter temperatures at the highest latitudes (Gray et al., 2009).

Canine babesiosis is transmitted by hard tick vector, *Dermacentor* (*D.*) *reticulatus* (Springer et al., 2024). However, this tick species has spread to numerous other previously undiscovered tick sites in Germany (Drehmann et al., 2020). These findings strongly imply that *D. reticulatus* has spread throughout Germany over the past three decades, especially in the country's East and Southwest. The presence of canine babesiosis in new regions of Switzerland, Netherlands, Hungary, and Germany offers additional support for the shift in distribution of *D. reticulatus* to new areas which were previously unknown for this tick specie (Ciuca et al., 2021). Increased host range and more fallow land due to warming climates can be the reasons that may be contributing to the recent expansion of *D. reticulatus* (Zając et al., 2021).

Immature stages of *Hyalomma* (*Hy.*) *marginatum* are now commonly seen on migratory birds traveling to temperate Europe from Africa (Földvári et al., 2022). That these immature ticks' stages are found on birds in Europe, indicates that reproduction may occur in Europe, and by implication, the potential northern extension and development of persistent populations of the tick in that continent is very significant (Földvári et al., 2022). This tick's life cycle is longer in Northern, colder areas, where immature stages of this tick are viable as late as June, while it is faster in Southern regions of its distribution range (Northern Africa) where larvae are viable as early as February (Hoogstraal et al., 1961). There are two distinct population clusters based on climate needs, according to a survey of the documented tick's distribution. One cluster stretches into Turkey and the Middle East from the species' Northern ecological limit in the Balkans (about 44°N latitude). The other one is limited to Western Spain and Africa (North of the Sahara). A temperature-related restricting variable for these Northern groups of ticks is evident from an evaluation of the first cluster's climate niche. September to December temperature is essential for the emergence of long-term populations. Unfed adults may thrive during the next active season if temperatures are sufficiently high to permit molting prior to the cold months (Gray et al., 2009).

It is obvious that unless there is adequate long-range tick transportation by hosts into new areas of compatibility, shifts in the geographical trends in climatic suitability for tick species would not lead to abrupt shifts in geographical ranges. While migratory birds seem able to accomplish faster ("pulled") range expansions, domestic hosts can only transmit ticks across comparatively limited distances, resulting in a ("pushed") gradual expansion of ticks and tick-borne disease ranges (Ogden et al., 2013). Both nidicolous ticks (Gray et al., 2014) and exophilic ticks and the infective agents they transmit (Cohen et al., 2015) appear to be especially spread by migratory birds. Therefore, it is reasonable to assume that a changing climate will frequently lead to changes in geographical range. By examining the areas where seasonal synchrony of life cycle stages (larva and nymph) of *I. ricinus* activity is anticipated to take place, one study has determined how climatic change affects the spatial distribution of tick-borne pathogens in Europe (Randolph and Rogers, 2000).

In Scandinavia, *I. ricinus* has also been shown to have expanded Northward, reaching Sweden (66°N) and Norway (66.2°N) (Hvidsten et al., 2020). Although roe deer have likely contributed to tick reproduction and spread into northern latitudes, warming temperatures seems to be a significant element in these developments. *Amblyomma americanum* Linnaeus, 1758, and *D. variabilis* Say, 1821 in North America are two other species that appear to be spreading their range (Molaei et al., 2019). Additionally, the tick of the cattle industry, *R. microplus* Canestrini, 1888, is thought to be now more likely to re-emerge in the United States due to climate change (Osbrink et al., 2022).

The introduction of exotic species of ticks in novel regions may be facilitated by climate change. A few of these are *Hy. marginatum marginatum* Koch, 1844, which only recently emerged in France, and *Haemaphysalis (Ha.) longicornis* Neumann, 1901, which is native to Asia and is currently established in the USA (Raghavan et al., 2019). Both species are significant carriers of tick-borne infections, including Crimean-Congo hemorrhagic fever (*Hy. marginatum*), and Theileriosis (*Ha. longicornis*). However, other factors including its preference for domestic animals (cattle, horses, and dogs) may be more important in its range extension because *Ha. longicornis*, a parthenogenetic and bisexual species, can adjust to a broad spectrum of climate types (Hoogstraal et al., 1968).

The main vector of *Rickettsia conorii, Ehrlichia canis* and *B. canis* in dogs is *Rhipicephalus sanguineus.* In Northern temperate Europe, *Rh. sanguineus* does not appear to survive in the outside environment long enough to do considerable contribution in disease spread, despite the fact that it may develop in kennels in central and Northern European latitudes, perhaps leading to brief localized disease outbreaks. However, authorities in several nations, including the UK and Ireland, have made tick treatment a part of the pet passport program because to worries about its potential entry and establishment (Gray et al., 2009).

7.4 Major Tick-Borne Protozoal Diseases

Babesiosis is a tick-borne disease caused by infection with the intraerythrocytic *Babesia* species (Table 7.1) Babesiosis occurs in livestock, canids, wild animals as well as humans (Wu et al., 2025). In tropical and subtropical regions, this infection exists in endemic state in ruminants. Bovine babesiosis is caused by *B. divergens, B. bovis* and*B. bigemina. B. divergens* is the major causative agent of red water fever, which is another name for bovine babesiosis (Azhar et al., 2023). The hard ticks, *Ixodes* and *Rhipicephalus* spp. act as vectors for *Babesia* spp. Mostly *B. bigemina* has been found to be prevalent in bovines (Ozubek et al., 2020). In acute phase of the disease, animals suffer from high grade fever, severe anemia, red urine (hemoglobinuria), lack of appetite and fatigue. From 50 to 100% parasitemia has been reported in *B. bigemina* infections while in cases of *B. bovis,* parasitemia has been reported to be as high as 90% (Alvarez et al., 2020).

Table 7.1 Species of *Babesia* and their vectors in different animals

Babesia species	Animal infected	Tick vector	References
Babesia bigemina	Cattle	*Rhipicephalus* (*Boophilus*) spp.	Srionrod et al., (2022)
Babesia bovis		*Rhipicephalus* spp.	
Babesia divergen	Buffaloes	*Ixodes* spp.	
Babesia major		*Haemaphysalis* spp.	
Babesia beliceri		*Hyalomma* spp.	
Babesia motasia	Sheep and Goat	*Rhipicephalus* spp.	Erster et al., (2016)
Babesia ovis			
Babesia caballi	Horse	*Rhipicephalus* spp.	Tirosh-Levy et al., (2021)
Babesia equi		*Dermacenter* spp.	
		Hyalomma spp.	
B. perroncitoi	Pig	*Rhipicephalus* sp.	Avenant et al., (2021)
B. trautmanni			
B. felis	Cat	*Rhipicephalus* spp.	Penzhorn and Oosthuizen, (2020)
Babesia microti	Dog	*Haemophysalis leachi*	Zygner et al., (2023)
Babesia canis (Large)		*Rhipicephalus sanguineus*	
Babesia valpes		*Haemophysalis eliptica*	
Babesia vogeli		*Haemaphysalis bispinosa*	
Babesia rossi (Large)		*Hyalomma* spp.	
Babesia gabsoni (small)		*Dermacenter* spp.	
Babesia conradae		*Rhipicephalus sanguineus*	

Canine babesiosis is caused by either large forms of *Babesia,* like *B. vogeli, B. rossi* and B. *canis* (Zygner et al., 2023), or by small forms, such as *B. conradae, B. vulpes and B. gibsoni* (Solano-Gallego et al., 2016). *Dermacentor reticulatus, R. sanguineus* and *Ha. longicornis* are the tick vectors associated with canine babesiosis (Bajer and Dwużnik-Szarek, 2021). Although *Babesia* spp. cause disease in humans very rarely worldwide, cases of human babesiosis are emerging in Europe, United States and China, with species reported for human disease include *B. venatorum, B. microti, B. duncani and B. divergens* (Kumar et al., 2021).

From an economic point of view, bovine babesiosis is considered important, however *Babesia* species also affect other animals (equines, companion animals and small ruminants) and humans. Babesiosis puts severe financial burden on the global livestock industry (Azhar et al., 2023). The pathogenesis of disease depends on the development of anemia in the host, which occurs due to the lysis of the infected red blood cells (RBCs). In case of infection by *B. bovis* or *B. canis,* a cerebral form of the disease may occur if the infected RBCs lodge in brain capillaries (Alvarez et al., 2019). The pathogenesis is exacerbated due to increased production of signaling molecules like cytokines and immune-mediated RBCs destruction by host defense system leading to vasodilatation, intravascular coagulation and vascular stasis-like effects. The hosts which survive the infection act as carriers and can transfer the parasite to tick vectors, and carrier humans can be source of transmission of zoonotic *Babesia spp.* like *B. microti* either through blood transfusions or congenitally to other humans (Wu et al., 2017).

Anti-babesial drugs, vaccines, and tick control are the tools for controlling babesiosis (Santos et al., 2023). In babesiosis-endemic regions, disease is controlled by live vaccines, but these vaccines are not used on a wide scale due to reasons like incompetency against variable strains and the chances of other haemo-pathogen contamination (Bastos et al., 2023). Although, to lower the economic burden due to babesiosis, chemotherapy is considered the primary option for the control of the disease, but these parasites have been reported to develop resistance against most of the anti-babesial agents (Yamasaki et al., 2021).

The protozoan *Theileria* spp. is an intracellular haemoparasite which is responsible for tick borne disease in livestock resulting in significant socio-economic burden on farming. Bovine Tropical Theileriosis is caused by *T. annulata* (Liu et al., 2022), while *T. parva* causes East Coast Fever (Agina et al., 2020). The pathogenesis of this parasite involves cell immortalization, rapid multiplication and propagation of infected white blood cells, the phenotypes similar to cancer resulting in transformation of the host cells (Woods et al., 2021). During this lympho-proliferative infection, infected mammalian host cells disperse throughout the body leading to the swelling of the lymph nodes. Theileriosis has been found in Asia, southern Europe, the Middle East, North and sub-Saharan Africa infecting the large population of cattle and buffalo (Woods et al., 2021).

The *Theileria* parasite completes its life cycle depending on two different hosts *i.e.*, ticks of genus *Hyalomma* and the mammalian host. During blood feeding, ticks inject the sporozoites of the parasite into the mammalian host which infect the different subsets of white blood cells and convert into a schizont form while the piroplasm stage develop in the erythrocytes (Zeb et al., 2022). The molecular mechanisms through which the transformation of the mammalian host cells occurs in theileriosis consist of modifications in cell signaling pathways. These changes result in fluctuations in the host cell gene expression/activation due to effect on host epigenome by *Theileria* parasites (Morrison, 2009). The changes in the oncogenic signaling pathways (like JNK and IKK) and modulation of transcription factors (like AP-1, HIF-1α and c-Myc) in the host are induced by the infected white blood cells in theileriosis (Metheni et al., 2015). Tropical theileriosis is characterized by various clinical signs, such as fever, mild ocular and nasal discharge, loss of appetite, salivation, respiratory problems, enlargement of superficial lymph nodes, acute anemia, jaundice, and death due to asphyxia. The typical pathological findings include lymph node enlargement, spleen enlargement, pulmonary emphysema, and subcutaneous and intramuscular hemorrhages (Ma et al., 2020).

7.5 Transmission Dynamics and Host Interactions

The host associations and patterns of transmission of protozoan parasites are intricate and influenced by a number of variables, such as the host immune system, the surrounding environment, and the parasite species. In order to survive and successfully spread, protozoans have developed a variety of adaptations, frequently at the cost of the health of their host. For the development of efficient therapies and preventive measures for protozoan infections, a thorough comprehension of these interactions is required (Jalovecka et al., 2018).

The life cycles of causative agents of babesiosis and both examples of extremely complex protozoal infection cycles that have several phases in both the host as well as the tick vector. Controlling these infections is made more difficult by the intricacy of these cycles and the parasites' highly developed immune escape strategies. Tick control, immunization, and chemotherapy are only a few of the many components that make up an effective control strategy. Mitigation of these infections is still difficult, though, due to the complex nature of the life cycles and host interactions, particularly in areas where tick populations are high, and animals are frequently exposed to infection (Elsworth and Duraisingh, 2021).

The *Babesia* species have nearly identical life cycles with few significant variations. During blood feeding, *Babesia* infected tick injects the sporozoites to the mammalian host, where they enter the red blood cells (RBCs) and start multiplication through an asexual reproduction process. *Babesia* species lack a pre-erythrocytic life span phase, in contrast to other closely associated apicomplexan, *Theileria* species that also multiply in their host's RBCs. *Babesia* spp. repeatedly replicate, and on destruction of infected RBCs, the emerging merozoites invade the other RBCs, this process goes on. A small number of parasites will dedicate themselves to the development of gametocytes, which can be taken up by other ticks. Upon ingestion, the gametocytes leave the RBCs and develop into gametes in the midgut. The male and female gametes fuse to form zygote. This zygote subsequently infiltrates the tick gut epithelial cells and divides through meiosis to produce kinetes (Jalovecka et al., 2018). These kinetes can spread to tick larvae by invading different tick tissues such as the ovaries and undergoing additional asexual reproduction, with the exception of *B. microti*, which mostly infiltrates nephrocytes (Howell et al., 2007). The salivary glands get the kinetes that develop in the tick gut cells and convert to sporozoites which are transmitted to the next mammalian host (Fig. 7.2) (Weber and Friedhoff, 1979). The signaling pathways and environmental stimuli that underpin *Babesia* spp. intricate life cycles have been examined (Fig. 7.2). An outline for the signaling pathways of *Babesia* spp. during their entire life cycle has been constructed. The signaling pathways of *Babesia* spp. are still poorly understood in comparison to the more thoroughly researched other apicomplexan parasites (*Toxoplasma gondii* and *Plasmodium* spp.). Up until now, all functional investigations have used small compounds to test signaling pathways. The number of genomes, transcriptomes, in vitro and in vivo culture systems, and genetic systems accessible for the research of *Babesia* species has recently increased. To comprehend both the molecular and cellular biology of these parasites and to support the development of innovative treatments and vaccines, it will be crucial to make use of and broaden these recent developments (Siddiqui et al., 2020).

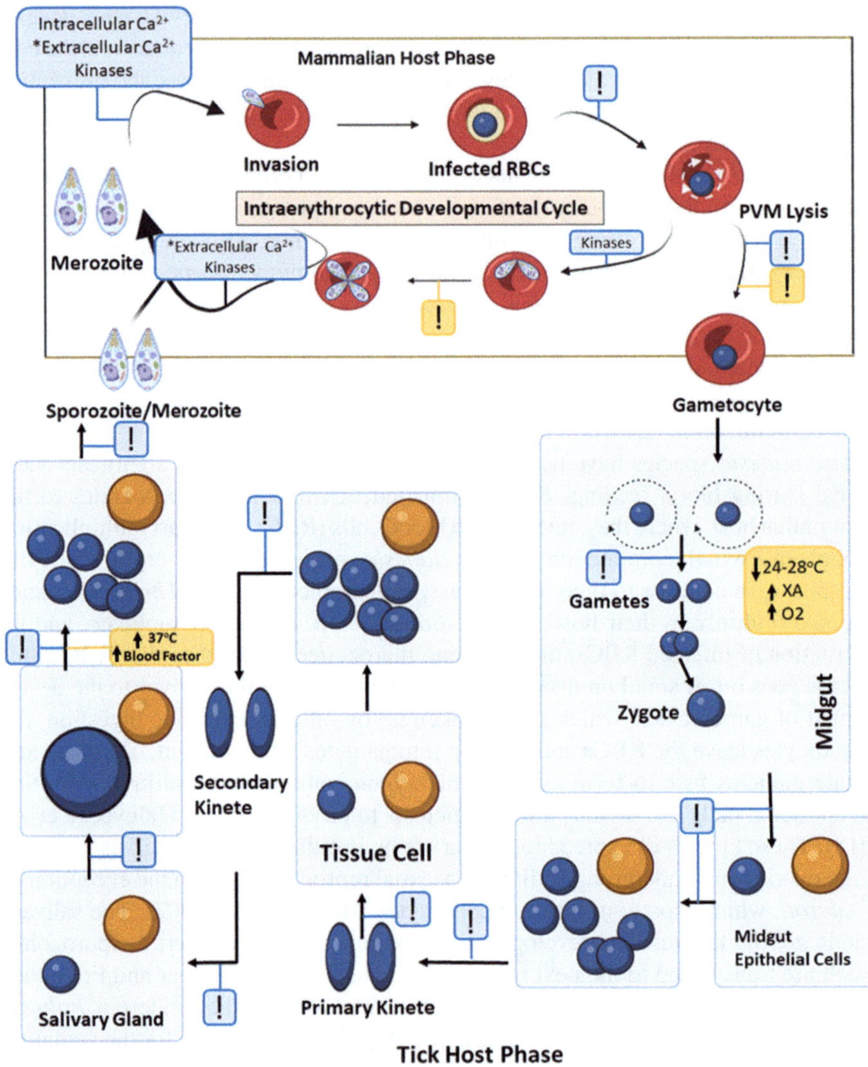

Fig. 7.2 A schematic diagram of the life cycle of *Babesia* spp., showing the required signaling pathways. Blue and yellow blocks indicate the required signaling molecules and external stimuli, respectively, by the protozoa at that phase in its life span. "*" Denotes the required extracellular Ca^{2+} by the *Babesia* at the time of invasion and exit from RBCs. "!" Indicates the points in the life cycle of the parasite at which signaling pathways and environmental stimuli were expected as explained in various studies for other apicomplexans but this information is lacking in case of *Babesia* spp. (XA: Xanthurenic acid)

Theileria spp. are the only piroplasms that typically undergo asexual reproduction (schizogony) in leukocytes (Liu et al., 2022) (Fig. 7.3). Following sporozoite invasion of leukocytes, schizogony begins, leading to the generation of merozoites, which then undergo merogony to reproduce. Schizogony may cause the leukocytes to undergo neoplastic shift, after which they will continue to divide continuously and this happens in case of *T. taurotragi*, *T. parva*, *T. lestoquardi* and *T. annulata* (Schreeg et al., 2016). Merozoite invasion of erythrocytes by *Theileria* species initiates the asexual replication (merogony) in the cell. The intraerythrocytic parasites then grow into trophozoites, which then asexually generate new merozoites. After the host erythrocytes burst, merozoites are liberated and infiltrate non infected red blood cells (Jalovecka et al., 2016). The appearance of gametocytes in the erythrocytes initiates the sexual reproductive phase (gamogony) of the parasite. It has been reported that the sexual multiplication phase of *Theileria* is similar to *Babesia* and gametocytes exist in the bloodstream (Lempereur et al., 2017). Gametocytes transform into gametes after tick blood absorption, and these gametes develop in the tick's midgut lumen. After fertilization, a zygote is produced that enters tick epithelial cells through the peritrophic network (Bishop et al., 2004). Here, the zygote divides through meiosis to produce kinetes, which are then discharged into the hemolymph. Primary kinetes infiltrate the salivary glands immediately (*Theileria* spp.), whereas kinetes undergo two rounds of asexual proliferation in different tick tissues in case of *Babesia* parasites before the secondary kinetes enter the tick's salivary glands. Following kinete invasion of salivary glands of tick, sporogony begins, resulting in the formation of the sporont (Yano et al., 2005). During tick molting, the sporont develops into sporoblast, an intracellular multinucleated meshwork. Sporozoites are secreted into tick saliva as a result of the parasite sporoblast maturing, which begins when the tick attaches to the host (Jalovecka et al., 2018).

7.6 Recent Trends and Outbreaks

There are around 100 described *Babesia* spp., and they cause disease in a variety of mammalian hosts. Of these, 18 of them seriously harm domestic animals, such as sheep, cats, goats, dogs, pigs and cattle, horses. For *B. bovis*, *B. major*, *B. divergens*, and *B. bigemina*, cattle serve as the main and reservoir hosts. Mostly found in tropical and subtropical regions, especially in Central and South America, Africa, parts of southern Europe, Australia and Asia, the two most common *Babesia* spp. are *B. bovis* and *B. bigemina.* Despite sharing the same geographic region, *B. bovis* and *B. bigemina* have somewhat diverse distributions. In Africa, *B. bigemina* is more prevalent than *B. bovis.* The infection and sero-prevalence of bovine babesiosis around globe are given in Table 7.2 as summarized by Jacob et al. (2020).

Theileria orientalis was initially believed to be comparatively benign (Yokoyama et al., 2012), in contrast to the extremely pathogenic *T. parva* and *T. annulata* that affect cattle in Africa, Europe, and Asia (Agina et al., 2020). Though outbreaks of diseases are primarily observed in regions of North America (Oakes et al.,

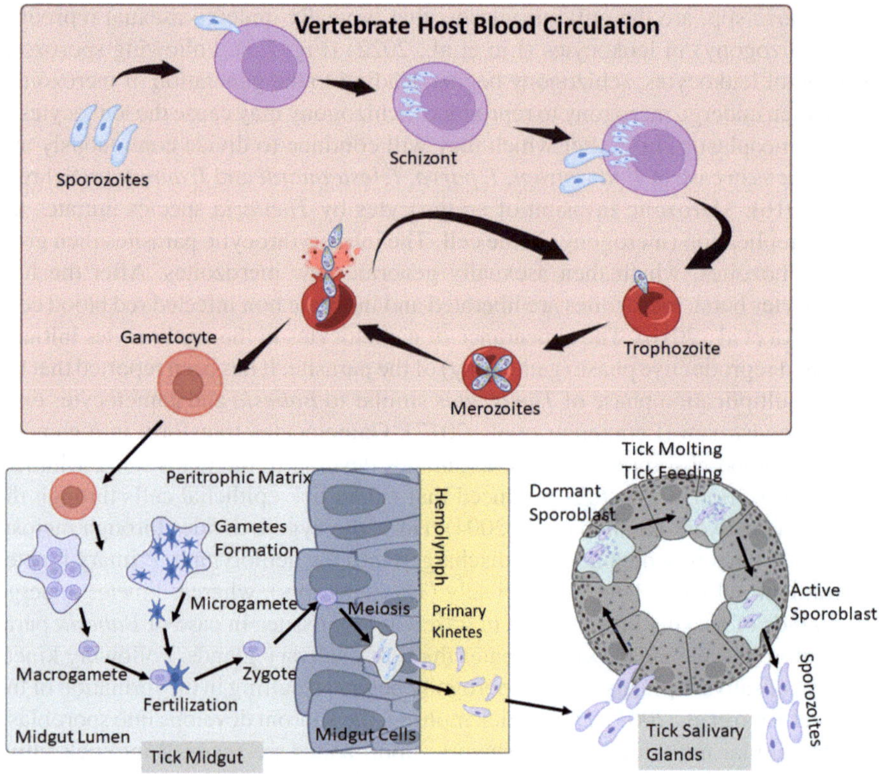

Fig. 7.3 Transmission cycle of *Theileria* spp

2019), Oceania (El-Alfy et al., 2025), and Asia (Kim et al., 2017). *T. orientalis* has been found in a number of continents like Africa (Gebrekidan et al., 2016) and Europe (Gomes et al., 2013). Although, 10 different genotypes of the *T. orientalis* have been found according to sequences of the main piroplasm surface protein gene, outbreaks are usually linked to a pathogenic genotype, the Ikeda genotype (or Type 2). Despite the fact that the majority of genotypes are thought to be harmless, outbreaks of acute theileriosis linked to the Ikeda genotype have increased recently, with outbreaks occurring in New Zealand, Australia, and the United States (Pulford et al., 2016).

In a study conducted in the Intermediate region of Sri Lanka, *T. orientalis* and *T. annulata* have been reported for the first time in dairy cattle with 55.0% of the theileriosis prevalence (Subhasinghe et al., 2024). In Pakistan, a study reported the 3.20% overall theileriosis prevalence in cattle, sheep and goat in year 2017–2018 (Mohsin et al., 2022). In a study conducted in the Portugal from 2018–2019, 10.8% of the theileriosis prevalence was reported in two breeds of cattle which was lower than recorded in previous studies (Valente et al., 2023). Molecular prevalence of 42.81% in cattle was reported in north-eastern China from 2017–2019 (Jia et al., 2020).

Table 7.2 Global recent trend in infection and sero-prevalence of bovine babesiosis

Continent	Country	Infection I Sero-prevalence
Asia	Bangladesh	03% I15%
	Indonesia	70% I 63%
	Japan	35% I not available
	China	19% I25%
	Iran	11% I not available
	Turkey	07% I 33%
	Vietnam	33% I77%
Africa	Egypt	16% I 29%
	Kenya	51% I 29%
	Nigeria	12% I 76%
	Tanzania	18% I 35%
	South Africa	71% I 64%
	Zambia	16% I 70%
North America	Costa Rica	03% I 59%
	Caribbean island	78% I not available
South America	Colombia	25% I 77%
	Argentina	42% I 58%
	Brazil	50% I 63%
Europe	Italy	02% I 35%
	Portugal	27% I 79%
	France	27% I 33%
Australia	Australia	50% I 70%

Numerous diseases spread by ticks have developed approaches to enduring infection in reservoir hosts. However, for the majority of infective agents (*Theileria, Anaplasma and Babesia spp.*) in the majority of host species, host defensive system typically leads to a successful host-to-tick dissemination and a comparatively brief period (days to weeks) of increased parasitemia (Ogden et al., 2021). Climate change will probably affect animal populations which act as reservoir host, while the effects will probably differ depending on the geographic location, the animal rearing system (domesticated animals), and species of animals like wildlife. Beyond some general trends, it may be challenging to forecast the elimination, incursions; and alterations in geographic range brought on by climate change that would impact specific species and communities (Tidman et al., 2021). According to some investigators, while expansions in biodiversity are beneficial (Keesing and Ostfeld, 2015), losses in host biodiversity are linked to an enhanced likelihood of tick-borne infections. It is highly controversial to predict to what degree shifts in host variety may affect the spread of tick-borne infective agents or may lead to more transmission or less (Levi et al., 2016).

One significant factor related to temperature influencing where and when dissemination can happen for infections, spread by arthropod vectors, both today and in the future due to climate change, may be the length of the extrinsic incubation period. However, the extrinsic incubation period is not the sole indicator of when and where tick-borne infection transmission cycles can occur because the infective agent develops in hard ticks while feeding on the host, at the body temperature of host, and considering spread generally occurs in an afterwards stage in life cycle (Ogden and Lindsay, 2016). Therefore, low temperatures restrict the spread of the tick-borne pathogens by primarily limiting the tick and reservoir host availability and abundance while climate change boosts the transmission by enhancing survival and distribution (Perumalsamy et al., 2024).

Spread of disease happens when human, vector, and reservoir activities converge, and vary depending on the infective agents and the environment. Modifications to the climate and the duration of the various seasons will have a direct impact on tick viability, activity, and growth. Modifications in development rates will lead to tick being readily accessible to various diapause windows, which are mostly influenced by the duration of the day, resulting in variations in seasonal activity (Gray, 2008). Climate change has an impact on vegetation which in turn result in an indirect influence of climate change on the abundance of infected ticks (Gilbert, 2021).

Furthermore, by influencing the viability and abundance of tick reservoir hosts (deer) and pathogen-maintenance hosts (rodents and birds), climate change will also indirectly impact the spread of tick-borne pathogens. Seasonal trends have an impact by affecting short-term human activities, and climate change could also impact risk of infection by modifying the long-term usage of land such as tourism or farming. Towards the geographic boundaries of both pathogen and vector distribution, climate impacts are more readily apparent. In addition to ecological environmental parameters, other factor including socioeconomic variables, ecosystems and biodiversity, human movement and colonization, bird patterns of migration, herd immunity and alterations to land-use and land-cover, can all have an impact on the extent of climate change's impacts in an endemic region. There is a complicated web of mechanisms that make it challenging to pinpoint the exact causes of fluctuations in disease incidence because some of these conditions are also impacted by climate change (Shivanna, 2022).

There are possibly three ways that climate change may impact the risk of babesiosis in human and animals. Firstly, because they are poikilothermic creatures, the ixodid ticks that carry babesiosis, and the *Babesia* protozoa in them can react promptly to variations in the environment. Secondly, both ticks and the vertebrate reservoirs hosts can have indirect impacts of climate change through its effects on vegetation, resulting in changes in their habitats and their ability to host food sources. Thirdly, human-induced climate change, particularly human behavior, as well as the care and handling of livestock reservoirs of *Babesia* spp., will influence the availability to the vectors and, consequently, the risk of infection (Gray and Ogden, 2021). A study conducted in sub-Saharan Africa shown that climate change has the influence on the range expansion of tick-borne disease, Theileriosis in cattle. This shift in range has been predicted due to fluctuation in rainfall and temperature (Olwoch et al., 2008).

7.7 Challenges in Disease Control and Management

Anti-protozoal drugs, vaccines, and tick control are the tools for controlling babesiosis and theileriosis (Santos et al., 2023). In babesiosis-endemic regions, disease is controlled by live vaccines, but these vaccines are not used on a wide scale due to reasons like incompetency against variable strains and the chances of other haemo-pathogen contamination (Bastos et al., 2023). Although, to lower the economic burden due to these tick-borne diseases, chemotherapy is considered the primary option for the control of the disease, but these parasites have been reported to develop resistance against most of the anti-babesial agents (Rizwan et al., 2021). Therefore, novel control strategies including discovery of new drug candidates and commercially effective vaccine keeping in view the changing climate have to be implemented.

7.8 Research Priorities for Understanding Climate Impacts

Using a One Health approach improves our ability to comprehend, identify, and attribute the consequences of climate change on ticks and tick-borne diseases. According to the One Health approach, animal, ecosystem and human health are interconnected and depend closely on one another, and comprehensive transdisciplinary approaches that take into account all three elements are necessary for either proactive or reactive solutions to risks to human well-being from zoonotic tick-borne diseases. The analyses and observations of the potential impacts of climate change on the incidence and degree of threat posed to humans or livestock by ticks and tick-borne diseases necessarily use straightforward measurable metrics like abundance/prevalence and presence/absence. Nonetheless, it is evident that climate change could impact extremely complicated tick lifespans and transmission cycles. Regulatory mechanisms and behaviour including developed host tolerance to infections and ticks may cause or lessen the effects (Ogden et al., 2019).

High temperatures and other environmental stresses might decrease the impact of host defenses on exposure to tick and pathogen population and hinder the development of acquired immunity by animal hosts (Bagath et al., 2019). In addition to affecting disease spread at the tick-host interface, density-associated developed resistance to ticks may restrict or perhaps overcome the rising tick abundance linked to a warming climate (Ogden et al., 2019). Predicting the potential development and resurgence of tick-borne infections is becoming more and more important for managing public and animal health. In order to account for the intricacy of the relationships required for climate effect forecasting, recognition, and attribution, a one health approach will be required throughout. These goals require a concerted research and monitoring effort that progressively connects in-silico studies with field and lab observations. However, long-term One Health empirical investigations that document as many facets of tick and tickborne disease ecology and epidemiology as feasible will serve as the cornerstone of this initiative. In order to detect alterations to occurrence patterns, and characterize them to climate change, these studies must: a) be Ideally placed along latitudinal, climatological, ecological, altitudinal transects; b) maintain for timescales sufficiently long to explore and consider drivers of changing patterns that are not related to climate change; and c) identify increases in addition to decreases in threat from tick and tickborne diseases (Ogden et al., 2021).

7.9 Conclusion

The effects of climate change on ticks and tick-borne protozoal infections in livestock represent a serious threat to global animal health and agricultural productivity. Warming temperatures and shifts in precipitation are causing tick vector habitats and behavior to alter, increasing the prevalence and geographic transmission of infections like babesiosis and theileriosis, which influence not only the health of livestock but also farmers' livelihoods, especially in areas where these infections were previously less prevalent. Outbreaks are becoming more frequent and intense, which is placing more strain on veterinary systems and necessitates for efficient disease management and prevention measures. Adopting an integrated strategy that incorporates better livestock health management techniques, tick control measures, and climate surveillance is essential. To lessen the effects of these infections and ensure the sustainability of livestock production in a changing climate, studies and advances in both veterinary medicine and climate science are crucial.

References

Abubakar, M., Perera, P. K., Iqbal, A., & Manzoor, S. (2018). Introductory chapter: Ticks and tick-borne pathogens. In *Ticks and Tick-Borne Pathogens*. IntechOpen.

Agina, O. A., Shaari, M. R., Isa, N. M. M., Ajat, M., Zamri-Saad, M., & Hamzah, H. (2020). Clinical pathology, immunopathology and advanced vaccine technology in bovine theileriosis: A review. *Pathogens, 9*(9), 697.

Alarcon-Chaidez, F. J. (2014). Salivary glands—structure, physiology and molecular biology. In D. E. Sonenshine & R. M. Rowe (Eds.), *The Biology of Ticks* (2nd ed., pp. 163–205). Oxford University Press.

Allen, M., Coninck, H. D., Dube, O. P., Hoegh-Guldberg, O., Jacob, D., Jiang, K., & Waterfield, T. (2018). Technical summary. In *Global warming of 1.5° C* (pp. 27–46). Intergovernmental Panel on Climate Change (IPCC).

Alvarez, J. A., Rojas, C., & Figueroa, J. V. (2019). Diagnostic tools for the identification of Babesia sp. in persistently infected cattle. *Pathogens, 8*(3), 143.

Alvarez, J. A., Rojas, C., & Figueroa, J. V. (2020). An overview of current knowledge on in vitro Babesia cultivation for production of live attenuated vaccines for bovine babesiosis in Mexico. *Frontiers in Veterinary Science, 7*, 364.

Avenant, A., Park, J. Y., Vorster, I., Mitchell, E. P., & Arenas-Gamboa, A. M. (2021). Porcine babesiosis caused by babesia sp. suis in a pot-bellied pig in South Africa. *Frontiers in Veterinary Science, 7*, 620462.

Azhar, M., Gadahi, J. A., Bhutto, B., Tunio, S., Vistro, W. A., Tunio, H., & Ram, T. (2023). Babesiosis: current status and future perspectives in Pakistan and chemotherapy used in livestock and pet animals. *Heliyon, 9*(6).

Bagath, M., Krishnan, G., Devaraj, C., Rashamol, V. P., Pragna, P., Lees, A. M., & Sejian, V. (2019). The impact of heat stress on the immune system in dairy cattle: A review. *Research in Veterinary Science, 126*, 94–102.

Bajer, A., & Dwużnik-Szarek, D. (2021). The specificity of Babesia-tick vector interactions: Recent advances and pitfalls in molecular and field studies. *Parasites & Vectors, 14*, 1–16.

Bastos, R. G., Capelli-Peixoto, J., Laughery, J. M., Suarez, C. E., & Ueti, M. W. (2023). Vaccination with an in vitro culture attenuated Babesia bovis strain safely protects highly susceptible adult cattle against acute bovine babesiosis. *Frontiers in Immunology, 14*, 1219913.

Belozerov, V. N., Fourie, L. J., & Kok, D. J. (2002). Photoperiodic control of developmental diapause in nymphs of prostriate ixodid ticks (Acari: Ixodidae). *Experimental & Applied Acarology, 28*, 163–168.

Berger, K. A., Ginsberg, H. S., Dugas, K. D., Hamel, L. H., & Mather, T. N. (2014). Adverse moisture events predict seasonal abundance of Lyme disease vector ticks (Ixodes scapularis). *Parasites & Vectors, 7*, 1–8.

Bishop, R., Musoke, A., Morzaria, S., Gardner, M., & Nene, V. (2004). Theileria: Intracellular protozoan parasites of wild and domestic ruminants transmitted by ixodid ticks. *Parasitology, 129*(S1), S271–S283.

Bouchard, C., Dibernardo, A., Koffi, J., Wood, H., Leighton, P. A., & Lindsay, L. R. (2019). N Increased risk of tick-borne diseases with climate and environmental changes. *Canada Communicable Disease Report, 45*(4), 83.

Burtis, J. C., Fahey, T. J., & Yavitt, J. B. (2019). Survival and energy use of Ixodes scapularis nymphs throughout their overwintering period. *Parasitology, 146*(6), 781–790.

Burtis, J. C., Sullivan, P., Levi, T., Oggenfuss, K., Fahey, T. J., & Ostfeld, R. S. (2016). The impact of temperature and precipitation on blacklegged tick activity and Lyme disease incidence in endemic and emerging regions. *Parasites & Vectors, 9*, 1–10.

Ceylan, O., Xuan, X., & Sevinc, F. (2021). Primary tick-borne protozoan and rickettsial infections of animals in Turkey. *Pathogens, 10*(2), 231.

Christensen, J. H., Hewitson, B. & Busuioc, A. (2007). Regional climate projections. In 'Climate change 2007: the physical science basis. Contribution of Working Group I to the Fourth Assessment Report of the Intergovernmental Panel on Climate Change'. (Eds S Solomon, D Qin, M Manning, Z Chen, M Marquis, KB Averyt, M Tignor, HL Miller) pp. 847–940.

Ciuca, L., Martinescu, G., Miron, L. D., Roman, C., Acatrinei, D., Cringoli, G., & Maurelli, M. P. (2021). Occurrence of Babesia species and co-infection with Hepatozoon canis in symptomatic dogs and in their ticks in Eastern Romania. *Pathogens, 10*(10), 1339.

Cohen, E. B., Auckland, L. D., Marra, P. P., & Hamer, S. A. (2015). Avian migrants facilitate invasions of neotropical ticks and tick-borne pathogens into the United States. *Applied and Environmental Microbiology, 81*(24), 8366–8378.

Dantas-Torres, F., & Otranto, D. (2022). Rhipicephalus sanguineus (Brown dog tick). *Trends in Parasitology, 38*(11), 993–994.

De Koning, H. P. (2020). The drugs of sleeping sickness: Their mechanisms of action and resistance, and a brief history. *Tropical Medicine and Infectious Disease, 5*(1), 14.

Diuk-Wasser, M. (2021). It's all in the timing: Effect of tick phenology on pathogen transmission dynamics. In P. Nuttall (Ed.), *Climate, Ticks and Disease* (pp. 283–292). CABI.

Drehmann, M., Springer, A., Lindau, A., Fachet, K., Mai, S., Thoma, D., & Strube, C. (2020). The spatial distribution of Dermacentor ticks (Ixodidae) in Germany—Evidence of a continuing spread of Dermacentor reticulatus. *Frontiers in Veterinary Science, 7*, Article 578220.

El-Alfy, E. S., Elseadawy, R., Saleh, S., Elmishmishy, B., Al-Kappany, Y., & Abbas, I. (2025). Genetic diversity and phylogeography of the global Theileria orientalis isolates inferred from MPSP gene sequences. *Parasitology International*, 103038.

Elmieh, N. (2022). The impacts of climate and land use change on tick-related risks. National Collaborating Centre for Environmental Health (NCCEH). (Accessed from: https://ncceh.ca/resources/evidence-reviews/impacts-climate-and-land-use-change-tick-related-risks)

Elsworth, B., & Duraisingh, M. T. (2021). A framework for signaling throughout the life cycle of Babesia species. *Molecular Microbiology, 115*(5), 882–890.

Erster, O., Roth, A., Wollkomirsky, R., Leibovich, B., Savitzky, I., Zamir, S., & Shkap, V. (2016). Quantitative analysis of Babesia ovis infection in sheep and ticks. *Veterinary Parasitology, 221*, 39–45.

Földvári, G., Szabó, É., Tóth, G. E., Lanszki, Z., Zana, B., Varga, Z., & Kemenesi, G. (2022). Emergence of Hyalomma marginatum and Hyalomma rufipes adults revealed by citizen science tick monitoring in Hungary. *Transboundary and Emerging Diseases, 69*(5), e2240–e2248.

Gebrekidan, H., Gasser, R. B., Baneth, G., Yasur-Landau, D., Nachum-Biala, Y., Hailu, A., & Jabbar, A. (2016). Molecular characterization of Theileria orientalis from cattle in Ethiopia. *Ticks and Tick-Borne Diseases, 7*(5), 742–747.

Ghafar, A., Abbas, T., Rehman, A., Sandhu, Z. U. D., Cabezas-Cruz, A., & Jabbar, A. (2020). Systematic review of ticks and tick-borne pathogens of small ruminants in Pakistan. *Pathogens, 9*(11), 937.

Gilbert, L. (2021). The impacts of climate change on ticks and tick-borne disease risk. *Annual Review of Entomology, 66*(1), 373–388.

Gomes, J., Soares, R., Santos, M., Santos-Gomes, G., Botelho, A., Amaro, A., & Inácio, J. (2013). Detection of Theileria and Babesia infections amongst asymptomatic cattle in Portugal. *Ticks and Tick-Borne Diseases, 4*(1–2), 148–151.

Gottlieb, Y., & Duron, O. (2021). Expected transitions in ticks and their heritable endosymbionts under environmental changes. In P. Nuttall (Ed.), *Climate, Ticks and Disease* (pp. 61–66). CABI.

Gray, J. S. (2008). Ixodes ricinus seasonal activity: Implications of global warming indicated by revisiting tick and weather data. *International Journal of Medical Microbiology, 298*, 19–24.

Gray, J. S., & Ogden, N. H. (2021). Ticks, human babesiosis and climate change. *Pathogens, 10*(11), 1430.

Gray, J. S., Dautel, H., Estrada-Peña, A., Kahl, O., & Lindgren, E. (2009). Effects of climate change on ticks and tick-borne diseases in Europe. *Interdisciplinary Perspectives on Infectious Diseases, 2009*(1), Article 593232.

Gray, J. S., Estrada-Peña, A., Vial, L., Sonenshine, D. E., & Roe, R. M. (2014). Ecology of nidicolous ticks. *Biology of Ticks, 2*, 39–60.

Gregory, N., Fernandez, M. P., & Diuk-Wasser, M. (2022). Risk of tick-borne pathogen spillover into urban yards in New York City. *Parasites & Vectors, 15*(1), 288.

Harley, C. D. (2011). Climate change, keystone predation, and biodiversity loss. *Science, 334*(6059), 1124–1127.

Hauser, G., Rais, O., Morán Cadenas, F., Gonseth, Y., Bouzelboudjen, M., & Gern, L. (2018). Influence of climatic factors on Ixodes ricinus nymph abundance and phenology over a long-term monthly observation in Switzerland (2000–2014). *Parasites & Vectors, 11*, 1–12.

Hoogstraal, H., Kaiser, M. N., Traylor, M. A., Gaber, S., & Guindy, E. (1961). Ticks (Ixodoidea) on birds migrating from Africa to Europe and Asia. *Bulletin of the World Health Organization, 24*(2), 197.

Hoogstraal, H., Roberts, F. H., Kohls, G. M., & Tipton, V. J. (1968). Review of haemaphysalis (kaiseriana) longicornis neumann (resurrected) of Australia, New Zealand, New Caledonia, Fiji, Japan, Korea, and Northeastern China and USSR, and its parthenogenetic and bisexual populations (Ixodea, Ixodidae). *The Journal of Parasitology*, 1197–1213.

Howell, J. M., Ueti, M. W., Palmer, G. H., Scoles, G. A., & Knowles, D. P. (2007). Persistently infected calves as reservoirs for acquisition and transovarial transmission of Babesia bovis by Rhipicephalus (Boophilus) microplus. *Journal of clinical microbiology, 45*(10), 3155–3159. https://www.ipcc.ch/report/ar6/syr/downloads/report/IPCC_AR6_SYR_SPM.pdf (Accessed online 26/12/2024).

Hvidsten, D., Frafjord, K., Gray, J. S., Henningsson, A. J., Jenkins, A., Kristiansen, B. E., & Wilhelmsson, P. (2020). The distribution limit of the common tick, Ixodes ricinus, and some associated pathogens in north-western Europe. *Ticks and Tick-Borne Diseases, 11*(4), Article 101388.

Intergovernmental Panel on Climate Change (IPCC). *Climate Change 2022: Impacts, Adaptation and Vulnerability* https://www.ipcc.ch/report/ar6/wg2/ (Accessed online 28/12/2024)

Jacob, S. S., Sengupta, P. P., Paramanandham, K., Suresh, K. P., Chamuah, J. K., Rudramurthy, G. R., & Roy, P. (2020). Bovine babesiosis: An insight into the global perspective on the disease distribution by systematic review and meta-analysis. *Veterinary Parasitology, 283*, Article 109136.

Jaenson, T. G., TäLleklint, L., Lundqvist, L., Olsen, B., Chirico, J. A. N., & Mejlon, H. (1994). Geographical distribution, host associations, and vector roles of ticks (Acari: Ixodidae, Argasidae) in Sweden. *Journal of Medical Entomology, 31*(2), 240–256.

Jalovecka, M., Bonsergent, C., Hajdusek, O., Kopacek, P., & Malandrin, L. (2016). Stimulation and quantification of Babesia divergens gametocytogenesis. *Parasites & Vectors, 9*, 1–11.

Jalovecka, M., Hajdusek, O., Sojka, D., Kopacek, P., & Malandrin, L. (2018). The complexity of piroplasms life cycles. *Frontiers in Cellular and Infection Microbiology, 8*, 248.

Jia, L., Zhao, S., Xie, S., Li, H., Wang, H., & Zhang, S. (2020). Molecular prevalence of Theileria infections in cattle in Yanbian, north-eastern China. *Parasite, 27*, 19.

Juyal, P. D., Singla, L. D., & Kaur, P. (2005). Management of surra due to Trypanosoma evansi in India: An overview. *Infectious Diseases of Domestic Animals and Zoonosis in India, 75*, 109–120.

Keesing, F., & Ostfeld, R. S. (2015). Is biodiversity good for your health? *Science, 349*(6245), 235–236.

Keesing, F., & Ostfeld, R. S. (2021). Impacts of biodiversity and biodiversity loss on zoonotic diseases. *Proceedings of the National Academy of Sciences, 118*(17), Article e2023540118.

Kim, S., Yu, D. H., Chae, J. B., Choi, K. S., Kim, H. C., Park, B. K., & Park, J. (2017). Pathogenic genotype of major piroplasm surface protein associated with anemia in Theileria orientalis infection in cattle. *Acta Veterinaria Scandinavica, 59*, 1–5.

Kumar, A., O'Bryan, J., & Krause, P. J. (2021). The global emergence of human babesiosis. *Pathogens, 10*(11), 1447.

Lee, H. (2023). AR6 synthesis report: climate change 2023. *Summary for Policymakers.*

Lempereur, L., Larcombe, S. D., Durrani, Z., Karagenc, T., Bilgic, H. B., Bakirci, S., & Shiels, B. (2017). Identification of candidate transmission-blocking antigen genes in Theileria annulata and related vector-borne apicomplexan parasites. *BMC Genomics, 18*, 1–14.

Levi, T., Keesing, F., Holt, R. D., Barfield, M., & Ostfeld, R. S. (2016). Quantifying dilution and amplification in a community of hosts for tick-borne pathogens. *Ecological Applications, 26*(2), 484–498.

Lindgren, E., Tälleklint, L., & Polfeldt, T. (2000). Impact of climatic change on the northern latitude limit and population density of the disease-transmitting European tick Ixodes ricinus. *Environmental Health Perspectives, 108*(2), 119–123.

Liu, J., Guan, G., & Yin, H. (2022). Theileria annulata. *Trends in Parasitology, 38*(3), 265–266.

Lowe, R., Gasparrini, A., Van Meerbeeck, C. J., Lippi, C. A., Mahon, R., Trotman, A. R., & Stewart-Ibarra, A. M. (2018). Nonlinear and delayed impacts of climate on dengue risk in Barbados: A modelling study. *PLoS Medicine, 15*(7), Article e1002613.

Ma, Q., Liu, J., Li, Z., Xiang, Q., Wang, J., Liu, A., & Luo, J. (2020). Clinical and pathological studies on cattle experimentally infected with Theileria annulata in China. *Pathogens, 9*(9), 727.

Mardani, A., Streimikiene, D., Cavallaro, F., Loganathan, N., & Khoshnoudi, M. (2019). Carbon dioxide (CO2) emissions and economic growth: A systematic review of two decades of research from 1995 to 2017. *Science of the Total Environment, 649*, 31–49.

Metheni, M., Lombès, A., Bouillaud, F., Batteux, F., & Langsley, G. (2015). HIF-1α induction, proliferation and glycolysis of T heileria-infected leukocytes. *Cellular Microbiology, 17*(4), 467–472.

Mohsin, M., Hameed, K., Kamal, M., Ali, A., Rafiq, N., Usman, T., & Yousafzai, G. J. (2022). Prevalence and risk factors assessment of theileriosis in livestock of Malakand Division, Pakistan. *Journal of the Saudi Society of Agricultural Sciences, 21*(4), 242–247.

Molaei, G., Little, E. A., Williams, S. C., & Stafford, K. C. (2019). Bracing for the worst—range expansion of the lone star tick in the northeastern United States. *New England Journal of Medicine, 381*(23), 2189–2192.

Morrison, W. I. (2009). Progress towards understanding the immunobiology of Theileria parasites. *Parasitology, 136*(12), 1415–1426.

Neelakanta, G., Sultana, H., Fish, D., Anderson, J. F., & Fikrig, E. (2010). Anaplasma phagocytophilum induces Ixodes scapularis ticks to express an antifreeze glycoprotein gene that enhances their survival in the cold. *The Journal of Clinical Investigation, 120*(9), 3179–3190.

Nyangiwe, N., Yawa, M., & Muchenje, V. (2018). Driving forces for changes in geographic range of cattle ticks (Acari: Ixodidae) in Africa: A review. *South African Journal of Animal Science, 48*(5), 829–841.

Oakes, V. J., Yabsley, M. J., Schwartz, D., LeRoith, T., Bissett, C., Broaddus, C., & Lahmers, K. K. (2019). Theileria orientalis Ikeda genotype in cattle, Virginia, USA. *Emerging Infectious Diseases, 25*(9), 1653.

Ochanda, H., Young, A. S., Mutugi, J. J., Mumo, J., & Omwoyo, P. L. (1988). The effect of temperature on the rate of transmission of Theileria parva parva infection to cattle by its tick vector. *Rhipicephalus Appendiculatus. Parasitology, 97*(2), 239–245.

O'Donoghue, P. (2017). Haemoprotozoa: Making biological sense of molecular phylogenies. *International Journal for Parasitology: Parasites and Wildlife, 6*(3), 241–256.

Ogden, N. H., & Lindsay, L. R. (2016). Effects of climate and climate change on vectors and vector-borne diseases: Ticks are different. *Trends in Parasitology, 32*(8), 646–656.

Ogden, N. H., Ben Beard, C., Ginsberg, H. S., & Tsao, J. I. (2021). Possible effects of climate change on ixodid ticks and the pathogens they transmit: Predictions and observations. *Journal of Medical Entomology, 58*(4), 1536–1545.

Ogden, N. H., Lindsay, L. R., Beauchamp, G., Charron, D., Maarouf, A., O'callaghan, C. J., & Barker, I. K. (2004). Investigation of relationships between temperature and developmental rates of tick Ixodes scapularis (Acari: Ixodidae) in the laboratory and field. *Journal of Medical Entomology, 41*(4), 622–633.

Ogden, N. H., Mechai, S., & Margos, G. (2013). Changing geographic ranges of ticks and tick-borne pathogens: Drivers, mechanisms and consequences for pathogen diversity. *Frontiers in Cellular and Infection Microbiology, 3*, 46.

Ogden, N. H., Wilson, J. R., Richardson, D. M., Hui, C., Davies, S. J., Kumschick, S., & Pulliam, J. R. (2019). Emerging infectious diseases and biological invasions: A call for a One Health collaboration in science and management. *Royal Society Open Science, 6*(3), Article 181577.

Olwoch, J. M., Reyers, B., Engelbrecht, F. A., & Erasmus, B. F. N. (2008). Climate change and the tick-borne disease, Theileriosis (East Coast fever) in sub-Saharan Africa. *Journal of Arid Environments, 72*(2), 108–120.

Osbrink, W. L., Thomas, D. B., Lohmeyer, K. H., & Temeyer, K. B. (2022). Climate change and alternative hosts complicate the eradication of cattle fever ticks (Acari: Ixodidae) in the southern United States, a review. *Annals of the Entomological Society of America, 115*(1), 39–55.

Ozubek, S., & Aktaş, M. (2019). Genetic diversity of Theileria orientalis from cattle in Turkey. *Comparative Immunology, Microbiology and Infectious Diseases, 65*, 132–136.

Ozubek, S., Bastos, R. G., Alzan, H. F., Inci, A., Aktas, M., & Suarez, C. E. (2020). Bovine babesiosis in Turkey: Impact, current gaps, and opportunities for intervention. *Pathogens, 9*(12), 1041.

Penzhorn, B. L., & Oosthuizen, M. C. (2020). Babesia species of domestic cats: Molecular characterization has opened pandora's box. *Frontiers in Veterinary Science, 7*, 134.

Perumalsamy, N., Sharma, R., Subramanian, M., & Nagarajan, S. A. (2024). Hard ticks as vectors: The emerging threat of tick-borne diseases in India. *Pathogens, 13*(7), 556.

Pulford, D. J., McFadden, A. M. J., Hamilton, J. S., & Donald, J. (2016). Investigation of the index case herd and identification of the genotypes of Theileria orientalis associated with outbreaks of bovine anaemia in New Zealand in 2012. *New Zealand Veterinary Journal, 64*(1), 21–28.

Raghavan, R. K., Barker, S. C., Cobos, M. E., Barker, D., Teo, E. J. M., Foley, D. H., & Peterson, A. T. (2019). Potential spatial distribution of the newly introduced long-horned tick, Haemaphysalis longicornis in North America. *Scientific Reports, 9*(1), 498.

Randolph, S. E., & Rogers, D. J. (2000). Fragile transmission cycles of tick-borne encephalitis virus may be disrupted by predicted climate change. *Proceedings of the Royal Society of London. Series B: Biological Sciences, 267*(1454), 1741–1744.

Rizwan, H. M., Abbas, H., Sajid, M. S., Maqbool, M., Jones, M. K., Ullah, M. I., & Ijaz, N. (2021). Drug resistance in protozoal infections. *Biochemistry of drug resistance*, 95–142.

Rodgers, S. E., Zolnik, C. P., & Mather, T. N. (2007). Duration of exposure to suboptimal atmospheric moisture affects nymphal blacklegged tick survival. *Journal of Medical Entomology, 44*(2), 372–375.

Santos, J. H. M., Siddle, H. V., Raza, A., Stanisic, D. I., Good, M. F., & Tabor, A. E. (2023). Exploring the landscape of Babesia bovis vaccines: Progress, challenges, and opportunities. *Parasites & Vectors, 16*(1), 274.

Schreeg, M. E., Marr, H. S., Tarigo, J. L., Cohn, L. A., Bird, D. M., Scholl, E. H., & Birkenheuer, A. J. (2016). Mitochondrial genome sequences and structures aid in the resolution of Piroplasmida phylogeny. *PLoS ONE, 11*(11), Article e0165702.

Shivanna, K. R. (2022). Climate change and its impact on biodiversity and human welfare. *Proceedings of the Indian National Science Academy, 88*(2), 160–171.

Siddiqui, G., Proellochs, N. I., & Cooke, B. M. (2020). Identification of essential exported Plasmodium falciparum protein kinases in malaria-infected red blood cells. *British Journal of Haematology, 188*(5), 774–783.

Šimo, L. A. D. I. S. L. A. V., Sonenshine, D. E., Park, Y., & Žitňan, D. (2013). Nervous and sensory systems. *Function, Genomics and Proteomics. In: Sonenshine, DE*, 309–367.

Solano-Gallego, L., Sainz, Á., Roura, X., Estrada-Peña, A., & Miró, G. (2016). A review of canine babesiosis: The European perspective. *Parasites & Vectors, 9*, 1–18.

Solomon, A., & Tanga, B. M. (2020). The First Investigation of Tick Vectors and Tick-Borne Diseases in Extensively Managed Cattle in Alle District. *Southwestern Ethiopia. Veterinary Medicine International, 2020*(1), 8862289.

Springer, A., Lindau, A., Probst, J., Fachet, K., Schäfer, I., Dobler, G., & Strube, C. (2024). Germany-wide Citizen Science study reveals spread of Babesia canis-infected Dermacentor reticulatus ticks by dogs travelling within the country. *Current Research in Parasitology & Vector-Borne Diseases, 6*, Article 100187.

Srionrod, N., Nooroong, P., Poolsawat, N., Minsakorn, S., Watthanadirek, A., Junsiri, W., & Anuracpreeda, P. (2022). Molecular characterization and genetic diversity of Babesia bovis and Babesia bigemina of cattle in Thailand. *Frontiers in Cellular and Infection Microbiology, 12*, 1065963.

Subhasinghe, J., Mahakapuge, T. A. N., Madusanka, K. S., Rajakaruna, R. S., Jabbar, A., & Perera, P. K. (2024). First report of oriental theileriosis in the Intermediate Zone, Sri Lanka: Is Theileria orientalis type 7 always apathogenic? *Veterinary Parasitology: Regional Studies and Reports, 55*, Article 101115.

TäLleklint, L., & Jaenson, T. G. (1998). Increasing geographical distribution and density of Ixodes ricinus (Acari: Ixodidae) in central and northern Sweden. *Journal of Medical Entomology, 35*(4), 521–526.

Thomson, M. C., & Stanberry, L. R. (2022). Climate change and vectorborne diseases. *New England Journal of Medicine, 387*(21), 1969–1978.

Tidman, R., Abela-Ridder, B., & de Castañeda, R. R. (2021). The impact of climate change on neglected tropical diseases: A systematic review. *Transactions of the Royal Society of Tropical Medicine and Hygiene, 115*(2), 147–168.

Tirosh-Levy, S., Mazuz, M. L., Savitsky, I., Pinkas, D., Gottlieb, Y., & Steinman, A. (2021). Serological and molecular prevalence of Babesia caballi in apparently healthy horses in Israel. *Pathogens, 10*(4), 445.

Turell, M. J., Cohnstaedt, L. W., & Wilson, W. C. (2020). Effect of environmental temperature on the ability of Culex tarsalis and Aedes taeniorhynchus (Diptera: Culicidae) to transmit Rift Valley fever virus. *Vector-Borne and Zoonotic Diseases, 20*(6), 454–460.

Valente, D., Dutra, A. P., Carolino, N., Gomes, J., Coelho, A. C., Espadinha, P., & Carolino, I. (2023). Prevalence and risk factors associated with Theileria annulata infection in two bovine Portuguese autochthonous breeds. *Pathogens, 12*(5), 669.

Watts, J. G., Playford, M. C., & Hickey, K. L. (2016). Theileria orientalis: A review. *New Zealand Veterinary Journal, 64*(1), 3–9.

Weber, G., & Friedhoff, K. (1979). Electron microscopic detection of initial and some subsequent developmental stages of Babesia bigemina in salivary glands of ticks.

Weiler, M., Duscher, G. G., Wetscher, M., & Walochnik, J. (2017). Tick abundance: A one year study on the impact of flood events along the banks of the river Danube, Austria. *Experimental and Applied Acarology, 71*, 151–157.

Woods, K., Perry, C., Brühlmann, F., & Olias, P. (2021). Theileria's strategies and effector mechanisms for host cell transformation: From invasion to immortalization. *Frontiers in Cell and Developmental Biology, 9*, Article 662805.

Wu, G., Zhou, X., Guo, F., Xu, J., Song, J., Jin, Z., & Yu, X. (2025). Babesia infection in cattle and dogs in Suizhou City, Hubei Province, China. *Infectious Medicine*, 100170.

Wu, J., Cao, J., Zhou, Y., Zhang, H., Gong, H., & Zhou, J. (2017). Evaluation on infectivity of Babesia microti to domestic animals and ticks outside the Ixodes genus. *Frontiers in Microbiology, 8*, 1915.

Yamasaki, M., Idaka, N., Abe, M., & Takiguchi, M. (2021). Reduced expression levels of heat shock protein 90 in a diminazene aceturate-resistant Babesia gibsoni isolate. *Experimental Parasitology, 221*, Article 108050.

Yano, Y., Saito-Ito, A., Anchalee, D., & Takada, N. (2005). Japanese Babesia microti cytologically detected in salivary glands of naturally infected tick Ixodes ovatus. *Microbiology and Immunology, 49*(10), 891–897.

Yokoyama, N., Sivakumar, T., Ota, N., Igarashi, I., Nakamura, Y., Yamashina, H., & Inokuma, H. (2012). Genetic diversity of Theileria orientalis in tick vectors detected in Hokkaido and Okinawa, Japan. *Infection, Genetics and Evolution, 12*(8), 1669–1675.

Zając, Z., Kulisz, J., Woźniak, A., Bartosik, K., & Khan, A. (2021). Seasonal activity of Dermacentor reticulatus ticks in the era of progressive climate change in eastern Poland. *Scientific Reports, 11*(1), 20382.

Zeb, J., Song, B., Aziz, M. U., Hussain, S., Zarin, R., & Sparagano, O. (2022). Diversity and distribution of Theileria species and their vectors in ruminants from India. *Pakistan and Bangladesh. Diversity, 14*(2), 82.

Zygner, W., Gójska-Zygner, O., Bartosik, J., Górski, P., Karabowicz, J., Kotomski, G., & Norbury, L. J. (2023). Canine babesiosis caused by large Babesia species: Global prevalence and risk factors—a review. *Animals, 13*(16), 2612.

Chapter 8
Tick-Borne Diseases in a Warming World: A Global Prospective

Maira Munir, Muhammad Usman, Aiman Maqsood, and Miranda Imeri

Abstract Within the past few decades, tick-borne diseases (TBDs) have gained increased recognition as an important problem for both human and veterinary medicine. The global increase in average temperature is likely to expand the range and the number of living ticks as well as increase the total number of TBDs. This chapter investigates how climate change affects TBDs ecology and explains how climate change initiates new diseases, what ecological and environmental factors influence disease risk, and where the gaps in surveillance and health systems are. Moreover, it addresses emerging adoption and mitigation of preventive measures such as integrated vector control, climate-smart public health policies, and new directions in the development of vaccines. One Health approach is important in the climate change human health system which is one response in countering the worst effects and disease in the modern world.

Keywords Global epidemiology · New geographic regions · Disease transmission dynamics · Ecological factors · Disease surveillance

M. Munir
Department of Parasitology, Faculty of Veterinary Sciences, University of Veterinary and Animal Sciences, Lahore, Pakistan

M. Usman (✉)
Department of Basic Sciences, KBCMA College of Veterinary and Animal Sciences, Narowal, Sub-Campus UVAS, Lahore, Pakistan
e-mail: usman.asif@uvas.edu.pk

A. Maqsood
Department of Zoology, Ramal International College, Zafarwal, District Narowal, Punjab, Pakistan

M. Imeri
Agronomy Department, University of Florida, Gainesville, FL, USA

H. M. Rizwan and M. S. Sajid (eds.), *Ticks in a Changing Climate*, Biometeorology 7,
https://doi.org/10.1007/978-3-032-12259-9_8

8.1 Introduction

The global surface temperature has increased by approximately 1.5 °C from pre-industrial times to now (Collins et al., 2018). Additionally, warming is not features uniformly as land regions have undergone some of the higher increases in temperature compared to the global average over several seasons. The Earth is experiencing warming due to the rise in the concentration of greenhouse gases such as carbon dioxide, methane, and nitrous oxide. These gases tend to be produced from anthropogenic activities (Abbass et al., 2022). Furthermore, climate change forecasts differ somewhat because there are still unknowns regarding the impact of clouds on global and regional warming (Bellouin et al., 2020).

Recognizing the effects of climate change on ticks and tick-borne diseases (TBDs) remains difficult because of the complex interplay of ecological and environmental factors (Gilbert, 2021). Ticks are poikilothermic ecto-parasites; that is, their body temperature changes in accordance with the external environment, which affects their metabolism, development, and behavior. Moreover, they're relatively long-lived arthropods; some species take over five years to fully mature into adults and spend a significant proportion of their life cycle off-host in the environment (Nuttall, 2022). These life history characteristics suggest that tick populations are vulnerable to environmental change. However, the slow spread of tick-borne pathogens when compared with the rapid spread of mosquito-borne diseases is staggering (Gilbert, 2021; Nuttall, 2022).

Ticks follow in the footsteps of mosquitoes in terms of spreading infectious diseases to both humans and animals (de la Fuente et al., 2023). There are around 800 species of ticks worldwide, all of which are obligate hematophagous parasites, meaning their existence and reproduction hinges completely on consuming blood (Dantas-Torres, 2015; Gray et. al., 2009). Excluding fish, ticks can inflict their parasitic nature on a wide variety of vertebrate hosts such as birds, reptiles, and mammals, which makes them remarkable carriers of diseases. In the tropical and temperate regions, ticks are responsible for a variety of bacterial, viral and protozoal pathogens that are extremely dangerous for human, livestock and wildlife health (Rochlin and Toledo, 2020).

It is now evident that new climatic conditions have affected tick population due to geographical movement, seasonal behaviour, and population numbers (Dantas-Torres, 2015; Gray et al., 2009). Increased temperature, higher humidity levels in some regions, and alterations in rainfall have all together made it possible for ticks to inhabit areas that were previously deemed unsuitable for them, including places with higher elevations and latitudinal regions (Porretta et al., 2013). These changes in the environment in conjunction with changes in host abundance and host habitat suitability are expected to increase TBD spread in the next decades (Maqbool et al., 2022; Perumalsamy et al., 2024).

As we know, climate change is becoming an issue that affects pretty much every aspect of our lives. In this chapter, we cover its consequences on the tick populations, how they expand geographically and the epidemiology of tick-borne diseases around the world. In this chapter, we analyze the effects of temperature changes, modified rainfall patterns, and other ecosystem changes on the survival of ticks, their hosts, and the associated transmission of diseases. Moreover, we highlight the implications for human health, livestock production, and biological diversity, focusing on the necessity of flexible disease monitoring and integrated epidemiological management approaches in an ever-increasing climate change context.

8.2 Global Epidemiology and Distribution of Tick-Borne Diseases

Ticks have the ability to pass on numerous pathogens such as bacteria, viruses, and protozoa which result in different diseases like Lyme disease, babesiosis, anaplasmosis, and tick borne encephalitis (Maqbool et al., 2022; Perumalsamy et al., 2024). The climatic conditions along with vegetation and host availability, determine the geographic range of the tick and the disease (Ogden et al., 2021).

The black-legged tick (*Ixodes scapularis*) is one of the primary vectors responsible for Lyme disease in North America. It is currently spreading to Canada where the average temperature and winter duration is increasing. The introduction of Lyme disease to new regions and the increase of cases in areas that previously did not report such incidences is due to the expansion of the disease (Ogden et al., 2006). In Europe, the expansion of the region that the castor bean tick (*Ixodes ricinus*) occupies has extended to the upper parts of Europe which has corresponded with an increase in cases of Lyme borreliosis and tick borne encephalitis (Medlock et al., 2013). In Asia, both the Siberian subtype of tick-borne encephalitis virus (TBEV) and the parasite are spread through the taiga tick (*Ixodes persulcatus*) and the ornate cow tick (*Dermacentor reticulatus*). Because of global warming, these ticks are spreading to new regions, and therefore raises the possibility of new TBDs emerging (Hayasaka et al., 2001; Süss, 2011).

TBDs are highly pronounced in Africa, especially in regions where humans and livestock live together. The tropical bont tick (*Amblyomma variegatum*) is responsible for spreading heartwater disease; this reduces the productivity of agricultural activities. Movement of cattle and changes in the environment has affected the range of this and other tick species in different parts of the continent (Githeko et al., 2000). On Australia's eastern coastline, both people and animals are at risk from paralysis ticks (*Ixodes holocyclus*). Climate change and land development may alter the geographical range and seasonal activity of this species raising public and veterinary health concerns (Gray et al., 2009). Brazil published reports of spotted fever from Colombian territory which is transmitted by *Amblyomma cajennense*. Agricultural expansion and deforestation has certainly changed the distribution of ticks over regions and the frequency of diseases within these regions (Polo et al., 2024). The epidemiology of TBDs is modified due to how climate change affects their survival, development, and host searching behavior. It is well known that higher temperatures tend to exacerbate the frequency of host seeking ticks which allows for an increased risk of the spread of pathogens (Ogden et al., 2021).

8.3 Impact of Climate Change on Tick-Borne Disease Ecology

Ticks are ectothermic, which means to some degree the development, survival, and activity of ticks depends on climatological factors like temperature and humidity. In moderate circumstances, warmer external temperature accelerates the tick's development because it takes a shorter period of time for a tick to complete its life cycle (Estrada-Peña et al., 2012). For example, the primary vector of Lyme disease in North America, *I. scapularis*, is known to reach higher population densities in regions that are warmer and have milder winters because this tick undergoes development more quickly (Ogden et al., 2021). Some European tick species, such as *I. ricinus*, have also increased the number of days active in a year during which humans and other vertebrates are most susceptible to tick bites (Medlock et al., 2013).

Adverse conditions that are more severe, such as heat and drought, can reduce tick population by desiccating them, especially in low moisture regions. Higher temperatures and more humidity on the other hand, improves survival ticks' chances and higher tick enduring condtions (Gilbert, 2021). However, less severe winters coupled with longer warm periods enable tick survival in places that were previously unsuitable. For instance, the expansion of *Dermacentor (D.) reticulatus* in Czech Republic (Siroký et al., 2011) and the northward expansion of *I. scapularis* into Canada is associated with climate change (Levy, 2017).

Moreover, climate change also affects the behavior and distribution of certain hosts, including reptiles, mammals, and birds. Change in the host's diet coupled with the extension of the habitat range increases the chances of migration which allows ticks and pathogens to invade novel ecosystems (Tardy et al., 2023). The increase in incidence of Lyme disease in North America is related to the growing population of white tailed deer, which is one of the principal reservoirs of *I. scapularis* (Kilpatrick et al., 2017). Likewise, migratory birds are important in dispersing ticks and TBDs to remote locations leading to the emergence of TBDs in previously disease-free areas (Hasle, 2013).

The primary driver of the disease ecologies of tick-borne pathogens is temperature and humidity. An increase in temperature tends to increase the replication of other pathogens in the tick vector enhancing infection rates. Higher temperature exposure is known to be correlated with greater prevalence rates of ticks infected with *Borrelia burgdorferi*, the agent that causes Lyme disease (Dantas-Torres et al., 2012). Furthermore, the TBEV is known to infect higher latitudes and altitudes due to warmer temperatures, escalating the potential for human infection in Europe and Asia (Randolph, 2008).

8.4 Emerging Tick-Borne Diseases in New Geographic Regions

The emergence and spread of TBDs in climate change, globalization, and modification of environment regions is largely due to human activity. Numerous species of ticks have moved northward and to higher elevations due to changed climate conditions. For instance, the *I. scapularis* tick has been migrating into higher latitudes of Canada because winters are getting milder and warm seasons are expanding, leading to increased human interaction with *Borrelia burgdorferi* (Leighton et al., 2012). Similarly, *I. ricinus* has moved further to the north of Europe and to higher altitudes of the Alps, which can be attributed to climate warming (Medlock et al., 2013).

More recently, ticks like *Hyalomma marginatum*, have expanded their range northward from Africa into Central and Southern Europe, which can be explained by warmer temperatures combined with shifts in bird migration patterns (Buczek et al., 2020). This tick species is particularly concerning due to its role in transmitting Crimean-Congo hemorrhagic fever virus (CCHFV), a zoonotic pathogen with high case fatality rate (Bah et al., 2022; Tazerji et al., 2024).

Emerging TBDs are increasingly being repoted in new geographic locations, particularly in North America and Europe, as climate and environmental changes facilitate vector range expansion (Bouchard et al., 2019; Rochlin and Toledo, 2020). In the U.S., a novel spirochete species, *Borrelia mayonii*, associated with Lyme disease, was detected in the upper Midwest, signifying changes tick ecology and host–pathogen interactions (Pritt et al., 2016). Furthermore, there have been increasing reports of *Rickettsia parkeri*, which causes spotted fever rickettsiosis, in the U.S. as *Amblyomma maculatum* ticks expand their range in the country (Paddock and Goddard, 2015). In Europe, human cases of babesiosis caused by protozoan parasite *Babesia venatorum* have been reported indicating the risk of such tick-borne protozoal infections in more temperate regions (Gray et al., 2010). *Anaplasma phagocytophilum*, the agent of human granulocytic anaplasmosis, was also reported in new parts of Scandinavia and the UK, and this continues to point to the impact of changing tick populations in those areas (Stuen et al., 2013). In Canada, newer TBD threats such as anaplasmosis, babesiosis, Pawassan virus and Borrelia miyamotoi disease are also emerging as I. scapularis spreads (Bouchard et al., 2019).

Wildlife hosts play a central role in the dispersal of ticks and the emergence of TBDs. Migratory birds are especially significant in the seasonal movement of tick vectors like *Ixodes ricinus* and *Hyalomma* spp. Because of this, new TBDs can be brought into non-endemic regions (Hasle, 2013). In a parallel manner, the increasing population of deer in suburban and rural localities facilitates the northerly migration of *I. scapularis* deer tick in North America, raising Lyme disease risks (Kilpatrick et al., 2017).

In Mongolia and other rapidly developing regions, environmental changes such as deforestation, urban expansion, and intensified agriculture alter natural tick-host relationships. These fragmented ecosystems increase contact between wildlife, livestock, and humans, creating new pathways for pathogen spillover (Medlock et al.,

2013). However, the detection and management of TBDs in newly affected areas present critical challenges. Surveillance efforts are hampered by underreporting, inconsistent diagnostic criteria, and limited capacity for monitoring non-human hosts (Madison-Antenucci et al., 2020; Rowan et al., 2023). Although imporved molecular and immunological detection techniques have enhanced diagnosis, TBDs remain underrecognized in many regions (Madison-Antenucci et al., 2020). Surveillance systems such as NNDSS and TickNET in the U.S. utilize passive, active, and sentinel methods, but still face issues with standardization and coverage (Rowan et al., 2023).

An adaptive surveillance framework has been propsed to address these limitations, advocating for integrated, intradiciolinary apporches across clinical medicine, veterinary scineces, ecology, and public health (Clow et al., 2019; Roth, 2012). This approach also aligns with future goals to strengthen surveillance networks and implement integfrated tick management strategies (Rowan et al., 2023). These emerging risks underscore the urgen need for surveillance refinement and proactive vector control in areas not previously impacted by TBDs (Estrada-Peña et al., 2014).

8.5 Climate-Driven Changes in Disease Transmission Dynamics

Ticks are highly dependent on temperature for their development, survival, and activities. Warmer conditions accelerate the growth of ticks, boost their reproduction, and decrease their molting time, eventually increasing their population density (Estrada-Peña et al., 2014). For instance, *I. scapularis*, is shifting its range northward due to less severe winter temperatures and longer growing seasons. This puts many previously unaffected areas at risk for transmission of *Borrelia burgdorferi* (Leighton et al., 2012). In the same fashion, *Ixodes ricinus*, one of the most important arthropods for the TBEV in Europe, has been discovered at higher elevations and latitudes as a result of climate change which permits the virus to survive in novel settings (Medlock et al., 2013). Increased temperatures also promote the reproduction of certain tick-borne pathogens, such as *Babesia microti* and *Anaplasma phagocytophilum*, which results in higher quantities of pathogens and increased rates of transmission efficiency (Gray et al., 2009).

Apart from temperature, tick survival and propagation of disease can be affected by moisture content and rainfall. For effective survival, ticks require humid conditions however, excess dryness interferes with their questing prowess (Ogden and Lindsay, 2016). Changes to the levels of precipitation which include higher amounts of rain in temperate areas as well as dry spells in dry regions, have an impact on tick populations and their expansion scope (Gage and Kosoy, 2005). For example, heavy rainfall favors tick abundant *Rhipicephalus sanguineus*, the vector of Mediterranean spotted fever, due to improvement in vegetative cover and favorable microhabitats for tick survival (Dantas-Torres, 2015). On the other hand, droughts in some areas tend to lower the tick population temporarily but change the concentration of host animals towards water sources, leading to concentrated tick-host interactions and localized outbreaks of TBDs (Ogden et al., 2021).

Changes in climate affects the range and number of hosts for ticks, such as wildlife, livestock, and even humans. In North America, deer populations have exploded into suburban and even urban areas due to warmer temperatures and habitat fragmentation, aiding the distribution of *Ixodes scapularis* and Lyme disease (Kilpatrick et al., 2017). Various bird species, particularly passerines like blackbirds, are known to carry ticks, with Ixodes ricinus and Hyalomma marginatum being the most commonly transported species in Europe (Buczek et al., 2020; Burnus et al., 2024). Birds can transport ticks across geographical barriers and introduce them to new areas, with documented cases of Ixodes scapularis being carried as far north as Slave Lake, Alberta, and Watson Lake, Yukon (Scott, 2016). The prevalence of ticks on birds varies depending on factors such as season, locality, and bird species (Hasle, 2013). In addition, the changes in the hosts' immune system responses, as well as pathogens' evolution due to climate changes could alter TBDs virulence and transmissibility factors. Some research indicates that higher temperatures may increase tick vector competence by changing their microbiota and immune system within the organism, which makes it easier to infect with pathogens (Estrada-Peña et al., 2014).

8.6 Ecological Factors Influencing Tick-Borne Disease Risks

Allan et al. (2003) indicate that anthropogenic changes to the landscape, especially deforestation, urban development, and agricultural activity, are key factors affecting the risk of TBD by modifying the distribution of tick habitats and host animals. Biodiversity reduction through habitat fragmentation is often accompanied by the decline in natural predators of ticks, and the increase in tick and reservoir host interactions (Ostfeld and Keesing, 2012). Areas where biodiversity is not under threat also show the so-called "dilution effect," meaning higher host diversity leads to reduced transmission of pathogens by ticks through diverting competent hosts (Ostfeld et al., 2006).

The dynamics of specific host populations are essential for the various life stages of ticks, and there is a corresponding higher-order relationship between the availability of hosts and the survival, as well as the disease transmission patterns, of ticks. For instance, deer, an abundant source of large mammalian hosts, are able to sustain tick reproduction, while smaller birds and rodents serve as reservoirs of pathogenic diseases (Ginsberg et al., 2021). The risks are apparently influenced when climate change, conservation programs, or other forms of habitat modifications result in increase or decrease of host populations. For example, the increase of rodent reservoirs owing to milder winters was connected with an increase of Lyme disease infections because rodents act as chief reservoirs for *Borrelia burgdorferi* (Schmidt and Ostfeld, 2001).

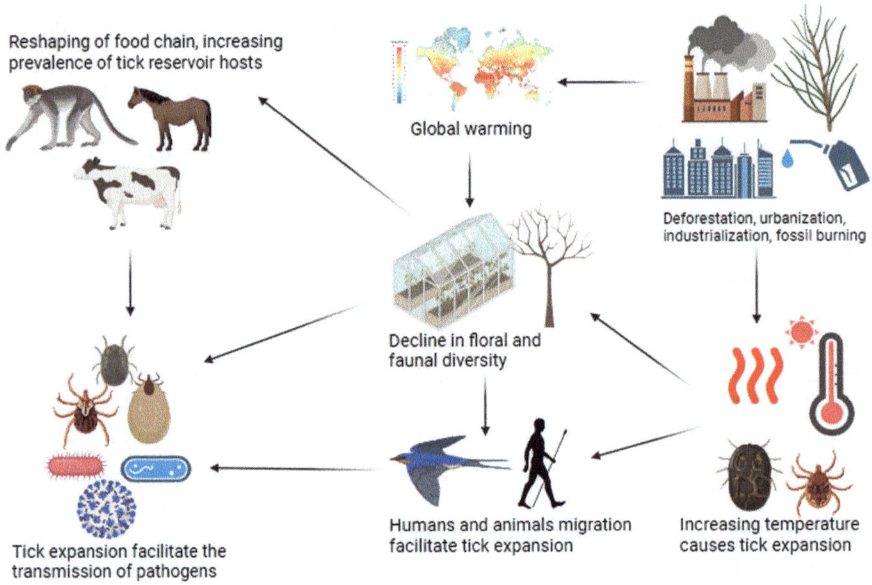

Fig. 8.1 Ecological factors affecting tick-borne diseases risks

With changes in land use, human travel, trade, and other economic activities have had a distinct role in the creation and circulation of the TBDs. Because of globalization, infected ticks together with hosts have moved continents, which escalates the risk of introducing diseases into non-endemic regions (Dantas-Torres, 2015). Urban development is also a reason for this as it brings people into contact with ticks by destroying the habitats of wildlife (Medlock et al., 2013). For instance, the fast growth of farming areas in Africa (Olwoch et al., 2008), and ecotourism and other outdoor activities have led to an increase in cases of TBDs in North America and Europe (Medlock et al., 2013; Randolph, 2008; Tardy et al., 2023). Ecological factors affecting influencing tick-borne diseases risk are given in Fig. 8.1.

8.7 Challenges and Opportunities for Disease Surveillance

The emergence of new pathogens, along with increased incidence and the expanding geographic distribution of TBDs, poses a growing public health concern (Eisen and Paddock, 2021). Predictive modeling for TBDs is complex, influenced by diverse variables such as environmental changes, host-vector dyniamics, and human behaviour. Effective surveillance is therefore crucial for timely diagnosis, rick assessment, and control measures. However, existing challneges. Including imited diagnostic capacity, underreporting, fragmented data systems, and ecological variability, continue to hinder comprehensive surveillance efforts (Rowan et al., 2023).

Nevertheless, there is an opportunity to improve surveillance through technological and interdisciplinary advancements. Novel tools such as machine learning-based risk mapping, genomic sequencing, remote sensing and mobile health platforms are expanding the toolkit for tracking emerging threats. Cross-sector collaboartions between public health, veterinary science, ecology, and data science are increasingly vital in building integrated surveillance networks that are responsive to the evolving nature of TBDs (Hulme et al., 2023; Zhang et al., 2024).

8.7.1 Challenges in Tick-Borne Disease Surveillance

One of the greatest issues in monitoring TBDs is underreporting from misdiagnosing the disease or lack of attention by providers (Schotthoefer et al., 2020). Many TBDs have nonspecific symptoms like fever, weakness, and myalgia, which are rather difficult to tell apart from other febrile illnesses (Maxwell et al., 2022). Also, some TBDs have long incubation periods, which subsequently results in people not recognizing or reporting them for a long time (Rochlin and Toledo, 2020). For instance, Lyme disease, is still considered to be underreported because of the wide range of clinical presentations and the relative unavailability of effective serological diagnostic tests (Mead, 2015). Anaplasmosis and babesiosis are, likewise, commonly misdiagnosed as viral infections, which creates gaps in the surveillance data and therefore underreporting (Tokarz and Lipkin, 2021). Different regions of the globe adopting heterogeneous standards regarding case definitions which further makes reporting more difficult (Dantas-Torres et al., 2012).

In lower and middle income countries, where monitoring and diagnostic resources are limited, surveillance systems for infectious diseases, particularly TBDs, are often inadequate, especially when new threats emerge (Fournet et al., 2018). Furthermore, more indigenous approaches such as passive case reporting do not describe the full scope of TBDs, especially in rural remotes regions where services are poor (Hart et al., 2022; Medlock et al., 2013). The collection of ticks and identification of pathogens in wildlife reservoirs are crucial in understanding the dynamics of risk associated with TBD (Ogden et al., 2021). Unfortunately, funding limitations, seasonal tick activity, and the general labor burden of field work poses significant challenges to effective and ongoing surveillance (Ginsberg et al., 2021).

The effects of climate change on tick dispersal patterns make surveillance more difficult (Ogden and Lindsay, 2016). Warmer temperatures along with an increase in rainfall creates new and favorable environments for the survival of ticks and their spread to other regions (Eisen et al., 2016). For example, the primary vector for Lyme disease, *Ixodes scapularis*, has been moving into Canada and northern U.S. states because milder winters have enabled ticks to survive winter months (Leighton et al., 2012). The flexibility of surveillance systems is limited in regards to including climate models and ecological monitoring. Despite ecological and policy changes, none have been made previously. The incorporation of climate data into TBD surveillance systems still has to be undertaken because of the difficulty in estimating vector-host–pathogen relationships in shifting environments (Estrada-Peña et al., 2014).

8.7.2 Opportunities to Enhance Tick-Borne Disease Surveillance

The development of new molecular diagnostic technologies, including high-throughput sequencing (HTS) and metagenomics, represents a new frontier for surveillance of TBDs (Sajid and Rizwan, 2024; Suminda et al., 2022). The combination of geographic information systems (GIS) and remote sensing technologies is revolutionizing the surveillance of TBD infections by facilitating monitoring of ticks and associated risks in real-time (Baldwin et al., 2022; Estrada-Peña, 2002). These technologies utilize satellite imagery and climatological data to estimate regions that are likely to have an excessive accumulation of ticks and an elevated probability of infection. For instance, GIS based models have successfully predicted the likelihood of Lyme disease in North America by using host distribution, land use, and temperature (Ozdenerol, 2015).

Another opportunity that has emerged for improving data collection revolves around engaging the general public in surveillance activities concerning TBDs via citizen science (Eisen and Eisen, 2021). In particular, smartphone applications and web services enable people to report their interactions with ticks, which helps track tick population changes as well as their movements. As an example, the Tick App in the U.S. and Canada allows participants to photograph and geo-locate ticks for imaging surveillance (Kache et al., 2023). In the same spirit, participatory surveillance activities like the European "Tick Map" project have been successful in tracking the tick distribution using the crowdsourced data from the public (Rubel et al., 2014).

For the effective surveillance of TBDs, the international links as well as the cross-departmental efforts must be undertaken (Eisen and Paddock, 2021). One Health approaches that combine human and animal health and environmental considerations provide a holistic framework to monitor and control TBDs (Johnson et al., 2022). International monitoring systems like the European Centre for Disease Prevention and Control (ECDC) and United States Center of Disease Control and Prevention (USCDC) assist with heading TBD surveillance activities in various regions of the globe (Medlock et al., 2013). Improving collaboration among public health professionals, veterinary specialists, ecologists, and computer scientists will enhance proactive measures in monitoring and controlling health threats. Opportunities to enhance tick-borne diseases surveillance are given below in Fig. 8.2.

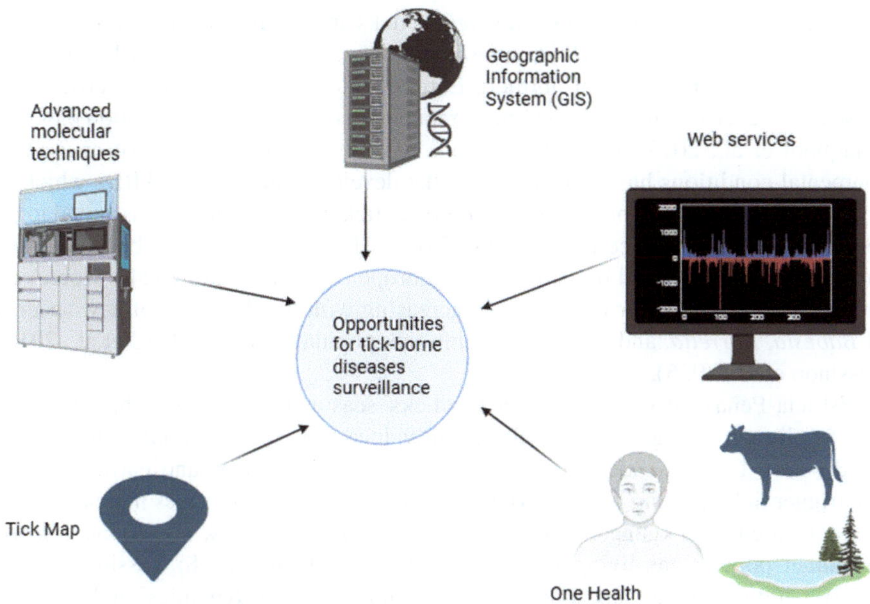

Fig. 8.2 Opportunities for enhancing tick-borne diseases surveillance

8.8 Public Health Implications of Climate-Driven Changes

Climate change is reshaping the global epidemiology of TBDs, with significant implications for public health systems. Global warming, changes in rainfall, and ne alteration of land has allowed for a greater rise in tick populations which has increased human contact with TBDs (Ogden and Lindsay, 2016). The global burden of TBDs is increasing, driven by the emergence of novel pathogens and the expansion of ecological suitable habitats for ticks (Eisen et al., 2016). Climate change and other external factors have already modified the public health infrastructure (e.g., surveillance system demands, diagnostic challenges) about vector ecology. Therefore, public health organizations are required to implement active response mechanisms as well as enhanced precautions such as effective monitoring, rapid diagnosis, and prevention methodologies.

For example, in the U.S., the CDC reported a 357% increase in reported Lyme disease cases from 1991 to 2021 in northern states, correlating with warming trends and northward expansion of tick populations (CDC, 2024). Conversely, many low- and middle-income countries in tropical regions report rising incidences of *Rickettsia* and Crimean-Congo hemorrhagic fever but often lack the surveillance capacity to track these diseases effectively (WHO, 2023; Yessinou et al., 2023).

The incidence of TBDs increases in those regions where ticks spread due to climate change (Medlock et al., 2013). Canada, Scandinavia, and even the higher altitude areas of Europe are reporting increased number of cases of Lyme disease because *Ixodes scapularis* and *Ixodes ricinus* ticks are moving northwards (Leighton et al., 2012; Picado et al., 2024; Semenza and Suk, 2018). Likewise, TBEV is being observed in newer regions of northern Europe and Asia because of the TBEV and these occurrences correlate with climate related alterations of tick migration patterns (Maqbool et al., 2022; Semenza and Suk, 2018). The development of new environmental conditions has also stimulated the development of novel TBDs which is another shocking development. The *Hyalomma* tick, for example, was assumed to be restricted to Africa and certain regions of Asia. However, it has now been detected in the Mediterranean and parts of Central Europe (Bernard et al., 2024). Moreover, the risk of zoonotic spillover events is increasing with the detection of new species of *Babesia*, *Borrelia*, and *Rickettsia* within tick populations (Bouchard et al., 2019; Yessinou et al., 2025).

Estrada-Peña et al. (2017) reported that ticks' seasonal activity range has extended due to climate change, which subsequently increases the potential transmission window for TBDs. Milder winters permit longer tick activity, and earlier springs can trigger tick questing behavior (Ogden et al., 2021). The increases in tick activity has coincided with rising reports of babesiosis, anaplasmosis, and Lyme disease in human populations living in temperate regions (Eisen, 2018; Yessinou et al., 2025). Climate change also impacts host-vector-pathogen dynamics, and worsens the already complex epidemiology of TBDs. Changes in density of some wildlife hosts, such as deer in suburban settings, results in increased tick populations and heightened human interactions (Ginsberg et a., 2021). Such ecological changes call for effective integrated public health measures to cope with risks presented by TBDs.

The geographical spread and increasing frequency of TBDs poses difficulties for disease diagnosis and proper monitoring. Various TBDs have non-specific flu-like symptoms which leads to misdiagnosis and underreporting being a common occurrence (Aucott and Seifter, 2011; Maxwell et al., 2022). Co-infections with several tick-borne pathogens present even greater challenges to accurate diagnosis, perplexing the clinical picture (Sanchez-Vicente and Tokarz, 2023). Public health agencies need to boost surveillance program capabilities and coverage in emerging hotspots for TBDs. Climate and ecology based GIS risk prediction shows that remote sensing technology is being used more commonly (Baldwin et al., 2022; Estrada-Peña, 2002). Yet, many areas, particularly in low- and middle-income countries, are still resource-poor and lack the requisite infrastructure to utilize these technologies. Another major issue is the unpreparedness of healthcare systems. Regions where TBDs are newly emerging will require healthcare providers who understand the pathology of these infections. Mitigating the public health consequences of climate change, particularly concerning TBDs, will require expanded laboratory diagnostic capacity, increased access to rapid molecular testing, and augmented TBD education (de la Fuent et al., 2023; Liao et al., 2024).

8.9 Strategies for Adaptation and Mitigation

The impact of climate change on TBDs is significant and require serious consideration to mitigation and adaptation strategies that reduces the impact of TBDs. Diseases caused by tick bites should be addressed by integrating vector control, mitigation of climate change policy, community comprehension, and public health surveillance (Caminade et al., 2019).

The control of ticks is central to the management of TBDs. Most common methods of tick control such as spraying acaricides do not work because of established resistance, non-target impacts, and environmental concern (Eisen and Dolan, 2016). Climate adaptive tick management policies include: ITM, host-targeted interventions, and genetic and microbial approaches. The ITM involves selective application of acaricide, biological control, and modification of the environment which can significantly lower the tick numbers with less environmental damage (Ginsberg et al., 2021). For example, controlled burns along with scrub management can be employed to reduce tick habitats, while *Metarhizium anisopliae* fungal pathogens could provide a chemical free method to acaricides (Rajput et al., 2024). Ticks have a close relationship with deer and rodents since they are the main hosts for larvae and adult ticks. According to Dolan et al. (2017) and Eisen et al. (2021), control measures such as the use of rodenticide bait boxes, managing the deer population, and applying tick collars on pets have proven beneficial in reducing tick burden. Shaw and Catteruccia (2019) reported that control of tick pathogens transmission using genetic engineering and on symbiotic microorganisms is still in progress. Tick-borne pathogens for instance can be neutralized by certain endosymbiotic bacteria called *Wolbachia* (Kolo and Raghavan, 2023; Narasimhan and Fikrig, 2015).

New surveillance systems are needed to efficiently detect and respond to emerging TBDs threats in a changing climate. Change in climate has impacted the spatial and temporal dispersion of ticks, which requires adaptive surveillance methods (Makwarela et al., 2025). To improve disease surveillance as well as early warning systems, strategic approaches such as integration of climate data with epidemiological monitoring, geospatial modeling and predictive mapping, as well as citizen science and community-based monitoring need to be implemented. Remote sensing and GIS are being used more frequently to predict areas prone to TBD risks based on human land use, climatic changes, and density of host population (Estrada-Pena, 2002; Baldwin et al., 2022). There is enhanced participation of the general public in tick surveillance activities through community tick collection programs and smartphone applications, which has improved the efficiency and scope of traditional surveillance methods (Kache et al., 2023). Climate based warning systems that rely on real-time observation of rainfall, temperature, and relative humidity incorporate epidemiological data to predict TBDs and provide effective control measures (Semenza and Suk, 2018).

Providing education to the public and altering their behaviors in order to reduce the effects of diseases is crucial to the TBDs mitigation plan. For public health education and the communication of risks, campaigns should emphasize tick removal and first aid training, personal protective practices, as well as risk communication for sensitive groups (Eisen and Eisen, 2021; Ginsberg et al., 2021). The public health challenges of TBDs within a transforming environment demands global action. Policymaking by governments, research centers, and intergovernmental bodies should give attention to integrate climate change into public health planning, strengthen One Health collaborations, and support research on climate-TBD interactions (Ogden and Lindsay, 2016; Semenza and Suk, 2018).

References

Abbass, K., Qasim, M. Z., Song, H., Murshed, M., Mahmood, H., & Younis, I. (2022). A review of the global climate change impacts, adaptation, and sustainable mitigation measures. *Environmental Science and Pollution Research, 29*(28), 42539–42559.

Allan, B. F., Keesing, F., & Ostfeld, R. S. (2003). Effect of forest fragmentation on Lyme disease risk. *Conservation Biology, 17*(1), 267–272.

Aucott, J. N., & Seifter, A. (2011). Misdiagnosis of early Lyme disease as the summer flu. *Orthopedic Reviews, 3*(2), Article e14.

Bah, M. T., Grosbois, V., Stachurski, F., Muñoz, F., Duhayon, M., Rakotoarivony, I., Appelgren, A., Calloix, C., Noguera, L., Mouillaud, T., & Andary, C. (2022). The Crimean-Congo haemorrhagic fever tick vector Hyalomma marginatum in the south of France: Modelling its distribution and determination of factors influencing its establishment in a newly invaded area. *Transboundary and Emerging Diseases, 69*(5), e2351–e2365.

Baldwin, H., Landesman, W. J., Borgmann-Winter, B., & Allen, D. (2022). A geographic information system approach to map tick exposure risk at a scale for public health intervention. *Journal of Medical Entomology, 59*(1), 162–172.

Bellouin, N., Quaas, J., Gryspeerdt, E., Kinne, S., Stier, P., Watson-Parris, D., Boucher, O., Carslaw, K. S., Christensen, M., Daniau, A. L., & Dufresne, J. L. (2020). Bounding global aerosol radiative forcing of climate change. *Reviews of Geophysics, 58*(1), e2019RG000660.

Bernard, C., Kukla, C. J., Rakotoarivony, I., Duhayon, M., Stachurski, F., Huber, K., Giupponi, C., Zortman, I., Holzmuller, P., Pollet, T., & Jeanneau, M. (2024). Detection of Crimean-Congo haemorrhagic fever virus in Hyalomma marginatum ticks, southern France, May 2022 and April 2023. *Eurosurveillance, 29*(6), 2400023.

Bouchard, C., Dibernardo, A., Koffi, J. K., Wood, H., Leighton, P. A., & Lindsay, L. R. (2019). N Increased risk of tick-borne diseases with climate and environmental changes. *Canada communicable disease report = Releve des maladies transmissibles au Canada, 45 4*, 83–89.

Buczek, A. M., Buczek, W., Buczek, A., & Bartosik, K. (2020). The potential role of migratory birds in the rapid spread of ticks and tick-borne pathogens in the changing climatic and environmental conditions in Europe. *International Journal of Environmental Research and Public Health, 17*(6), 2117.

Burnus, L., Wynn, J., Liedvogel, M., & Rollins, R. (2024). Beware of hitchhiking ticks? Clarifying the variable roles of bird families in tick movement along migratory routes. *Authorea Preprints.*

Caminade, C., McIntyre, K. M., & Jones, A. E. (2019). Impact of recent and future climate change on vector-borne diseases. *Annals of the New York Academy of Sciences, 1436*(1), 157–173.

CDC, Centers for Disease Control and Prevention. (2024). *Surveillance for Lyme disease after implementation of a revised case definition, United States, 2022.* Morbidity and Mortality Weekly Report, 73(6), 137–143. https://www.cdc.gov/mmwr/volumes/73/wr/mm7306a1.htm

Clow, K. M., Leighton, P. A., Pearl, D. L., & Jardine, C. M. (2019). A framework for adaptive surveillance of emerging tick-borne zoonoses. *One Health, 7*, Article 100083.

Collins, W. J., Webber, C. P., Cox, P. M., Huntingford, C., Lowe, J., Sitch, S., Chadburn, S. E., Comyn-Platt, E., Harper A. B., Hayman, G., & Powell, T. (2018). Increased importance of methane reduction for a 1.5 degree target. *Environmental Research Letters, 13*(5), 054003.

Dantas-Torres, F. (2015). Climate change, biodiversity, ticks and tick-borne diseases: The butterfly effect. *International Journal for Parasitology: Parasites and Wildlife, 4*(3), 452–461.

Dantas-Torres, F., Chomel, B. B., & Otranto, D. (2012). Ticks and tick-borne diseases: A One Health perspective. *Trends in Parasitology, 28*(10), 437–446.

de la Fuente, J., Estrada-Peña, A., Rafael, M., Almazán, C., Bermúdez, S., Abdelbaset, A. E., Kasaija, P. D., Kabi, F., Akande, F. A., Ajagbe, D. O., & Bamgbose, T. (2023). Perception of ticks and tick-borne diseases worldwide. *Pathogens, 12*(10), 1258.

Dolan, M. C., Schulze, T. L., Jordan, R. A., Schulze, C. J., Ullmann, A. J., Hojgaard, A., Williams, M. A., & Piesman, J. (2017). Evaluation of doxycycline-laden oral bait and topical fipronil delivered in a single bait box to control Ixodes scapularis (Acari: Ixodidae) and reduce Borrelia burgdorferi and Anaplasma phagocytophilum infection in small mammal reservoirs and host-seeking ticks. *Journal of Medical Entomology, 54*(2), 403–410.

Eisen, L. (2018). Pathogen transmission in relation to duration of attachment by Ixodes scapularis ticks. *Ticks and Tick-Borne Diseases, 9*(3), 535–542.

Eisen, L., & Dolan, M. C. (2016). Evidence for personal protective measures to reduce human contact with blacklegged ticks and for environmentally based control methods to suppress host-seeking blacklegged ticks and reduce infection with Lyme disease spirochetes in tick vectors and rodent reservoirs. *Journal of Medical Entomology, 53*(5), 1063–1092.

Eisen, L., & Eisen, R. J. (2021). Benefits and drawbacks of citizen science to complement traditional data gathering approaches for medically important hard ticks (Acari: Ixodidae) in the United States. *Journal of Medical Entomology, 58*(1), 1–9.

Eisen, R. J., & Paddock, C. D. (2021). Tick and tickborne pathogen surveillance as a public health tool in the United States. *Journal of Medical Entomology, 58*(4), 1490–1502.

Eisen, R. J., Eisen, L., & Beard, C. B. (2016). County-scale distribution of Ixodes scapularis and Ixodes pacificus (Acari: Ixodidae) in the continental United States. *Journal of Medical Entomology, 53*(2), 349–386.

Estrada-Peña, A. (2002). Increasing habitat suitability in the United States for the tick that transmits Lyme disease: A remote sensing approach. *Environmental Health Perspectives, 110*(7), 635–640.

Estrada-Peña, A., Ayllón, N., & De La Fuente, J. (2012). Impact of climate trends on tick-borne pathogen transmission. *Frontiers in Physiology, 3*, 64.

Estrada-Peña, A., Ostfeld, R. S., Peterson, A. T., Poulin, R., & de la Fuente, J. (2014). Effects of environmental change on zoonotic disease risk: An ecological primer. *Trends in Parasitology, 30*(4), 205–214.

Estrada-Peña, A., Roura, X., Sainz, A., Miró, G., & Solano-Gallego, L. (2017). Species of ticks and carried pathogens in owned dogs in Spain: results of a one-year national survey. *Ticks and Tick-Borne Diseases, 8*(4), 443–452.

Estrada-Peña, A., Venzal, J. M., Kocan, K. M., & Sonenshine, D. E. (2008). Overview: ticks as vectors of pathogens that cause disease in humans and animals.

Fournet, F., Jourdain, F., Bonnet, E., Degroote, S., & Ridde, V. (2018). Effective surveillance systems for vector-borne diseases in urban settings and translation of the data into action: A scoping review. *Infectious Diseases of Poverty, 7*, 1–14.

Gage, K. L., & Kosoy, M. Y. (2005). Natural history of plague: Perspectives from more than a century of research. *Annual Review of Entomology, 50*(1), 505–528.

Gilbert, L. (2021). The impacts of climate change on ticks and tick-borne disease risk. *Annual Review of Entomology, 66*(1), 373–388.

Ginsberg, H. S., Hickling, G. J., Burke, R. L., Ogden, N. H., Beati, L., LeBrun, R. A., Arsnoe, I. M., Gerhold, R., Han, S., Jackson, K., & Maestas, L. (2021). Why Lyme disease is common in the northern US, but rare in the south: The roles of host choice, host-seeking behavior, and tick density. *PLoS Biology, 19*(1), Article e3001066.

Githeko, A. K., Lindsay, S. W., Confalonieri, U. E., & Patz, J. A. (2000). Climate change and vector-borne diseases: A regional analysis. *Bulletin of the World Health Organization, 78*(9), 1136–1147.

Gray, J. S., Dautel, H., Estrada-Peña, A., Kahl, O., & Lindgren, E. (2009). Effects of climate change on ticks and tick-borne diseases in Europe. *Interdisciplinary Perspectives on Infectious Diseases, 2009*(1), Article 593232.

Gray, J., Zintl, A., Hildebrandt, A., Hunfeld, K. P., & Weiss, L. (2010). Zoonotic babesiosis: Overview of the disease and novel aspects of pathogen identity. *Ticks and Tick-Borne Diseases, 1*(1), 3–10.

Hart, C. E., Bhaskar, J. R., Reynolds, E., Hermance, M., Earl, M., Mahoney, M., Martinez, A., Petzlova, I., Esterly, A. T. Thangamani, S. (2022). Community engaged tick surveillance and tick MAP as a public health tool to track the emergence of ticks and tick-borne diseases in New York. *PLOS Global Public Health, 2*(6), e0000215.

Hasle, G. (2013). Transport of ixodid ticks and tick-borne pathogens by migratory birds. *Frontiers in Cellular and Infection Microbiology, 3*, 48.

Hayasaka, D., Ivanov, L., Leonova, G. N., Goto, A., Yoshii, K., Mizutani, T., Kariwa, H., & Takashima, I. (2001). Distribution and characterization of tick-borne encephalitis viruses from Siberia and far-eastern Asia. *Journal of General Virology, 82*(6), 1319–1328.

Hulme, P. E., Beggs, J. R., Binny, R. N., Bray, J. P., Cogger, N., Dhami, M. K., & Lockhart, P. J. (2023). Emerging advances in biosecurity to underpin human, animal, plant, and ecosystem health. *Iscience, 26*(9).

Johnson, N., Phipps, L. P., Hansford, K. M., Folly, A. J., Fooks, A. R., Medlock, J. M., & Mansfield, K. L. (2022). One Health approach to tick and tick-borne disease surveillance in the United Kingdom. *International Journal of Environmental Research and Public Health, 19*(10), 5833.

Kache, P. A., Bron, G. M., Zapata-Ramirez, S., Tsao, J. I., Bartholomay, L. C., Paskewitz, S. M., Diuk-Wasser, M. A., & del Pilar Fernandez, M. (2023). Evaluating spatial and temporal patterns of tick exposure in the United States using community science data submitted through a smartphone application. *Ticks and Tick-Borne Diseases, 14*(4), Article 102163.

Kilpatrick, A. M., Dobson, A. D., Levi, T., Salkeld, D. J., Swei, A., Ginsberg, H. S., Kjemtrup, A., Padgett, K. A., Jensen, P. M., Fish, D., & Ogden, N. H. (2017). Lyme disease ecology in a changing world: Consensus, uncertainty and critical gaps for improving control. *Philosophical Transactions of the Royal Society b: Biological Sciences, 372*(1722), 20160117.

Kolo, A. O., & Raghavan, R. (2023). Impact of endosymbionts on tick physiology and fitness. *Parasitology, 150*(10), 859–865.

Leighton, P. A., Koffi, J. K., Pelcat, Y., Lindsay, L. R., & Ogden, N. H. (2012). Predicting the speed of tick invasion: An empirical model of range expansion for the Lyme disease vector Ixodes scapularis in Canada. *Journal of Applied Ecology, 49*(2), 457–464.

Levy, S. (2017). Northern Trek: The spread of Ixodes scapularis into Canada. *Environmental Health Perspectives, 125*(7), Article 074002.

Liao, H., Lyon, C. J., Ying, B., & Hu, T. (2024). Climate change, its impact on emerging infectious diseases and new technologies to combat the challenge. *Emerging Microbes & Infections, 13*(1), 2356143.

Madison-Antenucci, S., Kramer, L. D., Gebhardt, L. L., & Kauffman, E. (2020). Emerging tick-borne diseases. *Clinical Microbiology Reviews, 33*(2), 10–1128.

Makwarela, T. G., Seoraj-Pillai, N., & Nangammbi, T. C. (2025). Tick control strategies: critical insights into chemical, biological, physical, and integrated approaches for effective hard tick management. *Veterinary Sciences, 12*(2), 114.

Maqbool, M., Sajid, M. S., Saqib, M., Anjum, F. R., Tayyab, M. H., Rizwan, H. M., Rashid, M. I., Rashid, I., Iqbal, A., Siddique, R. M., & Shamim, A. (2022). Potential mechanisms of transmission of tick-borne viruses at the virus-tick interface. *Frontiers in Microbiology, 13*, Article 846884.

Maxwell, S. P., Brooks, C., McNeely, C. L., & Thomas, K. C. (2022). Neurological pain, psychological symptoms, and diagnostic struggles among patients with tick-borne diseases. *In Healthcare, 10*(7), 1178.

Mead, P. S. (2015). Epidemiology of Lyme disease. *Infectious Disease Clinics, 29*(2), 187–210.

Medlock, J. M., Hansford, K. M., Bormane, A., Derdakova, M., Estrada-Peña, A., George, J. C., Golovljova, I., Jaenson, T. G., Jensen, J. K., Jensen, P. M., & Kazimirova, M. (2013). Driving forces for changes in geographical distribution of Ixodes ricinus ticks in Europe. *Parasites & Vectors, 6*, 1–11.

Narasimhan, S., & Fikrig, E. (2015). Tick microbiome: The force within. *Trends in Parasitology, 31*(7), 315–323.

Nuttall, P. A. (2022). Climate change impacts on ticks and tick-borne infections. *Biologia, 77*(6), 1503–1512.

Ogden, N. H., & Lindsay, L. R. (2016). Effects of climate and climate change on vectors and vector-borne diseases: Ticks are different. *Trends in Parasitology, 32*(8), 646–656.

Ogden, N. H., Ben Beard, C., Ginsberg, H. S., & Tsao, J. I. (2021). Possible effects of climate change on ixodid ticks and the pathogens they transmit: Predictions and observations. *Journal of Medical Entomology, 58*(4), 1536–1545.

Ogden, N. H., Maarouf, A., Barker, I. K., Bigras-Poulin, M., Lindsay, L. R., Morshed, M. G., O'callaghan, C. J., Ramay, F., Waltner-Toews, D., & Charron, D. F. (2006). Climate change and the potential for range expansion of the Lyme disease vector Ixodes scapularis in Canada. *International Journal for Parasitology, 36*(1), 63–70.

Olwoch, J. M., Reyers, B., Engelbrecht, F. A., & Erasmus, B. F. N. (2008). Climate change and the tick-borne disease, Theileriosis (East Coast fever) in sub-Saharan Africa. *Journal of Arid Environments, 72*(2), 108–120.

Ostfeld, R. S., & Keesing, F. (2012). Effects of host diversity on infectious disease. *Annual Review of Ecology, Evolution, and Systematics, 43*(1), 157–182.

Ostfeld, R. S., Canham, C. D., Oggenfuss, K., Winchcombe, R. J., & Keesing, F. (2006). Climate, deer, rodents, and acorns as determinants of variation in Lyme-disease risk. *PLoS Biology, 4*(6), Article e145.

Ozdenerol, E. (2015). GIS and remote sensing use in the exploration of Lyme disease epidemiology. *International Journal of Environmental Research and Public Health, 12*(12), 15182–15203.

Paddock, C. D., & Goddard, J. (2015). The evolving medical and veterinary importance of the Gulf Coast tick (Acari: Ixodidae). *Journal of Medical Entomology, 52*(2), 230–252.

Perumalsamy, N., Sharma, R., Subramanian, M., & Nagarajan, S. A. (2024). Hard ticks as vectors: The emerging threat of tick-borne diseases in India. *Pathogens, 13*(7), 556.

Picado, R., Baptista, C. J., Meneses, A., Legatti, S., Fonseca, J., & Belas, A. (2024). Lyme disease in companion animals: An updated state-of-art and current situation in Portugal. *Veterinary Research Communications, 48*(6), 3551–3561.

Polo, G., Gamarra, J., Robayo-Sánchez, L. N., Cortés-Vecino, J. A., & Ramírez-Hernández, A. (2024). Impact of climate change on the geographical distribution of ticks of public health significance in Colombia: Amblyomma ovale (Ixodida: Ixodidae), the Amblyomma maculatum (Ixodida: Ixodidae) complex and the Amblyomma cajennense (Ixodida: Ixodidae) complex. *Journal of Medical Entomology, 61*(6), 1489–1500.

Porretta, D., Mastrantonio, V., Amendolia, S., Gaiarsa, S., Epis, S., Genchi, C., Bandi, C., Otranto, D., & Urbanelli, S. (2013). Effects of global changes on the climatic niche of the tick Ixodes ricinus inferred by species distribution modelling. *Parasites & Vectors, 6*, 1–8.

Pritt, B. S., Mead, P. S., Johnson, D. K. H., Neitzel, D. F., Respicio-Kingry, L. B., Davis, J. P., Schiffman, E., Sloan, L. M., Schriefer, M. E., Replogle, A. J., & Paskewitz, S. M. (2016). Identification of a novel pathogenic Borrelia species causing Lyme borreliosis with unusually high spirochaetaemia: A descriptive study. *The Lancet Infectious Diseases, 16*(5), 556–564.

Rajput, M., Sajid, M. S., Rajput, N. A., George, D. R., Usman, M., Zeeshan, M., Iqbal, O., Bhutto, B., Atiq, M., Rizwan, H. M., & Daniel, I. K. (2024). Entomopathogenic Fungi as Alternatives to Chemical Acaricides: Challenges, Opportunities and Prospects for Sustainable Tick Control. *Insects, 15*(12), 1017.

Randolph, S. E. (2008). Tick-borne disease systems. *Rev Sci Tech off Int Epiz, 27*(2), 1–15.

Rochlin, I., & Toledo, A. (2020). Emerging tick-borne pathogens of public health importance: A mini-review. *Journal of Medical Microbiology, 69*(6), 781–791.

Roth, D. (2012). Surveillance for emerging infectious diseases: a Canadian perspective. National Collaborating Centre for Environmental Health.

Rowan, S., Mohseni, N., Chang, M., Burger, H., Peters, M., & Mir, S. (2023). From tick to test: A comprehensive review of tick-borne disease diagnostics and surveillance methods in the United States. *Life, 13*(10), 2048.

Rubel, F., Brugger, K., Monazahian, M., Habedank, B., Dautel, H., Leverenz, S., & Kahl, O. (2014). The first German map of georeferenced ixodid tick locations. *Parasites & Vectors, 7*, 1–5.

Sajid, M. S., & Rizwan, H. M. (2024). *Omics Approaches in Veterinary Parasitology: Diagnosis, Biomarkers, and Drug Development*. CRC Press.

Sanchez-Vicente, S., & Tokarz, R. (2023). Tick-borne co-infections: Challenges in molecular and serologic diagnoses. *Pathogens, 12*(11), 1371.

Schmidt, K. A., & Ostfeld, R. S. (2001). Biodiversity and the dilution effect in disease ecology. *Ecology, 82*(3), 609–619.

Schotthoefer, A., Stinebaugh, K., Martin, M., & Munoz-Zanzi, C. (2020). Tickborne disease awareness and protective practices among US Forest Service employees from the upper Midwest, USA. *BMC Public Health, 20*, 1–15.

Scott, J. D. (2016). Studies abound on how far north Ixodes scapularis ticks are transported by birds. *Ticks and Tick-Borne Diseases, 7*(2), 327–328.

Semenza, J. C., & Suk, J. E. (2018). Vector-borne diseases and climate change: a European perspective. *FEMS microbiology letters, 365*(2), fnx244.

Shaw, W. R., & Catteruccia, F. (2019). Vector biology meets disease control: Using basic research to fight vector-borne diseases. *Nature Microbiology, 4*(1), 20–34.

Široký, P., Kubelova, M., Bednář, M., Modrý, D., Hubálek, Z., & Tkadlec, E. (2011). The distribution and spreading pattern of Dermacentor reticulatus over its threshold area in the Czech Republic—how much is range of this vector expanding? *Veterinary Parasitology, 183*(1–2), 130–135.

Stuen, S., Granquist, E. G., & Silaghi, C. (2013). Anaplasma phagocytophilum—a widespread multi-host pathogen with highly adaptive strategies. *Frontiers in Cellular and Infection Microbiology, 3*, 31.

Suminda, G. G. D., Bhandari, S., Won, Y., Goutam, U., Pulicherla, K. K., Son, Y. O., & Ghosh, M. (2022). High-throughput sequencing technologies in the detection of livestock pathogens, diagnosis, and zoonotic surveillance. *Computational and Structural Biotechnology Journal, 20*, 5378–5392.

Süss, J. (2011). Tick-borne encephalitis 2010: Epidemiology, risk areas, and virus strains in Europe and Asia—an overview. *Ticks and Tick-Borne Diseases, 2*(1), 2–15.

Tardy, O., Acheson, E. S., Bouchard, C., Chamberland, É., Fortin, A., Ogden, N. H., & Leighton, P. A. (2023). Mechanistic movement models to predict geographic range expansions of ticks and tick-borne pathogens: Case studies with Ixodes scapularis and Amblyomma americanum in eastern North America. *Ticks and Tick-Borne Diseases, 14*(4), Article 102161.

Tazerji, S. S., Hajipour, P., Pshenichnaya, N., Rizwan, H. M., Abuova, G. N., & Duarte, P. M. (2024). *Crimean-Congo Haemorrhagic Fever. In The Handbook of Zoonotic Diseases of Goats*. CABI.

Tokarz, R., & Lipkin, W. I. (2021). Discovery and surveillance of tick-borne pathogens. *Journal of Medical Entomology, 58*(4), 1525–1535.

WHO, World Health Organization. (2023). *Crimean-Congo haemorrhagic fever*. https://www.who.int/news-room/fact-sheets/detail/crimean-congo-haemorrhagic-fever

Yessinou, R. E., Azihou, A. F., Yabi, J. A., Hounmanou, Y. G., Hounkpè, C. W., Gbaguidi-Haore, H., & Boko, M. (2023). Tick-borne diseases and gaps in surveillance systems in West Africa: A systematic review. *Parasite Epidemiology and Control, 22*

Yessinou, R. E., Koumassou, A., Galadima, H. B., Nanoukon-Ahigan, H., Farougou, S., & Pfeffer, M. (2025). Tick Diversity and Distribution of Pathogen in Ticks Collected from Wild Animals and Vegetation in Africa. *Pathogens, 14*(2), 116.

Zhai, P., Pörtner, H. O., Roberts, D., Skea, J., Shukla, P. R., Pirani, A., Moufouma-Okia, W., Péan, C., Pidcock, R., Connors, S., & Matthews, R. (2018). Global Warming of 1.5 C. An IPCC Special Report on the Impacts of Global Warming of 1.5 C above Pre-Industrial Levels and Related Global Greenhouse Gas Emission Pathways, in the Context of Strengthening the Global Response to the Threat of Climate Change. *Sustainable Development, and Efforts to Eradicate Poverty*, 32.

Zhang, L., Guo, W., & Lv, C. (2024). Modern technologies and solutions to enhance surveillance and response systems for emerging zoonotic diseases. *Science in One Health, 3*, Article 100061.

Chapter 9
Socio-Economic Consequences of Altered Tick-Borne Disease Patterns

Hafiz Muhammad Rizwan, Hizqeel Ahmed Muzaffar, Mohsin Raza, Mourad Ben Said, and Hanène Belkahia

Abstract Ticks and tick-borne diseases have remained an adverse challenge for poor communities. As compared to the estimated burden, the burden of tick-borne diseases is increasing worldwide. No region of the world is safe from these tick-borne diseases. But the most affected communities by these tick-borne diseases are those communities that have kept livestock only to earn their livelihoods. Health-care systems are feeling a high burden of tick-borne diseases. The world economy is suffering from the loss of billions of dollars annually in the form of loss of agricultural and livestock productivity. Moreover, the knowledge of poor communities regarding tick-borne diseases is very poor, which makes it difficult to implement control strategies effectively. Furthermore, recreational activities have become a greater source of the transmission of tick-borne diseases because tourists are at higher risk of coming in contact with ticks. Ultimately, people have to adopt personal protective behaviors and employers have to facilitate those employees who have been affected with tick-borne diseases for efficient management and control of tick-borne diseases. Public health professionals have to work effectively to inform people about these tick-borne diseases because a person's ability to understand their risks is highly dependent upon their knowledge.

H. M. Rizwan (✉)
Section of Parasitology, Department of Pathobiology, KBCMA College of Veterinary and Animal Sciences, Narowal, Sub-Campus UVAS, Lahore, Pakistan
e-mail: hm.rizwan@uvas.edu.pk

H. A. Muzaffar
Faculty of Veterinary Sciences, KBCMA College of Veterinary and Animal Sciences, Narowal, Sub-Campus UVAS, Lahore, Pakistan

M. Raza
Section of Physiology, Department of Basic Sciences, KBCMA College of Veterinary and Animal Sciences, Narowal, Sub-Campus UVAS, Lahore, Pakistan

M. B. Said
Department of Basic Sciences, Higher Institute of Biotechnology of Sidi Thabet, University of Manouba, Manouba, Tunisia

M. B. Said · H. Belkahia
Laboratory of Microbiology, National School of Veterinary Medicine of Sidi Thabet, University of Manouba, Manouba, Tunisia

H. M. Rizwan and M. S. Sajid (eds.), *Ticks in a Changing Climate*, Biometeorology 7,
https://doi.org/10.1007/978-3-032-12259-9_9

Keywords Economic burden · Socio-economic disparities · Public perception · Livestock productivity · Human health cost

9.1 Introduction

Among the economically important pathogens, ticks occupy an important position because of their prevalence supported by suitable climates (Kasaija et al., 2021). Tick-borne diseases are a major threat to public health. They are also a major threat to the economics of agricultural and equine industries. Tick-borne disease transmission is highly dependent upon ecological and environmental conditions, host-vector interaction, and people's behavior. To combat tick-borne diseases effectively, there should be a better understanding of social factors that promote the spread of diseases (Zortman et al., 2023). Tick-borne diseases cause severe economic losses in the form of mortality and morbidity of animals and are also a major threat to public health (Ramzan et al., 2021). Among the socio-economic factors responsible for the transmission of tick-borne diseases, the educational index is an important factor. Moreover, the importance of urban land area can't be neglected as it is also a valuable socio-economic factor regulating the transmission of tick-borne diseases (Zhang et al., 2022). The control of ticks and tick-borne diseases has a promising socio-economic impact on poor communities (Randela, 2007). In the forests where ticks are highly prevalent, the harvesting of wild foods for human consumption because of poverty and unemployment is the main socio-economic factor responsible for the transmission of tick-borne diseases. Data suggests that as the biotic and abiotic environmental conditions and socio-economic conditions are modified, the contact of humans with ticks could be increased (Šumilo et al., 2008). Cattle infected with ticks are a major concern for the farmers who keep these animals for livelihood. Moreover, the high prices of acaricides make it difficult for farmers to apply these on their animals (Adehan et al., 2018).

Although climate plays an important role in regulating the transmission of tick-borne diseases, many other factors are also key players. Some of these factors include epidemiological, socio-economic, and environmental factors (Parham et al., 2015). To understand the vulnerability of a person to tick-borne diseases, it is essential to understand a person's living habitat (Li et al., 2018). The spread of ticks has become a threat to health systems worldwide. People who spend their time outdoors are at greater risk of transmission of tick-borne diseases. This fact applies most to the tourists, who are easy targets for ticks, especially during climbing and hiking. Lack of surveillance, lack of resources, and lack of awareness make it difficult to control the transmission of ticks and tick-borne diseases (Klingelhöfer et al., 2022). Recreational places play an important role in the transmission of ticks and thus tick-borne diseases (Slunge et al., 2019a, 2019b). This chapter focuses on the socio-economic effect of the transmission of tick-borne diseases including the effects of these diseases on healthcare systems, economic burden, effects on livestock production, the role of tourism on the transmission of these diseases, and case studies of socio-economic

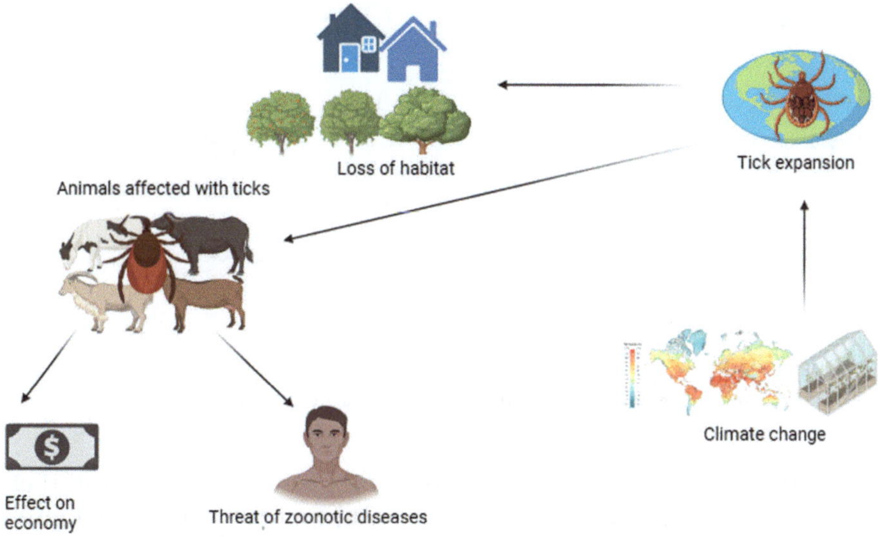

Fig. 9.1 Socio-economic consequences of tick-borne diseases

consequences in affected communities of the world are discussed. Socio-economic consequences of tick-borne diseases are given below in Fig. 9.1.

9.2 Economic Burden of Tick-Borne Diseases on Healthcare Systems

Tick-borne diseases are expanding rapidly worldwide. Different novel tick-borne diseases have been identified. So, their economic burden is increasing as compared to the estimated burden (Šmit & Postma, 2016). Tick-borne diseases impose unwanted stress on different healthcare systems including human and animal healthcare systems (Garcia et al., 2022). As the cases of tick-borne diseases increase within a healthcare reporting system, the burden on the economy also increases (Madison-Antenucci et al., 2020) (Fig. 9.2). While a tick-borne disease can be treated, untreated or misdiagnosed tick-borne disease leads to an increased burden on healthcare systems as in the case of Lyme disease (Mac et al., 2019). According to the reports, about 300,000 people are affected annually in the United States of America alone, leaving behind severe economic stress for the healthcare systems (Rochlin & Toledo, 2020). Similarly, Lyme disease has increased hospitalization of patients in Belgium (Geebelen, 2021). Moreover, the burden on healthcare systems has increased in Canada because of Lyme disease (Mac, 2022). Lyme disease requires proper lab tests (Lantos et al., 2013), serological tests (Wormser et al., 2013), antibiotics treatment (Shadick et al., 2001), other treatments (Müller et al., 2012), consultation (Joss

Fig. 9.2 Economic burden of tick-borne diseases on healthcare systems

et al., 2003), physician services (Henningsson et al., 2010), and hospital care (Lohr et al., 2015) for its proper management that require a healthcare system to work efficiently (Magid et al., 1992). It also requires the hospitalization of an affected patient to observe the effects of medication (Eckman et al., 1997). Among the total economic losses of a country caused by Lyme disease, healthcare costs constitute 48% alone (van den Wijngaard et al., 2017). Moreover, the vaccination cost for a tick-borne disease is also an important burden for a healthcare system (Hsia et al., 2002) (Fig. 9.2).

Tick-bone encephalitis imposes severe economic costs on healthcare systems (Avxentev et al., 2013). In Sweden, tick-borne encephalitis has caused the consumption of thousands of euros within the healthcare system (Slunge et al., 2022). Tick-borne encephalitis has caused healthcare systems to spend about € 400,000 per annum in Italy (Scaggiante et al., 2024). Because of the high treatment cost of a tick-borne disease, a healthcare system may also have to pay a part of the treatment cost (Hook et al., 2022). The mitigation of tick-borne diseases requires long-term surveillance that causes healthcare systems to spend a huge amount (Johnson et al., 2022). For the efficient control of tick-borne diseases, healthcare systems have to improve the surveillance and diagnostic capacity which will cause a healthcare system to spend a massive budget (Abdelbaset et al., 2022). In the regions where healthcare systems are not able to provide the costs for the management of tick-borne diseases, the establishment of the One Health concept is a major challenge (Charles et al., 2021). Mainly in rural areas, healthcare centers do have not much efficiency in diagnosing and reporting a tick-borne disease, leading to poor surveillance and unavailability of required data (Abdelbaset et al., 2023; de la Fuente et al., 2023) (Fig. 9.2).

9.3 Impact on Agricultural and Livestock Productivity

Because of global challenges, ticks, and tick-borne diseases are becoming more prevalent in domestic animals, leading to an economic loss of \$ 14–\$ 19 billion per year worldwide (Hussnain et al., 2021; Luan et al., 2023). Because of the poor economy of Africa, people there keep livestock to earn their livelihoods, but the prevalence of tick-borne diseases has caused a decline in livestock production, leading to further loss to the continent's economy (Heylen et al., 2023). In the continent of Africa, Uganda has suffered badly with ticks and tick-borne diseases. Tick-borne diseases cause an estimated loss of \$ 1.1 Billion per annum in Uganda alone (Kasaija et al., 2021). In India, the milk loss, leather loss, and treatment costs are calculated as \$ 595 million because of tick infestation and \$ 191 million because of tick-borne diseases, a cumulative loss of \$ 787 million (Singh et al., 2022). In Pakistan, under-resourced people keep small ruminants because of their fast growth (Khan et al., 2022), but their livelihood is now being affected by the tick infestation (Ghafar et al., 2021). More than 30 tick-borne diseases have now been identified in China, which has caused a massive drop in agricultural and livestock production (Li et al., 2020). Although Australia is isolated from other parts of the world, the introduction of domestic animals in Australia has caused the emergence of five tick species (Oskam et al., 2021), that are now a major threat to livestock production in Australia.

Tick-borne diseases have been increasing in the USA over the years (Rodino et al., 2020), affecting livestock production. Lyme disease is increasing in Canada (Ogden et al., 2022), becoming a risk for its economy. In South American countries, like Argentina, with an increase in the number of cattle ticks, the cattle industry is being adversely affected (Gilbert, 2021). Among several examples, the best example is *Rhipicephalus microplus* which is found in Asia, but now has migrated to Africa, Australia, and South America, and is now becoming a threat to the livestock industry in these countries (Johnson et al., 2022). FMD causes an estimated loss of \$ 3 billion in the United Kingdom (Folly et al., 2020), which is a tick-borne disease (George et al., 2002). Tikc-borne diseases, especially babesiosis, are becoming a threat to the livestock agricultural industry in many regions of Europe (Bajer et al., 2022). Direct and indirect losses due to ticks and tick-borne diseases have been summarized in Fig. 9.3.

9.4 Socio-Economic Disparities in Tick-Borne Diseases Risk

Unemployment and poverty cause an increase in the risk of tick-borne diseases (Šumilo et al., 2008). Tick-borne diseases lead to socio-economic transition. It leads to a decrease in the number of jobs as a result of which unemployment and poverty

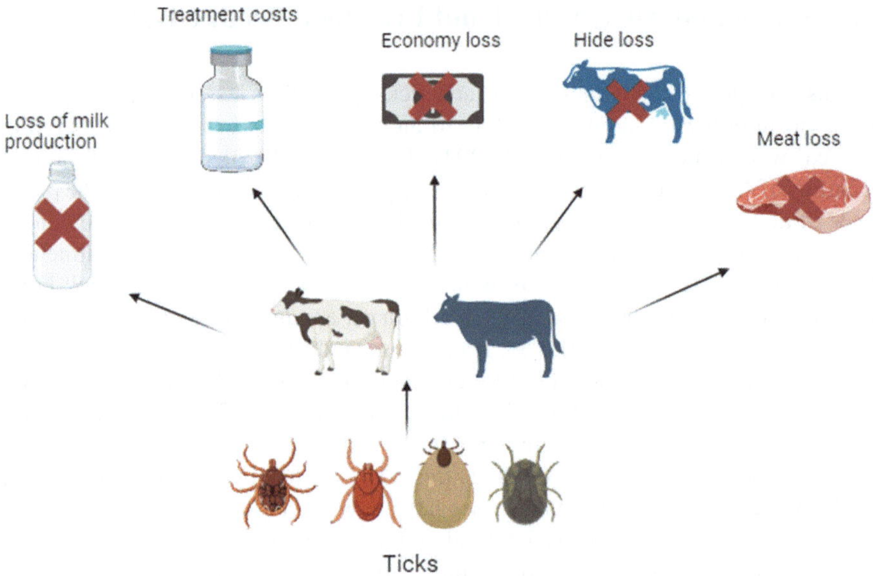

Fig. 9.3 Direct and indirect losses due to ticks and tick-borne diseases

increase. Because of this, economic inequality takes place (Randolph, 2008). Livestock farming plays a very important role in supporting the livelihoods of small-scale farmers (Namgyal et al., 2021). Tick-borne diseases are adversely affecting the livelihoods of marginalized and small-scale livestock owners in India and Africa (Minjauw & McLeod, 2003). The majority of those affected by these diseases reside in developing nations, where circumstances contribute to an increased risk of disease, particularly for underprivileged and marginalized groups such as rural Indigenous people, impoverished women, the elderly, and children (Hotez et al., 2020; Ortiz et al., 2021; Wilke et al., 2019). Tick-borne diseases are a major threat for livestock owners in Africa because most of the livestock owners there are keeping the livestock only to earn their livelihood (Sungirai et al., 2016). The other major concern regarding poor communities is that their knowledge regarding tick-borne diseases is very low (Katswara & Mukaratirwa, 2021), which leads to the uncontrolled spread of tick-borne diseases among these communities (Valente et al., 2015). Moreover, there is limited epidemiological data gathered from these communities that make control strategies difficult to implement (Simuunza et al., 2011) (Fig. 9.4).

As the human population is increasing, people are moving towards the cities to earn their livelihood. This kind of expansion causes a change in the natural structure, disturbing the natural habitat of several species. As a result, ticks have to migrate outside of their specific geographical areas, leading to the emergence of tick-borne diseases (Pfäffle et al., 2013). Human encroachment on wildlife habitats and deforestation has been proposed as a reason for the continuous increase in the incidence

Fig. 9.4 Factors contributing to increased risk of tick-borne diseases

of tick-borne diseases (Nchu et al., 2020). Tick management is also one of the problems of smallholder farmers. These farmers depend upon the acaricides provided by their governments. Improper use of these acaricides again leads to the development of resistance among the ticks against these acaricides, ultimately increasing the threat of tick-borne diseases (Muvhuringi et al., 2022). In short, we can say that due to barriers in the forms of culture, society, politics, and the health system, vulnerable groups such as indigenous populations and marginalized communities remain highly susceptible to the transmission of these diseases (Dell'Arciprete et al., 2014) (Fig. 9.4).

9.5 Tourism and Recreational Activities: Implications for Tick-Borne Disease Transmission

Tick bites increase when people come in contact with nature during outdoor activities including traveling, especially during the tick season (Haditsch & Kunze, 2013). Tick bites are the second most common threat for people visiting the parks (Brady et al., 2022). People travel to regions where ticks are in high population, and their bites can lead to the development of diseases that may be unfamiliar in their home countries (Gillingham et al., 2020). Ticks acquired in Africa, South America, and Australia

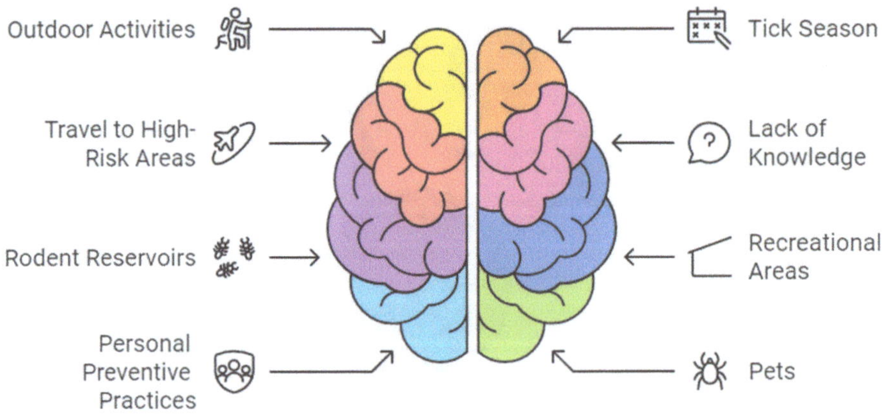

Fig. 9.5 Factors contributing to tick-borne transmission

now have been detected in Asia, Europe, the U.S.A., and New Zealand (Heath and Hardwick, 2011; McGarry, 2011; Pietzsch et al., 2014; Mathison et al., 2015; Pek et al., 2016; Molaei et al., 2018). Despite such kind of threat to the visitors, few visitors have sufficient knowledge about tick-borne diseases (Brady et al., 2024). In landscapes, rodents may be common because they give subadult ticks vital blood meals and act as natural reservoirs for numerous tick-borne illnesses, including Lyme borreliosis and tick-borne encephalitis (Vikentjeva et al., 2021). As the ticks occur mainly in forest areas, meadows, and tourist trails, the risk of contact with these ticks occurs when people visit these places for recreational purposes (Bartosik et al., 2008). Outdoor recreational areas situated in natural environments may carry a significant risk of tick infestation and tick-borne diseases (Fig. 9.5).

Adopting the personal preventive practices advised by the U.S. Centers for Disease Control and Prevention (CDC) is essential to lower the chance of tick-borne diseases when resting in environments in the absence of vaccinations. A few examples of personal protective actions are donning long sleeves and pants, tucking pants into socks, applying insect repellent to one's body and clothing, etc. Few individuals use these, despite their proven efficacy, thus they are not shielded from the danger of diseases carried by ticks (Omodior et al., 2015). To date, about 14 tick-borne diseases have been reported in international travelers. Notable diseases include Lyme disease, tick-borne relapsing fever, African tick bite fever, Mediterranean spotted fever (Eldin & Parola, 2018), rocky mountain spotted fever, tick typhus, anaplasmosis, ehrlichiosis, tularemia, babesiosis, tick-borne encephalitis, and tick paralysis (Jensenius et al., 2006). The other major factor to consider here is that people bring their pets, especially dogs and cats among them to recreational places. These companion animals also bring ticks and tick-borne diseases to non-endemic regions (Buczek & Buczek, 2020). However, the exact data regarding the transmission of tick-borne diseases is still unknown, and the majority of the cases remain undiagnosed (Haditsch & Kunze, 2013). As it is expected that recreational activities cannot be reduced, the only solution remains proper immunization, adapting personal protective behavior, and acquiring proper knowledge about the tick-borne diseases present in these recreational places (Fig. 9.5).

9.6 Influence on Labor Productivity and Workforce Health

A person's ability to work highly depends upon his health. Tick-borne diseases affect a person's health, thus limiting his ability to work. Companies have to allow their affected workers to take rest and manage fatigue. Stress should be minimized for an affected person (Leśniowska and Olejniczak, 2020). A person affected by a tick-borne disease has to be supported by the public and should be provided with disability benefits (Rochlin & Toledo, 2020). Since there is evidence linking rising air temperatures with the present or potential expansion of vector habitat, workers working outdoors are more vulnerable to vector-borne diseases. The evidence mostly relies on climate-driven model simulations for an evaluation of the influence on labor productivity (Levi et al., 2018). Public health professionals should be knowledgeable about the various tick species that are present within the region as well as the diseases they spread because they are on the front lines of tick-borne diseases in high-risk places (Rochlin & Toledo, 2020) (Fig. 9.6).

Climate change is believed to be causing climate-sensitive vector-borne diseases, which will influence occupational and public health. However, expansion may be influenced by additional natural and man-made factors such as increased travel and trade, reforestation and deforestation, altered land use, and urbanization. Outdoor workers like farmers, foresters, landscapers, groundskeepers, gardeners, painters,

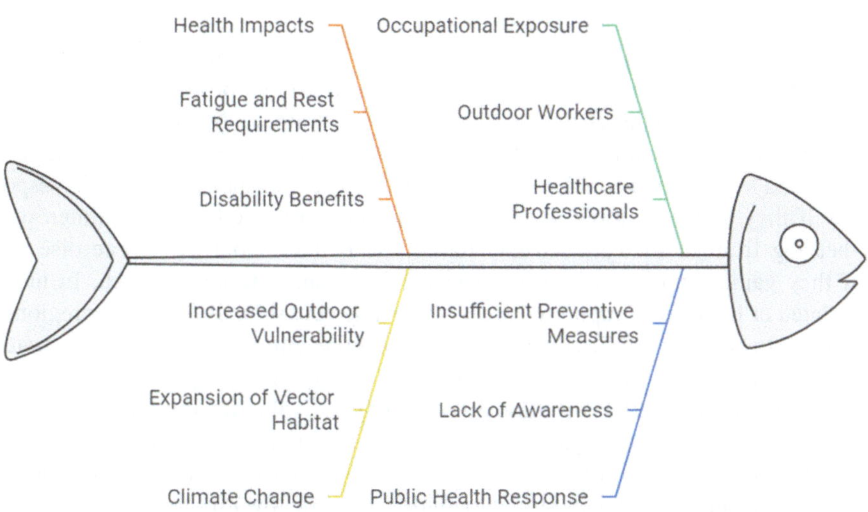

Fig. 9.6 Tick-borne diseases affecting labor productivity and workforce health

roofers, pavers, construction workers, laborers, mechanics, and any other individual who spends time outside represent the occupational category most affected by vector-borne diseases. Agriculture, forestry, fishing, building, mining, transportation, warehousing, utilities, and service sectors are among the industries that employ outdoor workers. Healthcare professionals and emergency responders who interact with infected patients run the danger of being exposed as well. Workers may be exposed to vector-borne diseases in places and during times where transmission was previously impossible as a result of climate change. Furthermore, adjustments to daily work schedules brought on by rising temperatures such as longer breaks during the day and more labor at dawn and dusk may coincide with the times of day when disease vectors are most active, raising the possibility of disease transmission (Vonesch et al., 2016). Tick-borne diseases are becoming more and more dangerous to human health, as tick spread continues to grow (Fig. 9.6).

Although there is little chance of contracting an illness carried by ticks after being bitten, the effects on health can be severe. Tick-related hazards are challenging for laypeople to evaluate due to these and other features of these diseases, and perceived risk might differ significantly from real risk (Slunge et al., 2019a, 2019b). It is also important to keep in mind that health practitioners, especially in endemic areas, have a poor level of knowledge of tick-borne diseases. This is valid for both CCHF in Eastern Europe and Lyme disease in North America. Surprisingly, considering the disease's infectious nature, 22% of Turkish healthcare professionals neglect to take any preventative action when interacting with CCHF patients (Mowbray et al., 2012).

9.7 Public Perception and Behavioral Responses to Tick-Borne Diseases Risks

Experts on tick-borne diseases may not be able to inform the public about scientific advancements if they are unable to communicate with them properly. The public is ultimately meant to benefit from the development of strategies by the medical, veterinary, and scientific communities to tackle novel or emerging tick-borne diseases. However, a lot of experts on tick-borne diseases may avoid speaking in public, especially if they believe a topic is politicized and the target audience is uninterested in hearing from them. Patients may have false beliefs about tick-borne diseases that they gained through television, social media, and internet research. In more populated or financially supported areas, there may occasionally be local or regional initiatives to enhance public education on tick bite prevention. However, these strategies frequently fall short in effectively reaching underprivileged populations, places where technology-based distribution is not an option. Evaluating the general public's knowledge of tick bite mitigation, resistance to insecticide use, ability to identify symptoms of tick-borne infections, and need for veterinary or medical consultation are all significant. It is also important to guide the public toward scientists who study tick identification (Sakamoto, 2018) (Fig. 9.7).

Fig. 9.7 Public perception and adoption of tick-borne disease prevention measures

The adoption of preventive measures is mostly dependent on people being aware of the risk of tick-borne diseases in the areas where they reside and travel. There is a shortage of information regarding public awareness of tick-borne diseases and their usage of preventative measures. Furthermore, several other significant tick-borne disease factors, such as the frequency of tick exposure and the behavior of seeking medical attention, have not been yet measured (Hook et al., 2015). Personal precautions that reduce the chance of tick contact, landscape changes or acaricidal treatments aimed at controlling tick populations on belongings, and community initiatives like treating landscape public parks are some strategies for preventing tick-borne diseases (Hinckley et al., 2016). Simple and affordable personal preventative measures include checking for ticks, applying insect repellent, and taking a bath after being in or near tick habitat (Niesobecki et al., 2019). The adoption and implementation of any preventative measure may be impacted by an individual's awareness and perspective of his likelihood of acquiring a disease carried by ticks. Current data on knowledge, perceptions, and application of the majority of feasible personal and property-based preventive strategies are lacking (Gould et al., 2008). Regular implementation of these preventive behaviors remains poor for people of endemic locations, even after decades of teaching about certain of these precautions, particularly tick checks and self-application of insect repellent (Connally et al., 2009). Compared to people with higher education levels, those with lower education levels are more likely to conduct a tick check. People with less education may be inclined to pursue professions that need them to go outside, including yard maintenance or landscaping, which means they will need to do tick checks more frequently. Higher earners and educated people can select more costly alternatives like property improvements, such as landscaping and lawn care. In the same way, those with lower earnings are more likely than those with higher incomes to apply insect repellent to themselves. Maybe those who have treated their yards believe they don't need to use personal insect repellent. As an alternative, those who don't usually use insect repellant could spray their yard (Niesobecki et al., 2019) (Fig. 9.7).

The concern and fear that livestock owners have regarding tick-borne diseases in their animals can be partially explained by their assessments of the perceived personal significance of tick presence, the possible negative effects on their daily lives, and their ability to cope by using various coping mechanisms (Johansson et al., 2020). People are more likely to conduct tick checks if they believe tick-borne diseases are very common in their neighborhood, if they or their family member have already received a tick-borne disease diagnosis, or if they think that tick exposure is likely near their house, cabin, or vacation home. In addition, the perceived risk of finding ticks around the house, lodge, or vacation home is linked to property-based insect management techniques. Additionally, people are more inclined to take preventative action if they have seen material from state health authorities. Forgetting, being unaware, and being worried about safety are the most often mentioned obstacles to utilizing any of these techniques (Beck et al., 2022). For those attempting to lower the prevalence of tick-borne diseases, increasing the public's adoption of preventive behaviors is still a significant problem (Mowbray et al., 2012) (Fig. 9.7).

9.8 Case Studies of Socio-Economic Consequences in Affected Communities

1. In addition to evaluating the socioeconomic effects of the dipping program, a study looks into whether the government's efforts to control tick-borne diseases and tick populations in South Africa's Northern Province are economically justified. It was predicted in a cross-sectional study conducted in the Northern Province's Venda region among 125 small-scale farmers. The cost–benefit analysis showed a ratio of 0.8 at a mortality rate of 4%, which is sensitive to changes in the mortality rate. This ratio suggests that there is no economic justification for the government to regulate tick populations and diseases carried by ticks. However, it might be argued that government funding for cattle dipping is still warranted due to the economic nature of the service it offers and the substantial socio-economic benefits to low-income people (Randela, 2007) (Fig. 9.8).
2. Although tick-borne diseases have a considerable financial impact, tick-borne diseases have little effect on the overall financial success of cattle ranching businesses. Tick control accounted for a significant amount of the economic costs. Due to ranchers' improvement of their cattle herds with tick-borne disease-susceptible dairy breeds, which led to high tick-borne disease mortality rates, the economic cost of tick-borne diseases was rising as ranches' herd sizes increased. Conversely, pastoralists raising indigenous cattle were benefiting from economies of scale, since the financial burden suffered from tick-borne diseases decreased as herd size increased (Ocaido et al., 2009).

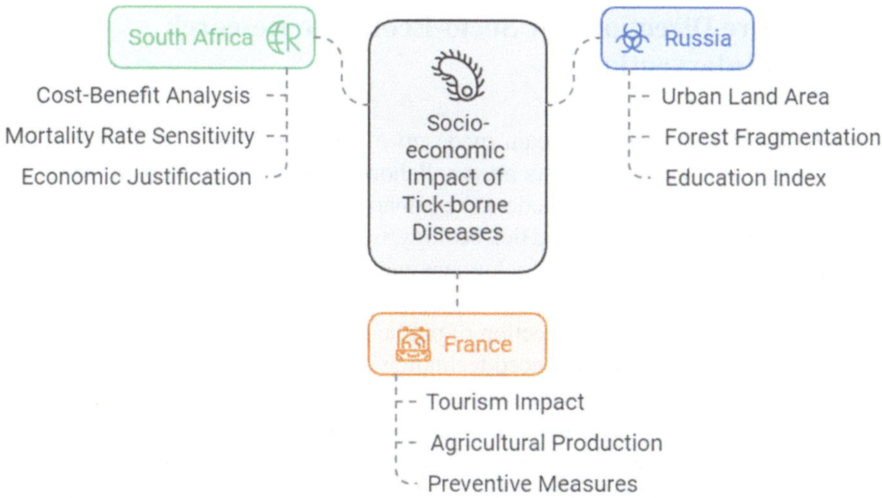

Fig. 9.8 Socio-economic consequences in affected communities in some countries

3. Tick-borne encephalitis (TBE) is an important tick-borne disease that Russia has been suffering from. A study conducted by Zhang et al., 2022 suggested that urban land area, forest fragmentation, and education index are the key socio-economic factors affecting the transmission of TBE in Russia (Zhang et al., 2022) (Fig. 9.8).
4. Due to its different landscapes and contrasting temperatures, the Occitanie region in southern France makes an excellent case study for examining the risk of tick-borne diseases within a socio-economic framework. Occitanie is a popular destination for vacationers and nature lovers because of its numerous regional and national parks, enormous mountain ranges, and Mediterranean beaches. The risk of tick-borne illnesses, however, might discourage future travelers. Occitanie, the region that produces the most ovine agriculture in France, is also a significant producer of agricultural products, accounting for almost 70% of the nation's milk supply and possessing a protected geographical designation for its well-known cheese. Tick bite prevention, together with early detection, management, and control strategies, are crucial to preventing serious health and economic implications, as tick-borne diseases impact all livestock species that are significant to the agriculture sector. Occitanie, however, has a major role in the use of acaricides as a preventive control measure for tick bites and diseases carried by ticks because it is the leading region for organic farming (Fig. 9.8). Therefore, while considering effective and regionally-adapted prevention, management, and control measures for tick bites and tick-borne diseases, important socio-economic stakes have to be taken into account (Zortman et al., 2023).

9.9 Future Directions for Socio-Economic Research and Intervention

An interdisciplinary research team made up of microbiologists, epidemiologists, agronomists, and social scientists must collaborate with local players in the region (such as farmers, livestock breeders, veterinarians, and environmental education associations) to better understand tick ecology, tick presence, and the general state of complex tick-borne disease risks for humans and animals in the area due to the socioeconomic concerns surrounding present and potentially arising tick-borne diseases (Zortman et al., 2023). The protection of public health depends on the development of highly skilled labor and advanced technology. It is significantly more difficult to raise awareness about the shifting danger of tick-borne diseases to stop the entrance of foreign infections because this is typically an unavoidable byproduct of increased trade and travel, and most often occurs by accident. Developing active surveillance of fatal infectious pathogens through the integration of veterinary and human health skills may be crucial in mitigating the incidence of diseases transmitted by ticks. History indicates that the effective management of diseases carried by ticks necessitates early detection, quick response, funding mobilization for the development of technical skills, and occasionally the use of harsh social measures (Khan, 2015). The complexity of tick-borne disease transmission dynamics, including the impact of larger societal contexts in which pathogen transmission occurs, should ideally be considered by the next generation of risk assessment techniques (Parham et al., 2015).

References

Abdelbaset, A. E., Nonaka, N., & Nakao, R. (2022). Tick-borne diseases in Egypt: A one health perspective. *One Health, 15*, Article 100443.

Abdelbaset, A. E., Kwak, M. L., Nonaka, N., & Nakao, R. (2023). Human-biting ticks and zoonotic tick-borne pathogens in North Africa: Diversity, distribution, and trans-Mediterranean public health challenges. *One Health, 16*, Article 100547.

Adehan, S. B., Adakal, H., Gbinwoua, D., Yokossi, D., Zoungrana, S., Toé, P., Ouedraogo, M., Gbaguidi, A. M., Adoligbé, C., Fandohan, A. B., & Hounmanou, G. (2018). West African cattle farmers' perception of tick-borne diseases. *EcoHealth, 15*, 437–449.

Avxentev, N., Avxentyeva, M., Platonov, A., Kolyasnikova, N., Gridneva, K., Dolgin, V., Titkov, A., & Derkach, E. V. (2013). Evaluating economic burden of tick-borne encephalitis. *Evidence from Russia. Value in Health, 16*(7), A347–A348.

Bajer, A., Beck, A., Beck, R., Behnke, J. M., Dwużnik-Szarek, D., Eichenberger, R. M., Farkas, R., Fuehrer, H. P., Heddergott, M., Jokelainen, P., & Leschnik, M. (2022). Babesiosis in Southeastern, Central and Northeastern Europe: An emerging and re-emerging tick-borne disease of humans and animals. *Microorganisms, 10*(5), 945.

Bartosik, K., Kubrak, T., Olszewski, T., Jung, M.. & Buczek, A. (2008). Prevention of tick bites and protection against tick-borne diseases in south-eastern Poland. *Annals of Agricultural and Environmental Medicine, 15*(2).

Beck, A., Bjork, J., Biggerstaff, B. J., Eisen, L., Eisen, R., Foster, E., Signs, K., Tsao, J. I., Kough, E., Peterson, M., & Schiffman, E. (2022). Knowledge, attitudes, and behaviors regarding tick-borne disease prevention in Lyme disease-endemic areas of the Upper Midwest. *United States. Ticks and Tick-Borne Diseases, 13*(3), Article 101925.

Brady, R. M., Lemieux, C. J., & Doherty, S. T. (2022). Linking visitor perceptions and behaviours related to ticks and lyme disease to risk management strategies in a protected areas context. *Journal of Outdoor Recreation and Tourism, 39*, Article 100515.

Brady, R. M., Lemieux, C. J., & Doherty, S. T. (2024). Ticks and lyme disease in natural areas: A segmentation analysis of visitor perceptions of risk and preferred communication strategies. *Journal of Outdoor Recreation and Tourism, 47*, Article 100794.

Buczek, A., & Buczek, W. (2020). Importation of ticks on companion animals and the risk of spread of tick-borne diseases to non-endemic regions in Europe. *Animals, 11*(1), 6.

Charles, R. A., Bermúdez, S., Banović, P., Alvarez, D. O., Díaz-Sánchez, A. A., Corona-González, B., Etter, E. M. C., Rodríguez González, I., Ghafar, A., Jabbar, A., & Moutailler, S. (2021). Ticks and tick-borne diseases in Central America and the Caribbean: A one health Perspective. *Pathogens, 10*(10), 1273.

Connally, N. P., Durante, A. J., Yousey-Hindes, K. M., Meek, J. I., Nelson, R. S., & Heimer, R. (2009). Peridomestic Lyme disease prevention: Results of a population-based case–control study. *American Journal of Preventive Medicine, 37*(3), 201–206.

de la Fuente, J., Estrada-Peña, A., Rafael, M., Almazán, C., Bermúdez, S., Abdelbaset, A. E., Kasaija, P. D., Kabi, F., Akande, F. A., Ajagbe, D. O., & Bamgbose, T. (2023). Perception of ticks and tick-borne diseases worldwide. *Pathogens, 12*(10), 1258.

Dell'Arciprete, A., Braunstein, J., Touris, C., Dinardi, G., Llovet, I., & Sosa-Estani, S. (2014). Cultural barriers to effective communication between Indigenous communities and health care providers in Northern Argentina: An anthropological contribution to Chagas disease prevention and control. *International Journal for Equity in Health, 13*, 1–10.

Eckman, M. H., Steere, A. C., Kalish, R. A., & Pauker, S. G. (1997). Cost effectiveness of oral as compared with intravenous antibiotic therapy for patients with early Lyme disease or Lyme arthritis. *New England Journal of Medicine, 337*(5), 357–364.

Eldin, C., & Parola, P. (2018). Update on tick-borne bacterial diseases in travelers. *Current Infectious Disease Reports, 20*, 1–9.

Folly, A. J., Dorey-Robinson, D., Hernández-Triana, L. M., Phipps, L. P., & Johnson, N. (2020). Emerging threats to animals in the United Kingdom by arthropod-borne diseases. *Frontiers in Veterinary Science, 7*, 20.

Garcia, K., Weakley, M., Do, T., & Mir, S. (2022). Current and future molecular diagnostics of tick-borne diseases in cattle. *Veterinary Sciences, 9*(5), 241.

Geebelen, L. (2021). Tick-borne diseases in Belgium: The incidence and economic burden of Lyme borreliosis.

George, J. E., Davey, R. B., & Pound, J. M. (2002). Introduced ticks and tick-borne diseases: The threat and approaches to eradication. *Veterinary Clinics: Food Animal Practice, 18*(3), 401–416.

Ghafar, A., Gasser, R. B., Abbas, T., Rehman, A., Gauci, C. G., & Jabbar, A. (2021). Ticks and tick-borne diseases of bovines in a smallholder livestock context: The Pakistani example. *Advances in Parasitology, 114*, 167–244.

Gilbert, L. (2021). The impacts of climate change on ticks and tick-borne disease risk. *Annual Review of Entomology, 66*(1), 373–388.

Gillingham, E. L., Cull, B., Pietzsch, M. E., Phipps, L. P., Medlock, J. M., & Hansford, K. (2020). The unexpected holiday souvenir: The public health risk to UK travellers from ticks acquired overseas. *International Journal of Environmental Research and Public Health, 17*(21), 7957.

Gould, L. H., Nelson, R. S., Griffith, K. S., Hayes, E. B., Piesman, J., Mead, P. S., & Cartter, M. L. (2008). Knowledge, attitudes, and behaviors regarding Lyme disease prevention among Connecticut residents, 1999–2004. *Vector-Borne and Zoonotic Diseases, 8*(6), 769–776.

Haditsch, M., & Kunze, U. (2013). Tick-borne encephalitis: A disease neglected by travel medicine. *Travel Medicine and Infectious Disease, 11*(5), 295–300.

Heath, A. C., & Hardwick, S. (2011). The role of humans in the importation of ticks to New Zealand: A threat to public health and biosecurity. *NZ Med. J, 124*, 67–82.

Henningsson, A. J., Malmvall, B. E., Ernerudh, J., Matussek, A., & Forsberg, P. (2010). Neuroborreliosis—an epidemiological, clinical and healthcare cost study from an endemic area in the south-east of Sweden. *Clinical Microbiology and Infection, 16*(8), 1245–1251.

Heylen, D. J., Kumsa, B., Kimbita, E., Frank, M. N., Muhanguzi, D., Jongejan, F., Adehan, S. B., Toure, A., Aboagye-Antwi, F., Ogo, N. I., & Juleff, N. (2023). Tick-borne pathogens and body condition of cattle in smallholder rural livestock production systems in East and West Africa. *Parasites & Vectors, 16*(1), 117.

Hinckley, A. F., Meek, J. I., Ray, J. A., Niesobecki, S. A., Connally, N. P., Feldman, K. A., Jones, E. H., Backenson, P. B., White, J. L., Lukacik, G., & Kay, A. B. (2016). Effectiveness of residential acaricides to prevent Lyme and other tick-borne diseases in humans. *The Journal of Infectious Diseases, 214*(2), 182–188.

Hook, S. A., Nelson, C. A., & Mead, P. S. (2015). US public's experience with ticks and tick-borne diseases: Results from national HealthStyles surveys. *Ticks and Tick-Borne Diseases, 6*(4), 483–488.

Hook, S. A., Jeon, S., Niesobecki, S. A., Hansen, A. P., Meek, J. I., Bjork, J. K., Dorr, F. M., Rutz, H. J., Feldman, K. A., White, J. L., & Backenson, P. B. (2022). Economic burden of reported Lyme disease in high-incidence areas, United States, 2014–2016. *Emerging Infectious Diseases, 28*(6), 1170.

Hotez, P. J., Damania, A., & Bottazzi, M. E. (2020). Central Latin America: Two decades of challenges in neglected tropical disease control. *PLoS Neglected Tropical Diseases, 14*(3), Article e0007962.

Hsia, E. C., Chung, J. B., Schwartz, J. S., & Albert, D. A. (2002). Cost-effectiveness analysis of the Lyme disease vaccine. *Arthritis & Rheumatism: Official Journal of the American College of Rheumatology, 46*(6), 1651–1660.

Hussain, S., Hussain, A., Ho, J., Li, J., George, D., Rehman, A., Zeb, J., & Sparagano, O. (2021). An epidemiological survey regarding ticks and tick-borne diseases among livestock owners in Punjab, Pakistan: A one health context. *Pathogens, 10*(3), 361.

Jensenius, M., Parola, P., & Raoult, D. (2006). Threats to international travellers posed by tick-borne diseases. *Travel Medicine and Infectious Disease, 4*(1), 4–13.

Johansson, M., Mysterud, A., & Flykt, A. (2020). Livestock owners' worry and fear of tick-borne diseases. *Parasites & Vectors, 13*, 1–11.

Johnson, N., Phipps, L. P., Hansford, K. M., Folly, A. J., Fooks, A. R., Medlock, J. M., & Mansfield, K. L. (2022). One Health approach to tick and tick-borne disease surveillance in the United Kingdom. *International Journal of Environmental Research and Public Health, 19*(10), 5833.

Joss, A. W. L., Davidson, M. M., Ho-Yen, D. O., & Ludbrook, A. (2003). Lyme disease—what is the cost for Scotland? *Public Health, 117*(4), 264–273.

Kasaija, P. D., Estrada-Peña, A., Contreras, M., Kirunda, H., & de la Fuente, J. (2021). Cattle ticks and tick-borne diseases: A review of Uganda's situation. *Ticks and Tick-Borne Diseases, 12*(5), Article 101756.

Katswara, T., & Mukaratirwa, S. (2021). Knowledge, attitudes and practices on African tick bite fever of rural livestock communities living in a livestock-wildlife interface area in the Eastern Cape Province of South Africa. *BMC Infectious Diseases, 21*(1), 497.

Khan, A., Muhammed, A. A., Nasreen, N., Iqbal, F., Cossio-Bayugar, R., ali Sha, S. S., Alanazi, A. D. & Zajac, Z. (2022). Tick-borne haemoparasitic diseases in small ruminants in Pakistan: Current knowledge and future perspectives. *Saudi Journal of Biological Sciences, 29*(4), pp.2014–2025.

Khan, M. A. H. N. A. (2015). Important vector-borne diseases with their zoonotic potential: present situation and future perspective. *Bangladesh Journal of Veterinary Medicine, 13*(2).

Klingelhöfer, D., Braun, M., Brüggmann, D., & Groneberg, D. A. (2022). Ticks in medical and parasitological research: Globally emerging risks require appropriate scientific awareness and action. *Travel Medicine and Infectious Disease, 50*, Article 102468.

Lantos, P. M., Brinkerhoff, R. J., Wormser, G. P., & Clemen, R. (2013). Empiric antibiotic treatment of erythema migrans-like skin lesions as a function of geography: A clinical and cost effectiveness modeling study. *Vector-Borne and Zoonotic Diseases, 13*(12), 877–883.

Leśniowska, J., & Olejniczak, M. (2020). Humanistic health management programs in anthropocene–a vector-borne disease case. *Organizacja i Zarządzanie: kwartalnik naukowy*, pp.59–73.

Levi, M., Kjellstrom, T., & Baldasseroni, A. (2018). Impact of climate change on occupational health and productivity: A systematic literature review focusing on workplace heat. *La Medicina Del Lavoro, 109*(3), 163.

Li, S., Juhász-Horváth, L., Trájer, A., Pintér, L., Rounsevell, M. D. A., & Harrison, P. A. (2018). Lifestyle, habitat and farmers' risk of exposure to tick bites in an endemic area of tick-borne diseases in Hungary. *Zoonoses and Public Health, 65*(1), e248–e253.

Li, Y., Wen, X., Li, M., Moumouni, P. F. A., Galon, E. M., Guo, Q., Rizk, M. A., Liu, M., Li, J., Ji, S., & Tumwebaze, M. A. (2020). Molecular detection of tick-borne pathogens harbored by ticks collected from livestock in the Xinjiang Uygur Autonomous Region. *China. Ticks and Tick-Borne Diseases, 11*(5), Article 101478.

Lohr, B., Müller, I., Mai, M., Norris, D. E., Schöffski, O., & Hunfeld, K. P. (2015). Epidemiology and cost of hospital care for Lyme borreliosis in Germany: Lessons from a health care utilization database analysis. *Ticks and Tick-Borne Diseases, 6*(1), 56–62.

Luan, Y., Gou, J., Zhong, D., Ma, L., Yin, C., Shu, M., Liu, G., & Lin, Q. (2023). The tick-borne pathogens: An overview of China's situation. *Acta Parasitologica, 68*(1), 1–20.

Mac, S., da Silva, S. R., & Sander, B. (2019). The economic burden of Lyme disease and the cost-effectiveness of Lyme disease interventions: A scoping review. *PLoS ONE, 14*(1), Article e0210280.

Mac, S. J. (2022). *Applying Health Technology Assessment Methodologies to Support Health Policy for Lyme Disease: Assessing Health and Economic Burden, and Cost-effectiveness of Potential Interventions* (Doctoral dissertation, University of Toronto (Canada).

Madison-Antenucci, S., Kramer, L. D., Gebhardt, L. L., & Kauffman, E. (2020). Emerging tick-borne diseases. *Clinical Microbiology Reviews, 33*(2), 10–1128.

Magid, D., Schwartz, B., Craft, J., & Schwartz, J. S. (1992). Prevention of Lyme disease after tick bites: A cost-effectiveness analysis. *New England Journal of Medicine, 327*(8), 534–541.

Mathison, B. A., Gerth, W. J., Pritt, B. S., & Baugh, S. (2015). Introduction of the exotic tick Hyalomma truncatum on a human with travel to Ethiopia: A case report. *Ticks and Tick-Borne Diseases, 6*(2), 152–154.

McGarry, J. W. (2011). Travel and disease vector ticks. *Travel Medicine and Infectious Disease, 9*(2), 49–59.

Minjauw, B., & McLeod, A. (2003). Tick-borne diseases and poverty: the impact of ticks and tick-borne diseases on the livelihoods of small-scale and marginal livestock owners in India and eastern and southern Africa.

Molaei, G., Andreadis, T. G., Anderson, J. F. & III, K. C. S. (2018). An exotic hitchhiker: a case report of importation into Connecticut from Africa of the human parasitizing tick, Hyalomma truncatum (Acari: Ixodidae). *The Journal of Parasitology, 104*(3), pp.302–305.

Mowbray, F., Amlôt, R., & Rubin, G. J. (2012). Ticking all the boxes? A systematic review of education and communication interventions to prevent tick-borne disease. *Vector-Borne and Zoonotic Diseases, 12*(9), 817–825.

Müller, I., Freitag, M. H., Poggensee, G., Scharnetzky, E., Straube, E., Schoerner, C., Hlobil, H., Hagedorn, H. J., Stanek, G., Schubert-Unkmeir, A., & Norris, D. E. (2012). Evaluating frequency, diagnostic quality, and cost of Lyme borreliosis testing in Germany: A retrospective model analysis. *Journal of Immunology Research, 2012*(1), Article 595427.

Muvhuringi, P.B., Murisa, R., Sylvester, D., Chigede, N., & Mafunga, K. (2022). Factors worsening tick borne diseases occurrence in rural communities. A case of Bindura district, Zimbabwe. *Cogent Food & Agriculture, 8*(1), p.2082058.

Namgyal, J., Tenzin, T., Checkley, S., Lysyk, T. J., Rinchen, S., Gurung, R. B., Dorjee, S., Couloigner, I., & Cork, S. C. (2021). A knowledge, attitudes, and practices study on ticks and tick-borne diseases in cattle among farmers in a selected area of eastern Bhutan. *PLoS ONE, 16*(2), Article e0247302.

Nchu, F., Nyangiwe, N., Muhanguzi, D., Nzalawahe, J., Nagagi, Y. P., Msalya, G., Joseph, N. A., Kimaro, E. G., Mollel, M., Temba, V., & Harouna, D. V. (2020). Development of a practical framework for sustainable surveillance and control of ticks and tick-borne diseases in Africa. *Veterinary World, 13*(9), 1910.

Niesobecki, S., Hansen, A., Rutz, H., Mehta, S., Feldman, K., Meek, J., Niccolai, L., Hook, S., & Hinckley, A. (2019). Knowledge, attitudes, and behaviors regarding tick-borne disease prevention in endemic areas. *Ticks and Tick-Borne Diseases, 10*(6), Article 101264.

Ocaido, M., Muwazi, R. T., & Opuda, J. A. (2009). Economic impact of ticks and tick-borne diseases on cattle production systems around Lake Mburo National Park in South Western Uganda. *Tropical Animal Health and Production, 41*, 731–739.

Ogden, N. H., Kotchi, S. O., Brazeau, S., Bouchard, C., White, J. C., Wulder, M. A., Davidson, A., Beaudoin, A., & Werle, D. (2022). Tick-borne diseases. *Earth observation, public health and one health: Activities, challenges and opportunities* (pp. 26–37). CABI.

Omodior, O., Pennington-Gray, L., & Donohoe, H. (2015). Efficacy of the Theory of Planned Behavior in Predicting the Intention to Engage in Tick-Borne Disease Personal Protective Behavior Amongst Visitors to an Outdoor Recreation Center. *Journal of Park & Recreation Administration, 33*(2).

Ortiz, D. I., Piche-Ovares, M., Romero-Vega, L. M., Wagman, J., & Troyo, A. (2021). The impact of deforestation, urbanization, and changing land use patterns on the ecology of mosquito and tick-borne diseases in Central America. *Insects, 13*(1), 20.

Oskam, C., Ronai, I., & Irwin, P. (2021). The emergence of tick-borne diseases in domestic animals in Australia. *Climate, ticks and disease* (pp. 424–429). CABI.

Parham, P. E., Waldock, J., Christophides, G. K., Hemming, D., Agusto, F., Evans, K. J., Fefferman, N., Gaff, H., Gumel, A., LaDeau, S., & Lenhart, S. (2015). Climate, environmental and socio-economic change: Weighing up the balance in vector-borne disease transmission. *Philosophical Transactions of the Royal Society B: Biological Sciences, 370*(1665), 20130551.

Pek, C. H., Cheong, C. S. J., Yap, Y. L., Doggett, S., Lim, T. C., Ong, W. C., & Lim, J. (2016). Rare cause of facial palsy: Case report of tick paralysis by Ixodes holocyclus imported by a patient travelling into Singapore from Australia. *The Journal of Emergency Medicine, 51*(5), e109–e114.

Pfäffle, M., Littwin, N., Muders, S. V., & Petney, T. N. (2013). The ecology of tick-borne diseases. *International Journal for Parasitology, 43*(12–13), 1059–1077.

Pietzsch, M., Hansford, K., Medlock, J., & Doggett, S. L. (2014). Australian paralysis tick imported on a traveller returning to the UK. *Travel Medicine and Infectious Disease, 12*, 196–197.

Ramzan, M., Murtaza, G., Sattar, A., Munawar, N., Ullah, A., Ejaz, A., Ayaz, F., Anwar, S., Jameel, K. and Kamran, F. (2021). Techniques for managing ticks and tick-borne diseases under changing climate; A review. *Egyptian Academic Journal of Biological Sciences, B. Zoology, 13*(1), pp.117–128.

Randela, R. (2007). Socio-economic impact of the control of ticks and tick-borne diseases in South Africa. *Impact of science on African agriculture and food security* (pp. 221–235). CABI.

Randolph, S. E. (2008). Tick-borne disease systems. *Rev Sci Tech off Int Epiz, 27*(2), 1–15.

Rochlin, I., & Toledo, A. (2020). Emerging tick-borne pathogens of public health importance: A mini-review. *Journal of Medical Microbiology, 69*(6), 781–791.

Rodino, K. G., Theel, E. S., & Pritt, B. S. (2020). Tick-borne diseases in the United States. *Clinical Chemistry, 66*(4), 537–548.

Sakamoto, J. M. (2018). Progress, challenges, and the role of public engagement to improve tick-borne disease literacy. *Current Opinion in Insect Science, 28*, 81–89.

Scaggiante, R., Guadagni, L., Orso, M., d'Angela, D., Carrieri, C., Polistena, B., Spandonaro, F., Bertoldi, I., Pilz, A., Schley, K., & Iantomasi, R. (2024). Incidence, healthcare resource

utilization and costs of hospitalized patients with tick-borne encephalitis (TBE) in Italy. *Ticks and Tick-Borne Diseases, 15*(6), Article 102372.
Shadick, N. A., Liang, M. H., Phillips, C. B., Fossel, K., & Kuntz, K. M. (2001). The cost-effectiveness of vaccination against Lyme disease. *Archives of Internal Medicine, 161*(4), 554–561.
Simuunza, M., Weir, W., Courcier, E., Tait, A., & Shiels, B. (2011). Epidemiological analysis of tick-borne diseases in Zambia. *Veterinary Parasitology, 175*(3–4), 331–342.
Singh, K., Kumar, S., Sharma, A. K., Jacob, S. S., RamVerma, M., Singh, N. K., Shakya, M., Sankar, M., & Ghosh, S. (2022). Economic impact of predominant ticks and tick-borne diseases on Indian dairy production systems. *Experimental Parasitology, 243*, Article 108408.
Slunge, D., Sterner, T., & Adamowicz, W. (2019a). Valuation when baselines are changing: Tick-borne disease risk and recreational choice. *Resource and Energy Economics, 58*, Article 101119.
Slunge, D., Jore, S., Krogfelt, K. A., Jepsen, M. T., & Boman, A. (2019b). Who is afraid of ticks and tick-borne diseases? Results from a cross-sectional survey in Scandinavia. *BMC Public Health, 19*, 1–13.
Slunge, D., Boman, A., & Studahl, M. (2022). Burden of tick-borne encephalitis, Sweden. *Emerging Infectious Diseases, 28*(2), 314.
Šmit, R., & Postma, M. J. (2016). Vaccines for tick-borne diseases and cost-effectiveness of vaccination: A public health challenge to reduce the diseases' burden. *Expert Review of Vaccines, 15*(1), 5–7.
Šumilo, D., Bormane, A., Asokliene, L., Vasilenko, V., Golovljova, I., Avsic-Zupanc, T., Hubalek, Z., & Randolph, S. E. (2008). Socio-economic factors in the differential upsurge of tick-borne encephalitis in central and Eastern Europe. *Reviews in Medical Virology, 18*(2), 81–95.
Sungirai, M., Moyo, D. Z., De Clercq, P., & Madder, M. (2016). Communal farmers' perceptions of tick-borne diseases affecting cattle and investigation of tick control methods practiced in Zimbabwe. *Ticks and Tick-Borne Diseases, 7*(1), 1–9.
Valente, S. L., Wemple, D., Ramos, S., Cashman, S. B., & Savageau, J. A. (2015). Preventive behaviors and knowledge of tick-borne illnesses: Results of a survey from an endemic area. *Journal of Public Health Management and Practice, 21*(3), E16–E23.
van den Wijngaard, C. C., Hofhuis, A., Wong, A., Harms, M. G., de Wit, G. A., Lugnér, A. K., Suijkerbuijk, A. W., Mangen, M. J. J., & van Pelt, W. (2017). The cost of Lyme borreliosis. *The European Journal of Public Health, 27*(3), 538–547.
Vikentjeva, M., Bragina, O., & Geller, J. (2021). Ticks and tick-borne pathogens in popular recreational areas in Tallinn, Estonia: an underestimated risk of tick-borne diseases.
Vonesch, N., D'Ovidio, M. C., Melis, P., Remoli, M. E., grazia Ciufolini, M., & Tomao, P. (2016). Climate change, vector-borne diseases and working population. *Annali dell'Istituto superiore di sanita, 52*(3), pp.397–405.
Wilke, A. B., Beier, J. C., & Benelli, G. (2019). Complexity of the relationship between global warming and urbanization–an obscure future for predicting increases in vector-borne infectious diseases. *Current Opinion in Insect Science, 35*, 1–9.
Wormser, G. P., Levin, A., Soman, S., Adenikinju, O., Longo, M. V., & Branda, J. A. (2013). Comparative cost-effectiveness of two-tiered testing strategies for serodiagnosis of Lyme disease with noncutaneous manifestations. *Journal of Clinical Microbiology, 51*(12), 4045–4049.
Zhang, L., Yang, L., Wang, L., Gu, L., Li, H., & Malkhazova, S. (2022). Natural and socioeconomic conditions influence tick-borne encephalitis cases in Russia. *Polish Journal of Environmental Studies, 32*(1), 427–437.
Zortman, I., de Garine-Wichatitsky, M., Arsevska, E., Dub, T., Van Bortel, W., Lefrançois, E., Vial, L., Pollet, T. and Binot, A., (2023). A social-ecological systems approach to tick bite and tick-borne disease risk management: Exploring collective action in the Occitanie region in southern France. *One Health*, p.100630.

Chapter 10
Adaptations Strategies for Tick Control in the Face of Climate Change

Ali Butt, Muhammad Sohail Sajid, Olivier Andre Sparagano, Muhammad Haziq Bajwa, and Muhammad Kamran

Abstract Ticks (Arthropoda: Acari), being poikilothermic, are susceptible to climatic and weather conditions. Climate change (CC) directly affects the distribution of ticks by boosting their survivability through increasing temperature. CC can also affect the hosts' resilience and ability to cope with tick infestations and tick-borne diseases (TBDs). Ticks due to their physiological adaptations have a greater ability to transmit diseases as compared to other arthropod vectors. Integrated tick management is a sustainable strategy that combines various control methods to reduce tick populations and the risks of TBDs. Vegetation management, application of acaricides, and reduction of susceptible host population significantly reduce the threat of ticks and TBDs. Biological control agents such as fungi, entomopathogenic nematodes, parasitoids, and predators can also be applied for tick suppression. Chemical control methods remain the corner stone for the developing countries; however, acaricidal resistance is also becoming a challenge. Adopting an integrated approach including tick surveillance, preventive management, biological control methods, and possibly immunization can be advisable. Biomarkers derived from tick saliva aid in effective modern surveillance of TBDs and their control. Periodic active and passive surveillance preferably through engaging the stakeholders involving participatory methods, can be effective for risk assessment of ticks and

A. Butt
Livestock and Dairy Development Department, Lahore, Pakistan

M. S. Sajid (✉)
Department of Parasitology, Faculty of Veterinary Sciences, University of Agriculture, Faisalabad, Pakistan
e-mail: drsohailuaf@hotmail.com; drsohailuaf@uaf.edu.pk

Climate Change Graduate Group, Institute of Soil and Environmental Sciences, University of Agriculture, Faisalabad, Pakistan

O. A. Sparagano
Department of Infection and Public Health, City University of Hong Kong, Hong Kong, China

UKMC, Manchester, UK

M. H. Bajwa · M. Kamran
Faculty of Veterinary Science, KBCMA College of Veterinary and Animal Sciences, Narowal, Sub-Campus UVAS, Lahore, Pakistan

H. M. Rizwan and M. S. Sajid (eds.), *Ticks in a Changing Climate*, Biometeorology 7,
https://doi.org/10.1007/978-3-032-12259-9_10

TBDs which can help planning appropriate mitigation and adaptation strategies in preview of the changing climate.

10.1 Introduction

Since pre-industrial times, the land surface temperature has risen by 1.5 °C (Allen et al., 2018). However, land surface warming has exceeded the world's average in some locations and seasons. This has been especially apparent in the Arctic. For example, in June 2020, at Siberian weather station, 38 °C was recorded, although the average temperature is 20 °C. The most significant changes have occurred during the Arctic winter; over the past 50 years, e.g., winter temperatures in western Canada have increased by approximately 3–4 °C (Černý et al., 2021). The main cause of Earth's increasing temperature is an increase in the concentrations of greenhouse gases (GHGs) brought on by human activity, mainly carbon dioxide, methane, and nitrous oxide (Myhre et al., 2014). Particulate pollution, particularly sulfate aerosols, which directly reflect sunlight and boost cloud reflectivity, partially offsets the warming effects of GHGs. The wide variation in future climate change (CC) projections—nearly twofold—largely stems from the uncertainty about how clouds will influence global and regional warming patterns (Bellouin et al., 2020). Some scientific reports proved that earth is warming due to anthropogenic activities, as seen by the increased frequency and intensity of heatwaves, droughts, and heavy precipitation events (Allen et al., 2018).

Determining whether or not ticks and tick-borne diseases (TBDs) are impacted by CC is even more uncertain. Environmental factors have a significant impact on the tick behavior and physiology since they are poikilothermic (those animals whose internal body temperature varies with the temperature of their environment). Ticks have a relatively long life cycle, Ticks spend the majority of their life cycle off-host (Apanaskevich et al., 2014). Water is essential for tick survival. *Ixodidae* ticks use hygroscopic secretions from their salivary glands to collect atmospheric water vapor (Bryer, 2021). Blood is also essential for tick survival because it provides nutrients for development and egg laying (Guizzo et al., 2022). While majority tick species feed on any vertebrate, they typically exhibit a preference related to whether they are endophilic or exophilic, specialists or generalists (Nuttall, 2022).

Individual tick survival, tick development time, and tick host-seeking activity are important to direct climatic and weather sensitivities. Because of these sensitivities, a warming climate may boost tick survivability, shorten tick life cycles, and extend tick activity seasons in some areas (Ogden et al., 2021). CC may also have an indirect impact on tick distribution and abundance through host changes (Peralbo-Moren et al., 2022). Changes in hosts brought on by CC are unlikely to have a significant effect on *Ixodes (I.) scapularis* and *I. ricinus*, which can reach higher levels of infestation and various species of ticks have been found co-feeding at the same time on one host. However, we may anticipate an important indirect influence on ticks if

CC negatively affects number of host populations, especially for more host-specific ticks like the long-legged bat tick *I. vespertilionis* (Gilbert, 2021).

Ticks after mosquitoes, far outperform all other arthropods in their ability to spread a wide range of pathogens, such as bacteria, viruses, and parasites (Singh, 2013). Among these, viral diseases spread by ticks have drawn increasing attention from the veterinary and medical communities (Vayssier-Taussat et al., 2015). The growing geographic range of tick species, outbreaks in previously unaffected areas brought on by migratory birds, changes in global climatic and socioeconomic conditions, and the lack of efficient control techniques are the main causes of this concern (Nepveu-Traversy et al., 2024). Enhancing extension services and educating farmers about tick control practices, using acaricides appropriately, developing and implementing vaccines to prevent TBDs, and monitoring and surveillance for TBDs and emerging acaricide resistance are all more sustainable and successful approaches for tick control (Kasaija et al., 2021).

10.2 Integrated ticks management approaches

Integrated tick management (ITM) represents a sustainable strategy that combines multiple control methods (Fig. 10.1) to reduce tick populations and the risks of TBDs (Makwarela et al., 2025). However, relatively few integrated management studies have been conducted, especially focusing on *I. scapularis* and *Amblyomma (A.) americanum* (Eisen & Dolan, 2016). Earlier tick management efforts in non-agricultural settings primarily concentrated on the use of acaricides, vegetation control, and exclusion of white-tailed deer to manage *A. americanum* populations (Stafford & Williams, 2017). A study in a recreational area demonstrated that employing these strategies either singly or in combination significantly suppressed tick abundance, with combinations being notably more effective than single approaches (Eisen & Stafford, 2021). Although deer exclusion fencing required high initial investment, it proved to be the most cost-effective method when evaluated over the projected ten-year lifespan of the structures (Ament et al., 2022). In contrast, annual vegetation management, particularly when combined with two rounds of acaricide applications, proved to be the most expensive option (Nath et al., 2018). Notably, neither acaricides nor vegetation management alone were sufficient to reduce *A. americanum* densities below economically significant thresholds (Villanueva, 2021).

Tick density played a critical role in determining the effectiveness of interventions. The highest control rates, between 92 and 99%, were achieved when all three strategies (use of acaricides, vegetation control, and exclusion of white-tailed deer) were combined, especially in areas with high tick densities (Makwarela et al., 2025). Due to the long life cycle of ticks and continual introduction of engorged sub-adults by birds and small mammals, deer exclusion alone was less effective for reducing nymphal and adult tick populations (Ferrari et al., 2023).

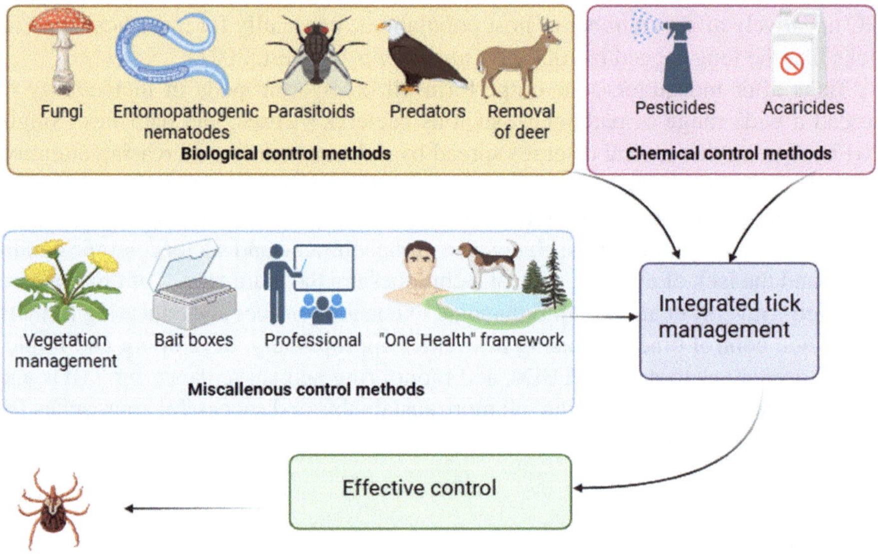

Fig. 10.1 Application of integrated tick management strategies to control ticks and tick-borne diseases

In Millstone, New Jersey, Schulze et al. (2017) implemented a combination of deer-focused interventions, rodent-targeted bait boxes containing fipronil, and granular deltamethrin barriers. These integrated strategies validated earlier computer simulation models designed to evaluate the population dynamics and management strategies of *I. scapularis*, *Dermacentor (D.) variabilis*, and *A. americanum* under varying environmental conditions (Mount et al., 1999). The results of the simulations suggested that the combined vegetation reduction with two applications of acaricides presented the greatest reduction in tick population over the short term (Stafford et al., 2017). Simulated outcomes aligned well with field observations, indicating that with full implementation, nymphal *A. americanum* populations could fall below tolerance thresholds within eight years, or even be eradicated if deer populations were entirely eliminated (White, 2019). Achieving long-term control, however, required repeated interventions, particularly multiple acaricide applications, followed by sustained ecological management strategies (Assouguem et al., 2022).

Similar findings were obtained using the LYMESIM model for *I. scapularis* (Mount et al., 1997). Strategies such as reducing deer density, clearing vegetation, and acaricide treatments for mice and deer were evaluated (Dolan et al., 2004). In small recreational or residential areas, vegetation management and area-wide acaricide application were found to be effective for short-term control, while treating deer proved to be the most cost-effective strategy for larger landscapes (Telford, 2017). Combining two methods often reduced densities of infected nymphs by more than 90%, below the predicted thresholds for disease transmission (Mandii et al.,

2021). Hayes et al. (1999) applied the LYMESIM model to a hypothetical residential community, demonstrating that, apart from the hypothetical optimal use of Lyme disease vaccine, treating deer was the most effective single intervention for reducing Lyme disease cases. Nonetheless, deer management presents considerable practical and logistical challenges. Field studies from Mohegan Island, Maine, showed that complete elimination of white-tailed deer led to a dramatic reduction in *I. scapularis* populations within three years, although some reintroduction via migratory birds was noted (Rand et al., 2010).

The development and application of ITM strategies have been significantly enhanced by recent advances in modeling tick population dynamics. Newer models incorporate updated data on tick-host interactions, the effectiveness of intervention combinations, human exposure factors, and the use of personal protective and environmental measures (Estrada-Peña et al., 2022). Although no human vaccines are widely available against most TBDs, integrated management approaches are considered the most effective strategies for controlling ticks and reducing disease transmission (Ghosh et al., 2007).

A successful ITM program must also include continuous education and training for forest rangers, ecologists, and the public and health professionals (Garg, 2017). A "One Health" approach, which integrates human, animal, and environment health, is crucial for managing tick-associated diseases. Since many tick-borne pathogens are zoonotic, affecting humans, livestock, wildlife, and companion animals, communication and collaboration between physicians and veterinarians are vital (Dantas-Torres et al., 2012). Although the One Health concept is not new, it has gained renewed relevance in addressing vector-borne and zoonotic diseases (Shaheeh, 2022). Control measures targeting ticks on companion animals should be an integral part of any ITM strategy. Pets not only serve as tick hosts but also act as sentinels for human disease risk. For instance, in Arizona, during a Rocky Mountain spotted fever outbreak among an American Indian community, a community-based program successfully suppressed *Rhipicephalus (R.) sanguineus* populations by promoting animal care practices and using long-acting tick collars (Drexler et al., 2014).

Despite the increasing threat of TBDs, there remains a substantial gap in medical knowledge. A knowledge, attitude, and behavior (KAB) survey among primary care physicians revealed limited awareness regarding tick-bite management, diagnostic criteria, and empirical treatment of Lyme disease (Magri et al., 2002). Furthermore, medical and veterinary professionals often receive insufficient training in tick identification (Dantas-Torres et al., 2012). To effectively address the growing challenges of tick-associated diseases, multidisciplinary expertise are essential. Entomologists, vector ecologists, public health officials, modelers, pest management professionals, and veterinarians must collaborate to develop and implement ITM strategies. Effective management targets various aspects, including host-seeking ticks, ticks feeding on reservoir animals, pathogen prevalence in ticks and hosts, and direct human-tick interactions (Eisen & Gray, 2016).

Challenges remain, including gaps in understanding human behavior related to tick exposure, the ecology of tick-pathogen systems, cost-effectiveness of interventions, and public acceptance of available tools (Eisen & Padddock, 2021). While some measures, such as acaricide application and vegetation management are accessible to private homeowners. Others, such as host-targeted technologies and deer population management, may require broader community involvement and investment (Černý et al., 2020). Several new technologies show promise for the future. For example, an oral Lyme disease vaccine targeting rodents is nearing regulatory approval and could soon become a valuable tool in disease control (Richer et al., 2014). However, many interventions remain experimental or underutilized. Ultimately, cost-sharing among households and municipalities may become necessary to implement area-wide tick control programs effectively. Despite emerging technologies, residential tick management is likely to continue depending heavily on the area-wide application of acaricides, with cost remaining a critical consideration (Stafford et al., 2017).

10.3 Biological Control Agents for Tick Suppression

With respect to changing climatic conditions, biological control agents are advancing strategies for the sustainable suppression of ticks. Biological control began in the early twentieth century with dispatching multiple pathogens, parasitoids, and predators (Ostfeld et al., 2006) and has often been incorporated into Integrated Pest Management (IPM) programs (Baker et al., 2020). Biological control of ticks in the USA began with the importation of parasitoid wasps (Ostfeld et al., 2006). Along with other areas of Africa that had not received oxpecker birds previously, Botes (2019) stated that oxpeckers, who feed on ticks, were also reintroduced in those areas.

Microbial control has made significant progress, with the Bio-Pesticide Manual now listing numerous commercially available active ingredients derived from microorganisms (Chaudhary et al., 2024). Although viral agents have not yet been widely implemented for tick control, promising studies are emerging. One such example is the development of a recombinant baculovirus engineered to express a tick chitinase gene, demonstrating potential as a biological control strategy (Assenga et al., 2006).

10.3.1 Fungi

Among the biological agents, entomopathogenic fungi have shown considerable promise. Over 700 species have been identified, although only about ten have been developed for insect control (Sharma et al., 2020). Fungi from the class Deuteromycetes (*Fungi Imperfecti*) are the most promising for tick control. Their spores adhere to the tick cuticle, germinate within 6–24 h post-infection, penetrate the cuticle within 2–5 days, and ultimately kill the tick by proliferating internally (Ahmad et al., 2019). Mortality is associated with massive fungal growth in the haemolymph and the production of toxins such as proteases and oxalic acid, which destroy internal tissues, including the Malpighian tubules and intestinal walls (Ma et al., 2024). A review on entomopathogenic fungi against arthropods was published by Rajput et al. (2024).

10.3.2 Entomopathogenic Nematodes (EPNs)

EPNs, particularly from the families Heterorhabditidae and Steinernematidae, are parasitic to a variety of insect species. Their infective juveniles enter the host through natural openings, release symbiotic bacteria, and kill the host within 24–72 h (Mabena, 2023). The nematodes multiply inside the host, and thousands of new infective juveniles emerge within few days (Vashisth et al., 2013). EPNs are remarkably adaptable to different soil conditions, making them suitable for varied climates against insect pests (Matuska-Łyżwa et al., 2024).

10.3.3 Parasitoids

Most parasitoids belong to the order Hymenoptera are utilized in pest management. They have been reported for over two-thirds of successful biological control cases (Taylor et al., 2012). However, only a few parasitoids specifically target ticks. Notably, *Ixodiphagus* species, belonging to the family Encyrtidae, are known to parasitize ticks (Ramos et al., 2023).

10.3.4 Predators

General predators such as ants, beetles, and birds also contribute to natural tick suppression (Lefèvre et al., 2022). However, because their populations do not directly depend on tick abundance, their impact on tick populations is typically limited (Samish, 2004). Manipulating predator populations for tick control could have unintended consequences on non-target species in natural ecosystems (Quadros et al., 2020).

While biological control agents present a valuable tool for tick management, especially under the pressures of climate change, their application must be carefully integrated into broader ecological and pest management strategies to ensure effectiveness and ecological balance (Makwarela et al., 2025).

10.4 Chemical Control Strategies: Pesticides and Acaricides

Ectoparasites significantly impact the productivity of domestic livestock worldwide (Muhammad et al., 2021). The extent of their effect is influenced by multiple factors, including host availability, climatic conditions, early diagnosis, the adaptive responses of hosts and the effectiveness of intervention strategies (Wall & Ellse, 2011). Among the major ectoparasites of livestock are arthropods such as lice, flies, fleas, mites and ticks (Akintunde & Adegoke, 2023). These parasites cause harm either directly through tissue damage and blood-feeding or indirectly by irritation that disrupts grazing and reproduction or by transmitting diseases (Behar et al., 2023).

Currently, the livestock sector faces critical challenges, including the need to enhance productivity to meet the growing global demand for protein. At the same time, reducing reliance on chemical treatments is essential to meet market preferences for low-chemical or chemical-free products. It is also important to minimize occupational and environmental exposure to pesticides (Davis, 2012). Despite the implementation of integrated control programs for many livestock parasites, chemical treatment remains a cornerstone in the management of ectoparasites (Lifschitz et al., 2024).

The main chemical groups used for ectoparasite control include organophosphates, synthetic pyrethroids, macrocyclic lactones, carbamates, phenylpyrazoles, neonicotinoids, and various growth regulator compounds (Singh & Randhawa, 2013). These chemicals are administered through different methods such as spot-on or pour-on formulations, dips, sprays, insecticidal ear tags, systemic injectables, oral drenches or feed additives. These chemicals are also use to treating the environment or livestock housing to control ticks (Lavy et al., 2022).

A major challenge in chemical control is the development of resistance among ectoparasite populations, which undermines the efficacy of available products (Küntüz et al., 2023). Moreover, many new chemical classes cannot be incorporated due to the expensive costs that come with developing and registering them. These new challenges are made worse by climate change, which ranges many ectoparasites into new areas and regions while also increasing their impact (Bett et al., 2017).

The epidemiological importance of ticks is only second to mosquitoes amongst blood-sucking arthropods. (Mullen & Durden, 2009). Ticks are parasitic on a wide range of vertebrate hosts and are responsible for transmitting several diseases, including East Coast and Texas cattle fevers (Gray et al., 2019). TBDs affect approximately 80% of the global cattle population, particularly in tropical and subtropical regions (Cooke et al., 2020). Ticks cause substantial indirect economic losses through disease transmission and direct losses through blood-feeding activities that lead to wounds, udder damage, reduced animal performance, and, in severe cases, death of calves deprived of adequate nutrition from heavily infested dams (Yadav & Upadhyay, 2024).

10.5 Novel Technologies for Tick Surveillance and Monitoring

As ticks continue to expand their geographical range globally (Omazic et al., 2023), effective surveillance and monitoring strategies are essential to assess their impact on human and domestic animal populations (Fig. 10.2). One possible mechanism for the introduction of ticks into new areas is through migratory birds, which has been suggested in the recent emergence of *Hyalomma* ticks (important vectors of Crimean-Congo hemorrhagic fever virus) in southern France (Bernard et al., 2024; Vial et al., 2016). Meanwhile, traditional control methods focusing on chemical pesticides or environmental interventions are showing decreasing efficacy, largely due to the evolution of pesticide resistance and changes in land use patterns (Wimms et al., 2023). These challenges underscore the urgent necessity for modern technological innovations and international collaboration in surveillance, diagnostic, and control efforts (Tiffin et al., 2022).

10.5.1 *Tick Biomarkers*

Biomarkers derived from tick saliva have emerged as promising tools for monitoring tick exposure and the threat posed by tick-borne pathogens (Dziedziech et al., 2024).

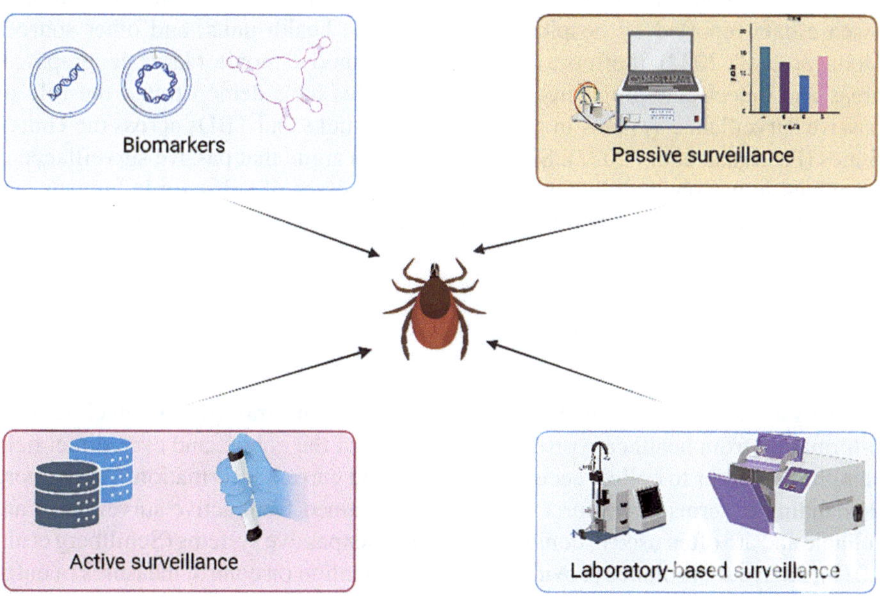

Fig. 10.2 Novel technologies for tick surveillance and monitoring

These biomarkers can improve understanding of host–vector contact, enhance evaluations of anti-vector interventions, and reveal shifts in vector–pathogen ecology (de Angeli Dutra et al., 2022). Importantly, tick exposure biomarkers could also aid in the diagnosis of TBDs in animal and humans. Many Lyme disease patients, for example, do not recall a tick bite. Self-reported exposure is often unreliable. Biomarkers can provide more accurate evidence of tick exposure (Hook et al., 2021). The idea of using antibodies against tick saliva components as exposure biomarkers was first proposed in 1990, when researchers found associations between anti-tick saliva antibodies, pet ownership, and Lyme disease seropositivity (Buezo Montero, 2021). Since then, numerous immunogenic tick salivary proteins have been identified (Table 10.1), including serine protease inhibitors, calreticulin (CRT), lipocalins, salivary proteins (Salps), metalloproteases, statins, and cement-like proteins (Ali et al., 2022; Olajiga et al., 2021). However, only a limited number of these molecules have been evaluated for their potential use as reliable biomarkers. Moreover, the immune response varies among candidates, with differences in antigenicity when tested with IgM or IgG. For example, antibodies against the Rs24p protein have been suggested as indicators of *R. sanguineus* exposure in dogs, but these antibodies were detectable only for a short period after infestation (Inokuma et al., 2000).

10.5.2 Passive Surveillance

Passive surveillance involves the routine collection, analysis, and dissemination of disease data reported by hospitals, clinics, public health units, and other sources (Johnson et al., 2022). It offers a cost-effective method to monitor large geographical areas and detect emerging threats. Various studies have demonstrated the role of passive surveillance systems in gathering data on ticks and TBDs across the United States (Pasternak et al., 2022). Siegel et al. (2023) argue that passive surveillance is beneficial in narrowing the gap between reported cases of a disease in humans and entomological field data, thereby enabling more effective public health responses.

10.5.3 Active Surveillance

Active surveillance is defined as the process involving trained personnel seeking information from healthcare providers, members of the public, and even direct field sampling in order to collect accurate, reliable, and current information. While more demanding in terms of resources, the results obtained from active surveillance are reliable and are often used in conjunction with other passive systems (Schillberg et al., 2018). Active surveillance provides critical information on control measures of entomological risk factors like tick density and pathogen prevalence. This is especially true for the nymphal stage ticks, which are severely under-represented in passive collections (Siegel et al., 2023).

Table 10.1 List of tick saliva components (full proteins or peptide regions) tested as biomarkers for tick bite exposure in animal sera

Name	Protein or peptide	Animal	Tick species	Outcome	References
Calreticulin (CRT)	Protein, peptide	Rabbits, humans, and mice	*Haemaphysalis longicornis, Dermacentor variabilis, Amblyomma americanum, Ixodes ricinus Ixodes scapularis,*	Recombinant AaCRT indicative of *A. americanum* and *D. variabilis* exposure in rabbits and humans; AaCRT and DvCRT found in rabbit serum after feeding	Alarcon-Chaidez et al. (2006); Hai et al. (2013); Zheng et al. (2022)
α-Gal modified proteins	Protein	Humans	*Ixodes holocyclus, Amblyomma americanum*	Tick proteins with α-Gal modifications from tick saliva illicit specific immune response in humans exposed to ticks	Villar et al. (2021)
Lipocalin	Peptide	Rabbits	*Ixodes ricinus*	Most antigenic peptide in rabbit serum	Bensaoud et al. (2022)
Microplusin				Moderate antigenic response in rabbit serum	
TIL domain protein					
Two Thyropin Domain protein				Antigenic response in rabbit serum	
Secreted protein				This peptide was not antigenic in rabbits	

10.5.4 Laboratory-Based Surveillance

Laboratory-based surveillance targets the detection and characterization of tick-borne pathogens associated with a clinical sample. In the USA, surveillance through laboratory testing has mostly focused on Lyme disease (Bisanzio et al., 2020). In areas with high incidence rates, switching to laboratory-based surveillance using the recommended two-tier testing algorithm may better standardize processes and lessen the workload for public health departments (Rowan et al., 2023). Moreover, in some studies, laboratory-confirmed cases were approximately 1.2 times more than the figures reported by the CDC, which points towards significant under-reporting by systems that rely on reporting frameworks. By focusing on laboratory-based methods, public health resources can be directed toward preventive strategies, novel intervention studies, and alternative data interrogation for a more comprehensive understanding of TBD epidemiology (Kugeler et al., 2022).

10.6 Community Engagement and Public Education Initiatives

To assist health professionals and governmental agencies dealing with public health and TBDs, it is important to understand what Knowledge, Attitude and Practices (KAPs) are achieved within the targeted populations. For instance, a study in China focusing on medical staff and their understanding of tick-borne rickettsial diseases highlighted only 55.78% (from 1,206 participants) had knowledge related to such diseases but reached much higher scores of 90.09% and 90.83% for attitudes and practices, respectively (Shao et al, 2024). A similar study with Italian Occupational health physicians found that the knowledge related to tick-borne encephalitis was around 58%, with only 19% of those professionals recommending TBE vaccination for at-risk workers (Riccò et al, 2020). 50% of Veterinarians in Ohio would be able to identify tick stages while only 9.6% knew that all tick life stages needed a blood meal (Eleftheriou et al., 2023). When it comes to farmers, some studies in Pakistan (Rajput et al, 2023) showed that only 42.5% (out of 240 respondents) were removing ticks manually and 57% were using acaricides but with variations in the usage frequency. A similar study by Hussain et al. (2021) showed 28.6% (out of 112 respondents) of Punjabi farmers removing ticks manually and 45.5% using acaricides regularly.

Community-based adaptation strategies emerged due to the increasing demand for more applicable science in responding to climate change challenges (Kirkby et al., 2018). Community-based adaptation (CBA) approaches focus on empowering communities through planning, coping, and implementing self-sustaining impact strategies while addressing climate change impacts (Reid et al., 2009). In the context of tick control, CBA emphasizes the integration of traditional, local, and scientific knowledge to better understand climate change impacts, adaptive strategies, and vulnerabilities (McNamara & Buggy, 2017). Research plays a vital role in supporting CBA efforts by expanding understanding through systematic investigation. It also facilitates capacity building, knowledge mobilization, empowerment, and training. While communities often possess valuable knowledge about their local environments, they may not have access to future climate projections or vulnerability assessments. This gap highlights the need for collaborative approaches that combine scientific insights with local expertise (Reid et al., 2009).

CBA projects range widely from attempts to trigger debates on the risks of climate change and building general resiliency to government and donor-funded projects that focus on the implementation and evaluation of specific adaptation strategies (Schipper et al., 2014; Tschakert et al., 2014). Importantly, there is a growing effort to scale up local-level initiatives to inform national and international adaptation programming (Schipper et al., 2014; Reid & Huq, 2014). While CBA has been predominantly applied in low- and middle-income countries, its relevance is increasingly recognized in high-income countries, particularly among Indigenous and resource-dependent communities (Ford, 2012).

The researchers were able to make particular contributions to tick surveillance and control in the context of climate change via community partnerships (Fig. 10.3). The traditional or public health based approaches to tick surveillance tend to be expensive because of repeated sampling at small scales (Glass, 2019). With the help of community members and molecular diagnostic tools, traditional surveillance approaches can be modified to make the process more viable (Peeling & McNerney, 2014). Initiatives that enable the general public to participate actively in research, like the community–academic based tick surveillance partnerships, enhance understanding of tick ecology while helping to solve surveillance data problems (Lewis et al., 2018). Seifert et al., 2016, for example, described how school partnerships effectively motivated students to practice tick bite prevention. In addition to education, surveillance data collected by citizens can improve the accuracy of models predicting seasonal, or even daily activity patterns of ticks, which is helpful for planning in public health (Garcia-Marti et al., 2017).

Fostering education, scientific collaboration, and providing local resources to communities enables them to cope with climate change challenges, as is evident from the case study on community-managed tick surveillance (Dambach et al., 2024).

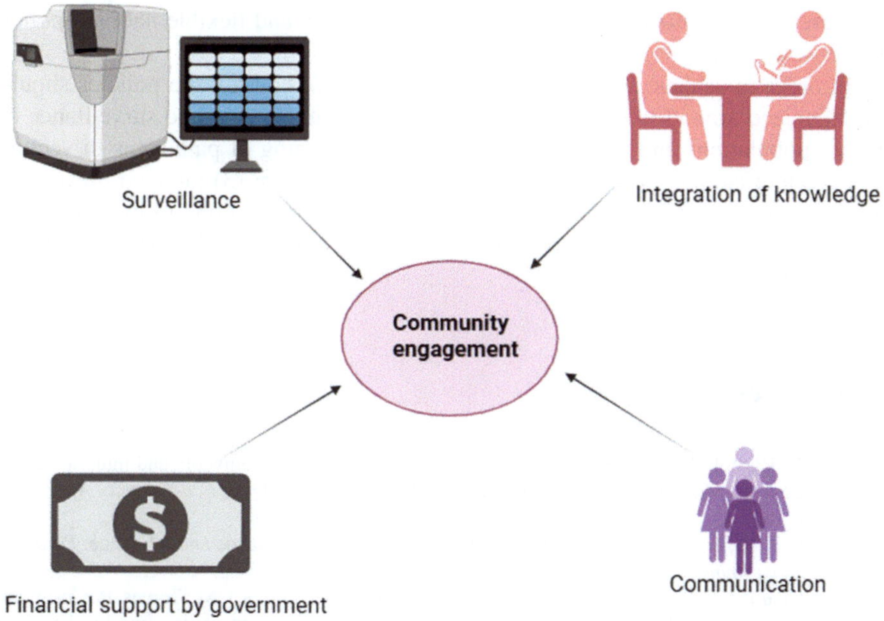

Fig. 10.3 Community-based adaptation approaches to control ticks and tick-borne diseases

10.7 Policy Implications and Regulatory Frameworks for Tick Control

As far as TBDs are concerned, climate change has transformed their distribution, seasonality, and intensity which creates the need to adopt effective policy and governance frameworks (Negev et al., 2015). In response to the worldwide challenge of vector-borne diseases, WHO initiated Global Vector Control Response 2017–2030 (GVCR). This initiative bears in mind that around 80% of the population worldwide is likely to be exposed to at least one VBD during their lifetime. Moreover, vector-borne diseases (VBDs) are responsible for over 700,000 deaths every year (WHO, 2017a). The GVCR focuses on integrated cross-sectoral approaches to improve surveillance and control, increase vector control efforts, and foster climate change adaptation to health vulnerabilities.

In alignment with the actions, WHO published the Framework for a National Vector Control Needs Assessment in 2017. This document assists evaluating the level of member countries' capabilities on VBD control providing highlights on the gaps that exist, thus aids in revising national policies in concordance with the GVCR's aims (WHO, 2017b).

The burden of VBDs including TBDs goes far beyond public health. These diseases have enormous socio-economic burdens by straining already stressed health care systems, deepening health inequalities, and negatively impacting critical economic sectors such as tourism, agriculture, and labor (Caminade et al., 2019). With climate change continuing to affect the transmission and geographic range of these diseases, there is an urgent need for effective and flexible national policy responses.

In order to achieve sustainable and comprehensive control, the policies should be organized around the following four pillars: (1) monitoring and surveillance of vector populations and their larval habitats; (2) screening of pathogens in vectors and sentinel animals or modern saliva capture techniques; (3) targeted prevention and control action; (4) ongoing assessment of the interventions to evaluate their effectiveness and determine appropriate subsequent strategies (Medlock et al., 2013; Estrada-Peña et al., 2017).

References

Ahmad, T., Rasool, A., Gull, S., Stephan, D., & Nabi, S. (2019). Ascomycota and integrated pest management. *Microbes for Sustainable Insect Pest Management: An Eco-Friendly Approach, 1*, 151–183.

Akintunde, J. K., & Adegoke, A. M. (2023). Ectoparasites host resistance and tolerance. *Parasitic Infections: Immune Responses and Therapeutics*, 124–140.

Alarcon-Chaidez, F., Ryan, R., Wikel, S., Dardick, K., Lawler, C., Foppa, I. M., Tomas, P., Cushman, A., Hsieh, A., Spielman, A., & Krause, P. J. (2006). Confirmation of tick bite by detection of antibody to Ixodes calreticulin salivary protein. *Clinical and Vaccine Immunology, 13*(11), 1217–1222.

Ali, A., Zeb, I., Alouffi, A., Zahid, H., Almutairi, M. M., Ayed Alshammari, F., Alrouji, M., Termignoni, C., Vaz Jr, I. D. S., & Tanaka, T. (2022). Host immune responses to salivary components-a critical facet of tick-host interactions. *Frontiers in Cellular and Infection Microbiology, 12*, Article 809052.

Allen, M., Coninck, H. D., Dube, O. P., Hoegh-Guldberg, O., Jacob, D., Jiang, K., Revi, A., Rogelj, J., Roy, J., Shindell, D., & Waterfield, T. (2018). Technical summary. In *Global warming of 1.5° C* 27–46. Intergovernmental Panel on Climate Change (IPCC).

Ament, R., Huijser, M., & May, D. (2022). *Animal Vehicle Collision Reduction and Habitat Connectivity Cost Effective Solutions-Final Report*. Nevada Department of Transportation.

Apanaskevich, D. A., Oliver, J. H., Sonenshine, D. E., & Roe, R. M. (2014). Life cycles and natural history of ticks. *Biology of Ticks, 1*, 59–73.

Assenga, S. P., You, M., Shy, C. H., Yamagishi, J., Sakaguchi, T., Zhou, J., Kibe, M. K., Xuan, X., & Fujisaki, K. (2006). The use of a recombinant baculovirus expressing a chitinase from the hard tick Haemaphysalis longicornis and its potential application as a bioacaricide for tick control. *Parasitology Research, 98*, 111–118.

Assouguem, A., Kara, M., Mechchate, H., Korkmaz, Y. B., Benmessaoud, S., Ramzi, A., Abdullah, K. R., Noman, O. M., Farah, A., & Lazraq, A. (2022). Current situation of Tetranychus urticae (Acari: Tetranychidae) in Northern Africa: The sustainable control methods and priorities for future research. *Sustainability, 14*(4), 2395.

Baker, B. P., Green, T. A., & Loker, A. J. (2020). Biological control and integrated pest management in organic and conventional systems. *Biological Control, 140*, Article 104095.

Behar, A., Yasur-Landau, D., & Leszkowicz-Mazuz, M. (2023). Vector-Borne Diseases in Ruminants. *Lester m. Shulman, 441*, 1–28.

Bellouin, N., Quaas, J., Gryspeerdt, E., Kinne, S., Stier, P., Watson-Parris, D., Boucher, O., Carslaw, K.S., Christensen, M., Daniau, A.L., & Stevens, B. (2020). Bounding global aerosol radiative forcing of climate change. *Reviews of Geophysics, 58*(1), e2019RG000660.

Bensaoud, C., Tenzer, S., Poplawski, A., Medina, J. M., Jmel, M. A., Voet, H., Mekki, I., Aparicio-Puerta, E., Cuveele, B., Distler, U., & Kotsyfakis, M. (2022). Quantitative proteomics analysis reveals core and variable tick salivary proteins at the tick-vertebrate host interface. *Molecular Ecology, 31*(15), 4162–4175.

Bernard, C., Kukla, C. J., Rakotoarivony, I., Duhayon, M., Stachurski, F., Huber, K., Giupponi, C., Zortman, I., Holzmuller, P., Pollet, T., & Vial, L. (2024). Detection of Crimean-Congo haemorrhagic fever virus in *Hyalomma marginatum* ticks, southern France, May 2022 and April 2023. *Eurosurveillance, 29*(6), 2400023.

Bett, B., Kiunga, P., Gachohi, J., Sindato, C., Mbotha, D., Robinson, T., Lindahl, J., & Grace, D. (2017). Effects of climate change on the occurrence and distribution of livestock diseases. *Preventive Veterinary Medicine, 137*, 119–129.

Bisanzio, D., Fernández, M. P., Martello, E., Reithinger, R., & Diuk-Wasser, M. A. (2020). Current and future spatiotemporal patterns of Lyme disease reporting in the Northeastern United States. *JAMA Network Open, 3*(3), e200319–e200319.

Botes, M. (2019). Foraging behaviour and health status of Red-billed Oxpeckers (*Buphagus erythrorhynchus)* in the Kruger National Park, South Africa.

Bryer, K. E. (2021). *Effects of Temperature and Humidity on the Mortality of the Tick Dermacentor Reticulatus in the UK* (Doctoral dissertation, University of Bristol).

Buezo Montero, S. (2021). Aedes albopictus 34k2. Salivary protein as epidemiological tool for the assesment of human exposure to the tiger mosquito.

Caminade, C., McIntyre, K. M., & Jones, A. E. (2019). Impact of recent and future climate change on vector-borne diseases. *Annals of the New York Academy of Sciences, 1436*(1), 157–173.

Černý, J., Elsterová, J., & Culler, L. (2021). Melting, melting pot-climate change and its impact on ticks and tick-borne pathogens in the Arctic. *Climate, ticks and disease* (pp. 460–468). CABI.

Černý, J., Lynn, G., Hrnková, J., Golovchenko, M., Rudenko, N., & Grubhoffer, L. (2020). Management options for Ixodes ricinus-associated pathogens: A review of prevention strategies. *International Journal of Environmental Research and Public Health, 17*(6), 1830.

Chaudhary, R., Nawaz, A., Khattak, Z., Butt, M. A., Fouillaud, M., Dufossé, L., Munir, M., ul Haq, I., & Mukhtar, H. (2024). Microbial bio-control agents: A comprehensive analysis on sustainable pest management in agriculture. *Journal of agriculture and food research*, 101421.

Cooke, R. F., Daigle, C. L., Moriel, P., Smith, S. B., Tedeschi, L. O., & Vendramini, J. M. (2020). Cattle adapted to tropical and subtropical environments: Social, nutritional, and carcass quality considerations. *Journal of Animal Science, 98*(2), skaa014.

Dambach, P., Louis, V. R., Standley, C. J., & Montenegro-Quiñonez, C. A. (2024). Beyond top-down: Community co-creation approaches for sustainable dengue vector control. *Global Health Action, 17*(1), 2426348.

Dantas-Torres, F., Chomel, B. B., & Otranto, D. (2012). Ticks and tick-borne diseases: A One Health perspective. *Trends in Parasitology, 28*(10), 437–446.

Davis, S. A. (2012). The movement for sustainable agricultural in Japan: A case study of post 3/11/2011 Organic Farmers in Saga, Fukuoka, Kagawa, and Hyogo Prefectures.

de Angeli Dutra, D., Poulin, R., & Ferreira, F. C. (2022). Evolutionary consequences of vector-borne transmission: How using vectors shapes host, vector and pathogen evolution. *Parasitology, 149*(13), 1667–1678.

Dolan, M. C., Maupin, G. O., Schneider, B. S., Denatale, C., Hamon, N., Cole, C., Zeidner, N. S., & Stafford, K. C. (2004). Control of immature Ixodes scapularis (Acari: Ixodidae) on rodent reservoirs of Borrelia burgdorferi in a residential community of southeastern Connecticut. *Journal of Medical Entomology, 41*(6), 1043–1054.

Drexler, N., Miller, M., Gerding, J., Todd, S., Adams, L., Dahlgren, F. S., Bryant, N., Weis, E., Herrick, K., Francies, J., & McQuiston, J. H. (2014). Community-based control of the brown dog tick in a region with high rates of Rocky Mountain spotted fever, 2012–2013. *PLoS ONE, 9*(12), Article e112368.

Dziedziech, A., Krupa, E., Persson, K. E., Paul, R., & Bonnet, S. (2024). Tick exposure biomarkers: A One Health approach to new tick surveillance tools. *Current Research in Parasitology & Vector-Borne Diseases, 6*, Article 100212.

Eisen, L., & Gray, J. S. (2016). 29. Lyme borreliosis prevention strategies: United States versus Europe. In *Ecology and prevention of Lyme borreliosis* (pp. 429–450). Wageningen Academic.

Eisen, L., & Dolan, M. C. (2016). Evidence for personal protective measures to reduce human contact with blacklegged ticks and for environmentally based control methods to suppress host-seeking blacklegged ticks and reduce infection with Lyme disease spirochetes in tick vectors and rodent reservoirs. *Journal of Medical Entomology, 53*(5), 1063–1092.

Eisen, L., & Stafford, K. C., III. (2021). Barriers to effective tick management and tick-bite prevention in the United States (Acari: Ixodidae). *Journal of Medical Entomology, 58*(4), 1588–1600.

Eisen, R. J., & Paddock, C. D. (2021). Tick and tickborne pathogen surveillance as a public health tool in the United States. *Journal of Medical Entomology, 58*(4), 1490–1502.

Eleftheriou, A., Swisher, S., Arruda, A., Berrian, A., & Pesapane, R. (2023). A survey of knowledge, attitudes, and practices of veterinary professionals regarding ticks and tick-borne diseases: Insights from Ohio, USA, *One Health, 17*(100592). https://doi.org/10.1016/j.onehlt.2023.100592.

Estrada-Peña, A., Gray, J. S., Kahl, O., Lane, R. S., & Nijhof, A. M. (2013). Research on the ecology of ticks and tick-borne pathogens—methodological principles and caveats. *Frontiers in Cellular and Infection Microbiology, 3*, 29.

Estrada-Peña, A., Mihalca, A., & Petney, T. N. (2017). *Ticks of Europe and North Africa*. Springer.

Estrada-Peña, A., Mallón, A. R., Bermúdez, S., De La Fuente, J., Domingos, A., García, M. P. E., Labruna, M. B., Merino, O., Mosqueda, J., Nava, S., Cruz, R. L., & Venzal, J. M. (2022). One health approach to identify research needs on Rhipicephalus microplus ticks in the Americas. *Pathogens, 11*(10), 1180.

Ferrari, G., Tagliapietra, V., Dagostin, F., Arnoldi, D., Inama, E., Rosso, F., Marini, G., & Rizzoli, A. (2023). Why account for biodiversity for mitigating tick-borne disease risk? Insights and perspectives from eastern Italian Alps. *JULIUS-KÜHN-ARCHIV, 473*, 97–98.

Ford, J. D. (2012). Indigenous health and climate change. *American Journal of Public Health, 102*(7), 1260–1266.

Garcia-Martí, I., Zurita-Milla, R., Van Vliet, A. J., & Takken, W. (2017). Modelling and mapping tick dynamics using volunteered observations. *International Journal of Health Geographics, 16*, 1–15.

Garg, B. S. (2017). Teaching/training and practice of public health/PSM/community medicine. *Indian Journal of Community Medicine, 42*(3), 127–130.

Ghosh, S. S. R. I. K. A. N. T. A., Azhahianambi, P. A. L. A. V. E. S. A. M., & Yadav, M. P. (2007). Upcoming and future strategies of tick control: A review. *Journal of Vector Borne Diseases, 44*(2), 79.

Gilbert, L. (2021). The impacts of climate change on ticks and tick-borne disease risk. *Annual Review of Entomology, 66*(1), 373–388.

Glass, C. S. (2019). *The relationships between livestock and human wealth, health, and wellbeing in a rural Maasai community of southwestern Kenya* (Doctoral dissertation, University of British Columbia).

Gray, J. S., Estrada-Peña, A., & Zintl, A. (2019). Vectors of babesiosis. *Annual Review of Entomology, 64*(1), 149–165.

Guizzo, M. G., Tirloni, L., Gonzalez, S. A., Farber, M. D., Braz, G., Parizi, L. F., Dedavid e Silva, L.A., da Silva Vaz Jr, I., & Oliveira, P. L. (2022). Coxiella endosymbiont of Rhipicephalus microplus modulates tick physiology with a major impact in blood feeding capacity. *Frontiers in Microbiology, 13*, 868575.

Hai, V. V., Pages, F., Boulanger, N., Audebert, S., Parola, P., & Almeras, L. (2013). Immunoproteomic identification of antigenic salivary biomarkers detected by Ixodes ricinus-exposed rabbit sera. *Ticks and Tick-Borne Diseases, 4*(5), 459–468.

Hayes, E. B., Maupin, G. O., Mount, G. A., & Piesman, J. (1999). Assessing the prevention effectiveness of local Lyme disease control. *Journal of Public Health Management and Practice, 5*(3), 84–92.

Hook, S. A., Nawrocki, C. C., Meek, J. I., Feldman, K. A., White, J. L., Connally, N. P., & Hinckley, A. F. (2021). Human-tick encounters as a measure of tickborne disease risk in Lyme disease endemic areas. *Zoonoses and Public Health, 68*(5), 384–392.

Hussain, S., Hussain, A., Ho, J., Li, J., George, D., Rehman, A., Zeb, J., & Sparagano, O. (2021). An epidemiological survey regarding ticks and tick-borne diseases among livestock owners in Punjab, Pakistan: A One Health Context. *Pathogens, 10*, 361. https://doi.org/10.3390/pathogens10030361

Inokuma, H., Ohno, K., & Onishi, T. (2000). Is the detection of anti-Rhipicephalus sanguineus (Rs24p) antibodies a valuable epidemiological tool of tick infestation in dogs? *Veterinary Research, 31*(3), 365–369.

Johnson, N., Phipps, L. P., Hansford, K. M., Folly, A. J., Fooks, A. R., Medlock, J. M., & Mansfield, K. L. (2022). One Health approach to tick and tick-borne disease surveillance in the United Kingdom. *International Journal of Environmental Research and Public Health, 19*(10), 5833.

Kasaija, P. D., Estrada-Peña, A., Contreras, M., Kirunda, H., & de la Fuente, J. (2021). Cattle ticks and tick-borne diseases: A review of Uganda's situation. *Ticks and Tick-Borne Diseases, 12*(5), Article 101756.

Kirkby, P., Williams, C., & Huq, S. (2018). Community-based adaptation (CBA): Adding conceptual clarity to the approach, and establishing its principles and challenges. *Climate and Development, 10*(7), 577–589.

Kugeler, K. J., Cervantes, K., Brown, C. M., Horiuchi, K., Schiffman, E., Lind, L., Barkley, J., Broyhill, J., Murphy, J., Crum, D., & Hinckley, A. F. (2022). Potential quantitative effect of a laboratory-based approach to Lyme disease surveillance in high-incidence states. *Zoonoses and Public Health, 69*(5), 451–457.

Küntüz, T., Güneş, Y., Sarı, A. B., & Keleş, O. Ü. (2023). Navigating the resistance: Current perspectives on ectoparasite control in veterinary medicine. *Journal of Istanbul Veterinary Sciences, 7*(2), 56–67.

Lavy, E., Kirmayer, D., Nudelman, Z., Orenshtein-Vilensky, L., Rowan, T. G., Shenderovich-Gefter, J., & Friedman, M. (2022). Aspects in controlled drug delivery for topical applications in veterinary medicine. *Veterinary and Animal Science, 15*, Article 100235.

Lefèvre, T., Sauvion, N., Almeida, R. P., Fournet, F., & Alout, H. (2022). The ecological significance of arthropod vectors of plant, animal, and human pathogens. *Trends in Parasitology, 38*(5), 404–418.

Lewis, J., Boudreau, C. R., Patterson, J. W., Bradet-Legris, J., & Lloyd, V. K. (2018). Citizen science and community engagement in tick surveillance—A Canadian case study. In *Healthcare*. MDPI.

Lifschitz, A., Nava, S., Miró, V., Canton, C., Alvarez, L., & Lanusse, C. (2024). Macrocyclic lactones and ectoparasites control in livestock: efficacy, drug resistance and therapeutic challenges. *International Journal for Parasitology: Drugs and Drug Resistance*, 100559.

Ma, M., Luo, J., Li, C., Eleftherianos, I., Zhang, W., & Xu, L. (2024). A life-and-death struggle: Interaction of insects with entomopathogenic fungi across various infection stages. *Frontiers in Immunology, 14*, 1329843.

Mabena, N. B. (2023). *Characterization of EPNs and their bacterial symbionts* (Master's thesis, University of the Witwatersrand, Johannesburg (South Africa).

Magri, J. M., Johnson, M. T., Herring, T. A., & Greenblatt, J. F. (2002). Lyme disease knowledge, beliefs, and practices of New Hampshire primary care physicians. *The Journal of the American Board of Family Practice, 15*(4), 277–284.

Makwarela, T. G., Seoraj-Pillai, N., & Nangammbi, T. C. (2025). Tick Control Strategies: Critical Insights into Chemical, Biological, Physical, and Integrated Approaches for Effective Hard Tick Management. *Veterinary Sciences, 12*(2), 114.

Mandli, J. T., Lee, X., Bron, G. M., & Paskewitz, S. M. (2021). Integrated tick management in south central Wisconsin: Impact of invasive vegetation removal and host-targeted acaricides on the density of questing Ixodes scapularis (Acari: Ixodidae) nymphs. *Journal of Medical Entomology, 58*(6), 2358–2367.

Matuska-Łyżwa, J., Duda, S., Nowak, D., & Kaca, W. (2024). Impact of Abiotic and Biotic Environmental Conditions on the Development and Infectivity of Entomopathogenic Nematodes in Agricultural Soils. *Insects, 15*(6), 421.

McNamara, K. E., & Buggy, L. (2017). Community-based climate change adaptation: A review of academic literature. *Local Environment, 22*(4), 443–460.

Medlock, J. M., Hansford, K. M., Bormane, A., Derdakova, M., Estrada-Peña, A., George, J. C., & Van Bortel, W. (2013). Driving forces for changes in geographical distribution of Ixodes ricinus ticks in Europe. *Parasites & Vectors, 6*, 1–11.

Mount, G. A., Haile, D. G., Barnard, D. R., & Daniels, E. (1999). Integrated management strategies for Amblyomma americanum (Acari: Ixodidae) in non-agricultural areas. *Experimental & Applied Acarology, 23*, 827–839.

Mount, G. A., Haile, D. G., & Daniels, E. (1997). Simulation of management strategies for the blacklegged tick (Acari: Ixodidae) and the Lyme disease spirochete, Borrelia burgdorferi. *Journal of Medical Entomology, 34*(6), 672–683.

Muhammad, A., Bashir, R., Mahmood, M., Afzal, M. S., Simsek, S., Awan, U. A., Khan, M. R., Ahmed, H., & Cao, J. (2021). Epidemiology of ectoparasites (ticks, lice, and mites) in the livestock of Pakistan: A review. *Frontiers in Veterinary Science, 8*, Article 780738.

Mullen, G. R., & Durden, L. A. (Eds.). (2009). *Medical and veterinary entomology*. Academic press.

Myhre, G., Shindell, D., Bréon, F. M., Collins, W., Fuglestvedt, J., Huang, J., Koch, D., Lamarque, J.F., Lee, D., Mendoza, B., & Zhang, H. (2014). Anthropogenic and natural radiative forcing. *Climate Change 2013-The Physical Science Basis*, 659–740.

Nath, S., Mandal, S., Pal, S., Jadhao, S., Ottalwar, N., & Sanyal, P. (2018). Impact and management of acaricide resistance-pertaining to sustainable control of ticks. *International Journal of Livestock Research, 8*(10), 46–60.

Negev, M., Paz, S., Clermont, A., Pri-Or, N. G., Shalom, U., Yeger, T., & Green, M. S. (2015). Impacts of climate change on vector borne diseases in the Mediterranean Basin—implications for preparedness and adaptation policy. *International Journal of Environmental Research and Public Health, 12*(6), 6745–6770.

Nepveu-Traversy, M. E., Fausther-Bovendo, H., & Babuadze, G. (2024). Human tick-borne diseases and advances in anti-tick vaccine approaches: A comprehensive review. *Vaccines, 12*(2), 141.

Nuttall, P. A. (2022). Climate change impacts on ticks and tick-borne infections. *Biologia, 77*(6), 1503–1512.

Ogden, N. H., Ben Beard, C., Ginsberg, H. S., & Tsao, J. I. (2021). Possible effects of climate change on ixodid ticks and the pathogens they transmit: Predictions and observations. *Journal of Medical Entomology, 58*(4), 1536–1545.

Olajiga, O., Holguin-Rocha, A. F., Rippee-Brooks, M., Eppler, M., Harris, S. L., & Londono-Renteria, B. (2021). Vertebrate responses against arthropod salivary proteins and their therapeutic potential. *Vaccines, 9*(4), 347.

Omazic, A., Han, S., Albihn, A., Ullman, K., Choklikitumnuey, P., Perissinotto, D., & Grandi, G. (2023). Ixodid tick species found in northern Sweden-Data from a frontier area. *Ticks and Tick-Borne Diseases, 14*(6), Article 102244.

Ostfeld, R. S., Price, A., Hornbostel, V. L., Benjamin, M. A., & Keesing, F. (2006). Controlling ticks and tick-borne zoonoses with biological and chemical agents. *BioScience, 56*(5), 383–394.

Pasternak, A. R., & Palli, S. R. (2022). Mapping distributions of the Lyme disease vector, Ixodes scapularis, and spirochete, Borrelia burgdorferi, in Kentucky using passive and active surveillance. *Ticks and Tick-Borne Diseases, 13*(2), Article 101885.

Peeling, R. W., & McNerney, R. (2014). Emerging technologies in point-of-care molecular diagnostics for resource-limited settings. *Expert Review of Molecular Diagnostics, 14*(5), 525–534.

Peralbo-Moreno, A., Baz-Flores, S., Cuadrado-Matías, R., Barroso, P., Triguero-Ocaña, R., Jiménez-Ruiz, S., Herraiz, C., Ruiz-Rodríguez, C., Acevedo, P., & Ruiz-Fons, F. (2022). Environmental factors driving fine-scale ixodid tick abundance patterns. *Science of the Total Environment, 853*, Article 158633.

Quadros, D. G., Johnson, T. L., Whitney, T. R., Oliver, J. D., & Oliva Chávez, A. S. (2020). Plant-derived natural compounds for tick pest control in livestock and wildlife: Pragmatism or Utopia? *Insects, 11*(8), 490.

Rajput, M., Sajid, M. S., Imran, M., Javed, M. T., & Sparagano, O. A. (2023). A participatory approach in assessing the knowledge, attitude, and practices (KAP) of stakeholders and livestock owners about ticks and tick-borne diseases from Sindh. *Pakistan Pathogens, 2023*(12), 800. https://doi.org/10.3390/pathogens12060800

Rajput, M., Sajid, M. S., Rajput, N. A., George, D. R., Usman, M., Zeeshan, M., Iqbal, O., Bhutto, B., Atiq, M., Rizwan, H. M., et al. (2024). Entomopathogenic fungi as alternatives to chemical acaricides: Challenges, opportunities and prospects for sustainable tick control. *Insects, 2024*(15), 1017. https://doi.org/10.3390/insects15121017

Ramos, R. A. N., de Macedo, L. O., Bezerra-Santos, M. A., de Carvalho, G. A., Verocai, G. G., & Otranto, D. (2023). The role of parasitoid wasps, Ixodiphagus spp.(Hymenoptera: Encyrtidae), in tick control. *Pathogens, 12*(5), 676.

Rand, P. W., Lacombe, E. H., Elias, S. P., Lubelczyk, C. B., Amand, T. S., & Smith, R. P. (2010). Trial of a minimal-risk botanical compound to control the vector tick of Lyme disease. *Journal of Medical Entomology, 47*(4), 695–698.

Reid, H. (Ed.). (2009). *Community-based adaptation to climate change* (Vol. 60). Iied.

Reid, H., & Huq, S. (2014). Mainstreaming community-based adaptation into national and local planning. *Climate and Development, 6*(4), 291–292.

Riccò, M., Gualerzi, G., Ranzieri, S., Ferraro, P., & Bragazzi, N. L. (2020). Knowledge, Attitudes, Practices (KAP) of Italian Occupational Physicians towards Tick Borne Encephalitis. *Trop. Med. Infect. Dis., 5*, 117. https://doi.org/10.3390/tropicalmed5030117

Richer, L. M., Brisson, D., Melo, R., Ostfeld, R. S., Zeidner, N., & Gomes-Solecki, M. (2014). Reservoir targeted vaccine against Borrelia burgdorferi: A new strategy to prevent Lyme disease transmission. *The Journal of Infectious Diseases, 209*(12), 1972–1980.

Rowan, S., Mohseni, N., Chang, M., Burger, H., Peters, M., & Mir, S. (2023). From tick to test: A comprehensive review of tick-borne disease diagnostics and surveillance methods in the United States. *Life, 13*(10), 2048.

Samish, M., Ginsberg, H., & Glazer, I. (2004). Biological control of ticks. *Parasitology, 129*(S1), S389–S403.

Schillberg, E., Lunny, D., Lindsay, L. R., Nelder, M. P., Russell, C., Mackie, M., Coats, D., Berry, A., & Young Hoon, K. N. (2018). Distribution of Ixodes scapularis in northwestern Ontario: Results from active and passive surveillance activities in the Northwestern Health Unit catchment area. *International Journal of Environmental Research and Public Health, 15*(10), 2225.

Schipper, E. L., Ayers, J., Reid, H., Huq, S., & Rahman, A. (2014). *Community-based adaptation to climate change: Scaling it up.* Routledge.

Schulze, T. L., Jordan, R. A., Williams, M., & Dolan, M. C. (2017). Evaluation of the SELECT tick control system (TCS), a host-targeted bait box, to reduce exposure to Ixodes scapularis (Acari: Ixodidae) in a Lyme disease endemic area of New Jersey. *Journal of Medical Entomology, 54*(4), 1019–1024.

Seifert, V. A., Wilson, S., Toivonen, S., Clarke, B., & Prunuske, A. (2016). Community partnership designed to promote Lyme disease prevention and engagement in citizen science. *Journal of Microbiology & Biology Education, 17*(1), 63–69.

Shaheen, M. N. (2022). The concept of one health applied to the problem of zoonotic diseases. *Reviews in Medical Virology, 32*(4), Article e2326.

Shao, M., Hu, P., Xu, P., Sun, J., Zhang, X., Zhang, D., Shen, Y., Gao, D., Zhang, W., Qin, W., & Lyu, Y. (2024). Knowledge, attitudes, and practices (KAPs) regarding tick-borne rickettsial disease among medical staff in endemic areas of China. *Peer J, 12*, Article e17562. https://doi.org/10.7717/peerj.17562

Sharma, A., Srivastava, A., Shukla, A. K., Srivastava, K., Srivastava, A. K., & Saxena, A. K. (2020). Entomopathogenic fungi: a potential source for biological control of insect pests. *Phytobiomes: current insights and future vistas*, 225–250.

Siegel, E. L., Lavoie, N., Xu, G., Brown, C. M., Ledizet, M., & Rich, S. M. (2023). Human-biting Ixodes scapularis submissions to a crowd-funded tick testing program correlate with the incidence of rare tick-borne disease: A seven-year retrospective study of anaplasmosis and babesiosis in Massachusetts. *Microorganisms, 11*(6), 1418.

Singh, S. K. (Ed.). (2013). *Viral infections and global change.* John Wiley & Sons.

Singh, N. K., & Randhawa, S. N. S. (2013). Chemotherapeutics for control and treatment of Ectoparasites in Companion animals. *Intas Polivet, 14*(2), 257–263.

Stafford, K. C., III., & Williams, S. C. (2017). Deer-targeted methods: A review of the use of topical acaricides for the control of ticks on white-tailed deer. *Journal of Integrated Pest Management, 8*(1), 19.

Stafford, K. C., III., Williams, S. C., & Molaei, G. (2017). Integrated pest management in controlling ticks and tick-associated diseases. *Journal of Integrated Pest Management, 8*(1), 28.

Taylor, P. B., Duan, J. J., Fuester, R. W., Hoddle, M., & Van Driesche, R. (2012). Parasitoid guilds of Agrilus woodborers (Coleoptera: Buprestidae): their diversity and potential for use in biological control. *Psyche: A Journal of Entomology, 2012*(1), 813929.

Telford, S. R., III. (2017). Deer reduction is a cornerstone of integrated deer tick management. *Journal of Integrated Pest Management, 8*(1), 25.

Tiffin, H. S., Rajotte, E. G., Sakamoto, J. M., & Machtinger, E. T. (2022). Tick control in a connected world: Challenges, solutions, and public policy from a United States border perspective. *Tropical Medicine and Infectious Disease, 7*(11), 388.

Tschakert, P., Dietrich, K., Tamminga, K., Prins, E., Shaffer, J., Liwenga, E., & Asiedu, A. (2014). Learning and envisioning under climatic uncertainty: An African experience. *Environment and Planning A, 46*(5), 1049–1068.

Vashisth, S., Chandel, Y. S., & Sharma, P. K. (2013). Entomopathogenic nematodes-A review. *Agricultural Reviews, 34*(3), 163–175.

Vayssier-Taussat, M., Cosson, J. F., Degeilh, B., Eloit, M., Fontanet, A., Moutailler, S., Raoult, D., Sellal, E., Ungeheuer, M. N., & Zylbermann, P. (2015). How a multidisciplinary 'One Health'approach can combat the tick-borne pathogen threat in Europe. *Future Microbiology, 10*(5), 809–818.

Vial, L., Stachurski, F., Leblond, A., Huber, K., Vourc'h, G., René-Martellet, M., Desjardins, I., Balança, G., Grosbois, V., Pradier, S., & Estrada-Peña, A. (2016). Strong evidence for the presence of the tick Hyalomma marginatum Koch, 1844 in southern continental France. *Ticks and Tick-borne Diseases, 7*(6), 1162–1167.

Villanueva, L. (2021). *Efficacy of Commercial Acaricide with Insect Growth Regulators Against Amblyomma americanum (Acari: Ixodidae) and Haemaphysalis longicornis (Acari: Ixodidae)* (Master's thesis, Texas A&M University-Kingsville).

Villar, M., Pacheco, I., Mateos-Hernández, L., Cabezas-Cruz, A., Tabor, A. E., Rodríguez-Valle, M., Mulenga, A., Kocan, K. M., Blouin, E. F., & de La Fuente, J. (2021). Characterization of tick salivary gland and saliva alphagalactome reveals candidate alpha-gal syndrome disease biomarkers. *Expert Review of Proteomics, 18*(12), 1099–1116.

Wall, R., & Ellse, L. S. (2011). Climate change and livestock parasites: Integrated management of sheep blowfly strike in a warmer environment. *Global Change Biology, 17*(5), 1770–1777.

White, A. L. (2019). *Tick control methods for Amblyomma americanum in Virginia: applications and modeling* (Doctoral dissertation, Old Dominion University).

Wimms, C., Aljundi, E., & Halsey, S. J. (2023). Regional dynamics of tick vectors of human disease. *Current Opinion in Insect Science, 55*, Article 101006.

World Health Organization. (2017a). *Global vector control response 2017–2030.* World Health Organization.

World Health Organization. (2017b). *Framework for a national vector control needs assessment* (p. 48). WHO.

Yadav, N., & Upadhyay, R. K. (2024). Status of Economic Losses and Health Impact of the Tick Bites on Farm, Field and Dairy Animals. *Acta Scientific Medical Sci*ences, *8*.

Zheng, W., Hu, H., Jiang, J., Sun, X., Fu, R., Tao, H., & Chen, S. (2022). Haemaphysalis longicornis calreticulin is not an effective molecular tool for tick bite diagnosis and disruption of tick infestations. *Veterinary Parasitology, 309*, Article 109775.

Chapter 11
Community Health Resilience: Addressing Tick-Borne Diseases

Syeda Hina Shah, Mahvish Rajput, Surrya Khanam, Osama B. Mohammed, and Shumaila Naz

Abstract This chapter examines the relationship between climate change and the increasing frequency of tick-borne diseases (TBDs) and highlights the importance of community health resiliency in addressing these matters. The chapter reviews new developments in the serological and molecular diagnostics of tick-borne pathogens in relation to traditional microscopy and culture methods, discussing their advantages and shortcomings. The chapter analyzes how socio-economic gaps, specifically within low-income groups, result in a lack of healthcare access and disproportionate treatment delays, leading to more advanced disease stages. Empowering communities through health education and targeted outreach programs tailored to the newly identified vulnerable populations is a critical component of the discussion. The chapter emphasizes the absence of proactive public health approaches to chronic problems in climate change and resilience-focused burdened public health policies regarding TBDs that need to be addressed. Equitable social strategies that encompass social justice and occupational health are recommended to mobilize community capacity to respond to TBDs.

S. H. Shah · S. Naz (✉)
Department of Biological Sciences, National University of Medical Sciences, Rawalpindi, Pakistan
e-mail: shumaila.naz@numspak.edu.pk

S. H. Shah
e-mail: hinahamadani.76@gmail.com

M. Rajput
Department of Parasitology, Faculty of Veterinary Science, University of Agriculture Faisalabad, Faisalabad, Pakistan
e-mail: drmehvish.aslam@gmail.com

S. Khanam
Department of Zoology, Women University Swabi, Swabi, Pakistan
e-mail: surryiamalik@gmail.com

O. B. Mohammed
Department of Zoology, College of Science, King Saud University, Riyadh, Saudi Arabia
e-mail: omohammed@ksu.edu.sa

H. M. Rizwan and M. S. Sajid (eds.), *Ticks in a Changing Climate*, Biometeorology 7,
https://doi.org/10.1007/978-3-032-12259-9_11

Keywords Community-based surveillance · Public education · Access to health care services · Socio-economic factors · Policy recommendations · Risk communication

11.1 Introduction

The impact of climate change on the transmission of ticks and tick-borne diseases (TBDs) is garnering concern and appears to be detrimental to public health (Adepoju et al., 2023). These changes create a need for new strategies in community health systems and intervention planning (Pinero de Plaza et al., 2023). An important oversight in the literature is the social and behavioral determinants that heighten the risk of human interaction with ticks and TBDs (Fang et al., 2024). Following studies must incorporate psycho-behavioral factors to examine how knowledge of climate-driven tick risks modifies the behavior, motivation, and preparedness of the public (Bouchard et al., 2019).

Any changes in the natural environment, especially temperature or precipitation, have a drastic impact on the risk landscape for TBDs. Meteorological factors have the potential to change the distribution and, in addition, the quantity of both the tick vectors and their animal reservoirs. This will subsequently influence the dynamics of disease transmission (Randolph, 2013). For example, increased temperatures tend to enhance the geographic distribution of potential animal reservoirs like white-footed mice which go beyond serving as primary hosts for the immature stages of ticks to act as infection reservoirs (Rocha et al., 2022). Adult female ticks are dependent on larger mammals such as deer for blood meals which are necessary for reproduction. Climate change is likely to increase the number of deer which will increase the period of host-tick interactions. This means that there are likely to be more ticks and their reproductive success and distribution is guaranteed to increase (Levi et al., 2012).

Ticks are known to become active in early spring when there is some snow cover and continue to be active until the fall. People are predicted to spend more time outdoors as the weather gets warmer, increasing their chances of coming into contact with ticks (Gilbert, 2021). People's contact with ticks is further facilitated by milder winters that cause later snowfall (Gray & Ogden, 2021). On the other hand, outdoor human activity and tick activity may temporarily lessen during extreme summer heat and drought, slightly reducing the risk of exposure (Estrada-Peña et al., 2014). Regardless, the overall reduction in the duration of winter and high activity periods for ticks significantly increases the potential for infection (Nuttall, 2022).

The geographical distribution of ticks correlates to certain environmental preferences. In particular, microhabitat factors like humidity, temperature, and soil type are essential for their survival, reproduction, and developing new populations (Estrada-Peña et al., 2014). These ecological changes impact the patterns of human risk and community susceptibility to TBDs (Gilbert, 2021). There is a strong historical evidence that supports the relationship between disease emergence and environmental changes. For example, agricultural land reforestation in the northeastern

United States during mid-twentieth century resulted in favorable conditions for the white-tailed deer population. Consequently, there was an increase in *Ixodes (I.) scapularis* ticks, and the emergence of Lyme disease which was first identified in the 1970s (Kugeler et al., 2011). Such events highlight the need for integrated strategies that consider ecological, behavioral, and climatic factors in building resilient communities capable of responding to future TBD threats.

11.2 Community-Based Surveillance Systems for Early Detection of TBDs

Community-based surveillance systems play a pivotal role in the early detection and containment of TBDs. These systems depend on local participation, accessible diagnostic tools, and timely communication to public health authorities. In both human and veterinary medicine, various diagnostic techniques outlined in Table 11.1 are employed to identify TBDs efficiently.

11.2.1 Traditional Diagnostic Techniques

Giemsa staining remains a conventional method for visualizing tick-borne pathogens such as *Anaplasma (An.) phagocytophilum*, *Ehrlichia* species, and *Babesia microti* in blood smears. This method facilitates rapid identification of intracellular infections (Killmaster & Levin, 2016). However, its sensitivity is limited compared to molecular methods like PCR. Moreover, recent approaches like TBD CapSeq have enhanced pathogen detection by identifying unknown agents such as *Ehrlichia, Rickettsia*, and *Borrelia (B.) miyamotoi* (Chung et al., 2022). The diagnostic accuracy of traditional methods is often influenced by the timing of sample collection and pathogen load. For instance, the sensitivity of detecting *Anaplasma* in blood smears is highest during the first week of illness when morulae are most likely visible, occurring in 25–75% of cases (Schotthoefer et al., 2013). Diagnosing babesiosis is particularly challenging when parasitemia levels are low, necessitating multiple smears for confirmation (Hussain et al., 2021).

Cultural methods of pathogen identification are considered the gold standard. But this method hardly implemented for TBDs because of high resource costs associated with cultivate methods and the slow growth nature of tick-borne pathogens (Marques 2015). Additionally, the contamination risk, biosafety needs, and absence of trained microbiologists make both microscopy and culture techniques unrealistic for routine TBD diagnosis in community settings at the field level.

Table 11.1 Diagnostic techniques to identify the tick-borne diseases

Diagnostic techniques	*TBDs*	References
Microscopy Examination and culture testing		
Microscopy (blood smear analysis)	Acute *Babesia* infection	Springer et al. (2021)
Pathogenic culture	Anaplasmosis and ehrlichiosis, *Rickettsia* spp.*Neoehrlichia mikurensi, B. burgdorferi s.l*	Schotthoefer et al. (2013), Springer et al. (2021)
Molecular diagnostic methods–nucleic acid amplification techniques (NAAT)		
Real-time quantitative PCR (qPCR)	*Borrelia miyamotoi* disease (BMD), acute anaplasmosis and babesiosis, Omsk hemorrhagic fever	Sanchez-Vicente & Tokarz (2023), CDC (2024)
NGS broad-panel system	*Ehrlichia* spp.,*B. burgdorferi s.s., A. phagocytophilum,* and *R. rickettsia* infection	Chung et al. (2022)
Loop-mediated isothermal amplification (LAMP)	TBEV and Severe fever with thrombocytopenia syndrome virus (SFTSV)	Gui et al. (2024)
Serological diagnostic techniques		
Rapid Immunochromatographic tests (ICT)	*Ehrlichia* spp., Lyme borreliosis antibodies against *B. burgdorferi s.l.,* and *Anaplasma* spp., in canine heartworm antigen	Richardson et al. (2023), Smit et al. (2015), Genchi et al. (2018)
Enzyme-linked Immunosorbent assay (ELISA)	*B. burgdorferi* s.l., Anti-*B. burgdorferi* s.l. antibodies in humans as well as in horses and dogs, NS1 of TBEV	Gerritzen & Brandt (2012)

11.2.2 Molecular Diagnostic Methods

The polymerase chain reaction (PCR) technique is preferred over traditional methods because of its higher sensitivity and specificity (Eikeland et al., 2024). Chan et al. (2013) also noted that PCR is very useful in diagnosing acute cases of babesiosis and anaplasmosis. However, for *B. burgdorferi* (Lyme disease)-sensitive specimens, PCR is usually low due to the small amounts of spirochetes present in the blood during the acute phase. Other sample types such as synovial fluid and skin lesions (erythema migrans) tend to have higher detection rates. Tagliafierro et al. (2022) argued that cerebrospinal fluid (CSF) testing in cases of neuroborreliosis is unreliable.

Springer et al. (2021) explains that qRT-PCR is more sensitive than standard PCR because it quantifies the target DNA and measures whether gene copies or Cycle threshold (Ct) values were produced (as in ct values are count of cycles). Digital PCR (dPCR) is another new technique, which improves sensitivity by subdividing reactions of the samples into many parts. This is particularly helpful in cases where

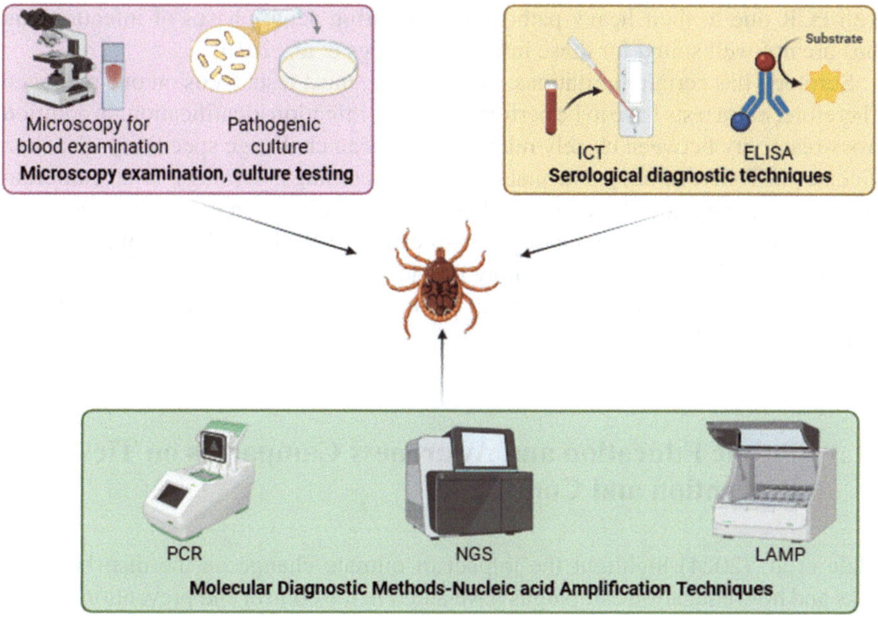

Fig. 11.1 Community-based surveillance systems for early detection of ticks and tick-borne diseases

B. burgdorferi sensu lato has been detected in patients where spirochete numbers were so low that they could not be detected (Das et al., 2020).

The Next Generation Sequencing (NGS) make it possible to diagnose different pathogens at once from a single sample, verifying to be an effective diagnostic tools (Sanchez-Vicente et al., 2022). However, NGS's have high implementation costs, less sensitive as compared to PCR, and absence of established cutoff values (Intirach et al., 2024; Jain et al., 2021). Another diagnostic tool like Loop-Mediated Isothermal Amplification (LAMP) have reported the successful detection of tick-borne pathogens in areas where PCR is unavailable (Chua et al., 2020). Community-based surveillance systems for early detection of TBDs of have been given in Fig. 11.1.

11.2.3 Serological Testing

The need to detect antibodies against TBD pathogens has catalyzed the development of serological assays such as enzyme-linked immunosorbent assay ELISA, indirect immunofluorescent assays, (IFA) and western blotting (Jain et al., 2021). Such techniques are helpful in the diagnosis of advanced Lyme borreliosis. Unfortunately, most cases of *Babesia* or *Ehrlichia* infections tend to be more efficiently diagnosed with PCR due to their heavy pathogen load during acute phases of infection, and thus are not well suited for these infections (Biggs et al., 2016).

Serology has certain limitations. For instance, most tests focus on one pathogen. Therefore, extra tests have to be performed for co-infection identification. In addition, cross-reactivity between closely related species can challenge specificity (Sanchez-Vicente & Tokarz, 2023). No matter the shortcomings, serology is an important method for community-based surveillance, particularly when it comes to analyzing exposure histories and conducting sero-prevalence studies (Hay et al., 2023). In combination with other tests, serology provides early detection, but must be paired with molecular methods due to the gap between infection and detectable antibodies (Cassedy et al., 2021).

11.3 Public Education and Awareness Campaigns on Tick Prevention and Control

Alale et al. (2024) highlight the impact of climate change on the distribution of ticks and how education campaigns have aided in the control and prevention of TBD exposure in affected areas. Community resilience in health is defined by Bardosh et al. (2017) as the capacity to provide a person with the educational resources necessary to defend against ticks and get medical help. The proactive geographical mapping of risk areas is one of the most important innovations in the monitoring of emerging hotspots (Beard et al., 2018). Available evidence suggests that new regions are now emerging with climate driven the onset of ticks and TBDs (Behar et al., 2023). In these areas, the general population as well as health care workers tend to be uninformed about the risks involved and how to efficiently prevent them, making education on the matter much more timely and effective (Bouchard et al., 2023).

It is important to ensure active community participation at the grassroots level. Thoroughly examined etiological guides and educational textbooks for region specific teaching aids like handouts, infographics, posters, and other media content must be prepared and made readily available to systematically integrating public health agencies (Vernon-Wilson et al., 2024). Targeted education for at-risk populations—such as forest workers, hikers, farmers, pet owners, and children—is particularly important (Thompson et al., 2014). These campaigns should highlight the importance of wearing personal protective equipment (PPE) such as long sleeves and pants, using EPA-approved tick repellents, and conducting regular tick checks after outdoor activities. Prompt removal of ticks is a key message to reduce the risk of infection transmission (Eisen, 2022).

Healthcare providers also play a pivotal role in educational outreach. By counseling patients on tick prevention during consultations, especially for those traveling to endemic or high-risk areas, clinicians can bridge the knowledge gap at the individual level (Valente et al., 2015). Education for healthcare workers themselves is equally important. Given the nonspecific symptoms of many TBDs—such as fever, headache, and muscle aches—clinicians should be trained to maintain a high index of suspicion and consider TBDs even in the absence of classical symptoms like a rash (Taba et al., 2017). The clinical challenges of TBDs are compounded by diagnostic limitations. As seen in Canada, the rise in certain TBDs has revealed gaps in laboratory testing and diagnostic confirmation (Rowan et al., 2023). Physicians must be aware that a negative test for diseases like Lyme does not rule out the possibility of other tick-borne infections, especially in patients presenting with compatible clinical symptoms (Aenishaenslin et al., 2017). Public campaigns should therefore also aim to educate on the importance of seeking medical evaluation following tick exposure, even in the absence of immediate symptoms.

In addition to individual-level behavior change, public awareness efforts must address broader ecological interventions (Golden & Earp, 2012). Campaigns can inform communities about the role of climate change in expanding tick habitats and encourage support for integrated tick management strategies, including landscaping practices that reduce tick habitats, host animal population control, and environmental monitoring (Tiffin et al., 2022). On the other hand, teaching campaigns should focus on the relations between humans, animals, and the environment within whose regions they work through the One Health system (Singh et al., 2024). These messages can be more effectively disseminated through school-based programs as well as through community workshops in better collaboration with local organizations and local media (Iman et al., 2011). Components of public education regarding TBDs have been summarized in Fig. 11.2.

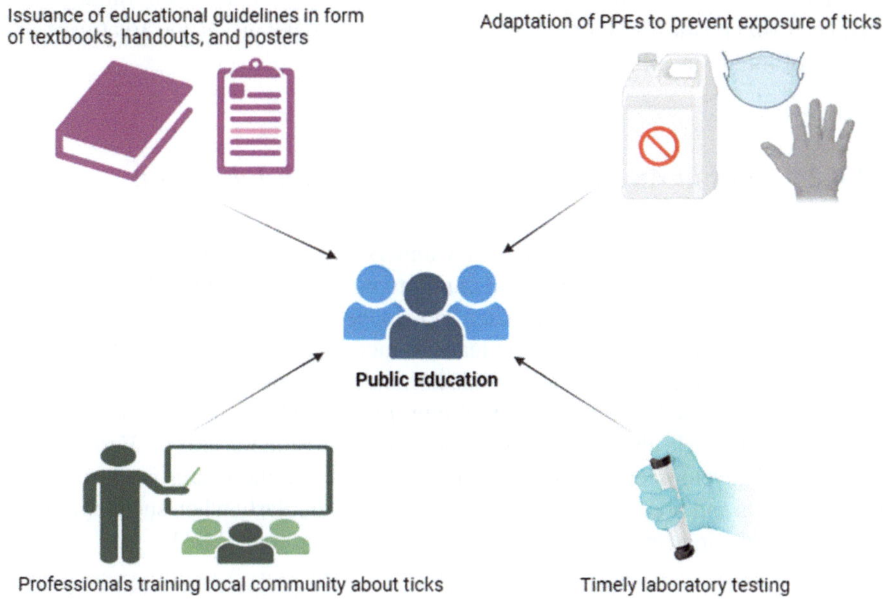

Fig. 11.2 Components of public education regarding TBDs

11.4 Access to Health Care Services for Diagnosis and Treatment of TBDs

The global prevalence of TBDs, inclusive of Lyme disease, anaplasmosis, babesiosis, and Rocky Mountain spotted fever is on the rise (Ramzan et al., 2021). When these infections are not diagnosed or left untreated, they can result in serious complications like capricious cardiac issues, chronic neurologic disorders, and even rheumatologic conditions (Pace & O'Reilly, 2020). There is, however, an uneven distribution of access to these healthcare services for timely diagnosis and treatment, particularly in agriculture and other remote areas that are more exposed to ticks (Norris-Roozmon, 2021).

11.4.1 Geographic Distribution and Healthcare Infrastructure

Ticks and TBDs are most common in remote rural areas that are forested as well as wooded since these regions are lightly populated and lack proper health infrastructure (Pfäffle et al., 2013). For example, rural communities of Midwestern and Northeastern areas of USA are more prone to Lyme disease due to increased risk of environmental exposure. On the other hand, availability of medical personnel and diagnostic equipment in these regions are very low. Due to unavailability of advanced diagnostics in these regions, people suffer delays in the treatment and control of infection (Maxwell et al., 2022; Schwartz, 2017).

11.4.2 Gaps in Awareness and Public Health Education

In regions where TBDs are remain poorly understood or emerging, a lack of professional or public awareness adds significantly to misdiagnosis and under-recognition (Vasic et al., 2022). The lack of exact knowledge about early symptoms, preventive behaviors, and risk factors often leads to increased disease burden, negligence, and delayed treatment. Public health education campaigns are crucial to encourage protective practices, like the wearing of appropriate clothing, use of tick repellents, and conducting tick checks after outdoor activities (Somers, 2021). Furthermore, professional development programs for healthcare practitioners improve the management of TBDs and precision of clinical diagnosis, particularly when such regions are undergoing changes in tick and disease patterns (Coyer et al., 2023).

11.4.3 Diagnostic Challenges and Socioeconomic Barriers

Maxwell et al. (2022) discussed how the non-specific presenting symptoms of fever, exhaustion, muscle pain, and even headache tend to pose the most diagnostic challenges in TBDs. And in parts where TBDs are emerging or just starting to appear, health professionals are likely to be unfamiliar with the presentation which makes them misdiagnose or delayed the treatment of infection (Norris-Roozmon, 2021). The CDC has noted the importance of clinical assessment and past history of symptoms in known tick exposure areas (CDC, 2024).

Socioeconomic status plays a critical role in determining access to healthcare services for individuals affected by TBDs. In many resource-limited settings, the availability of commonly prescribed antibiotics such as doxycycline and amoxicillin can be restricted, impeding timely treatment (Wright et al., 2012). Individuals living in rural or remote areas often face geographic and transportation challenges, with long distances to the nearest healthcare facilities or pharmacies delaying the initiation of care (Mac et al., 2019). Furthermore, patients suffering from post-treatment Lyme disease syndrome (PTLDS) require extended monitoring and support, which can impose significant logistical and financial burdens. These barriers frequently hinder access to consistent and affordable care, particularly for those without adequate insurance or social support networks (Wong et al., 2022).

11.5 Control and Prevention of Ticks and TBDs

11.5.1 Community Participation in the Control of TBDs

Achieving sustainable tick control requires active community involvement. Community participation in recognizing and removing potential tick habitats, such as tick friendly areas, and overgrown vegetation. Community participation should also boosts the effectiveness and sustainability of intervention approaches (Eastman et al., 2024). Community members who are educated about ticks and TBDs are more likely to practice personal protective measures, as well as implement behavioral and habitat alteration strategies (Slunge & Boman, 2018). Examples include promoting environmentally friendly pest control, leaf litter and refuse removal, landscape maintenance, and waste management (Tambo et al., 2017).

Communities that are fully empowered have actively demonstrated success in volunteer-based TBD control programs and surveillance. To increases the local capacity to manage disease risk, it is important to train the peoples about how to pinpoint tick-infested areas, document the case, and preparation of report (Bardosh et al., 2017). Additionally, proactive community advocacy, such as campaigning to reduce exposure zones or promoting cleaner environments also help in reduction in ticks (Gay, 2015). Custom vector control strategies based on participatory epidemiology research have historically been more effective in managing the socio-ecological realities of the affected communities (Rivera et al., 2023).

11.5.2 Creating Cooperation in Managing TBDs

As noted in (Tiffin et al., 2022), academic institutions and non-profit organizations alongside government agencies can pool their resources to study the biology of ticks, the epidemiology of the diseases, and put in place effective measures. Innovation is important in the advancement of tick control. Invention of new tools and technologies such as advanced repellents, vaccines, and diagnostic kits can greatly enhance disease prevention and treatment (Nchu et al., 2020). Additionally, outreach and public health access are important. Community groups need to partner with local governments and health departments to educate the public about the TBDs, their signs, and prevention strategies (Sakamoto et al., 2018).

Another area of collaboration that is equally important to focus on is the agriculture and livestock industry as they face the consequences of TBDs. Some parties, including farmers, veterinarians, and environmental scientists, have implemented coordinated efforts that can minimize tick infestations on livestock and therefore reduce the risk of zoonotic transmission to humans (Diuk-Wasser et al., 2021).

11.5.3 Capacity Building and Training Initiatives

The public and health care practitioners understanding surveillance, prevention, and response strategies greatly improves their ability to carry out risk management and mitigation associated with TBDs (Coderre-Ball et al., 2024). Citizen science, for instance, enables the public to assist in the identification of tick infested areas and the submission of tick samples (Lewis et al., 2018).

A common barrier in the control of TBD is the underestimation of the processes of tick identification or diagnosis, as well as lack of proper understanding of the clinical features of tick-borne infections. This is not helped by the fact that many practitioners have little formal instruction in these topics (Mattoon et al., 2021). Therefore, the design of these capacity-building programs to incorporate workshops, online lectures, and other practical teaching methods is crucial in improving diagnostic precision, increasing the rate of appropriate responses, and minimizing the sequelae of complications due to TBDs (Dantas-Torres et al., 2012).

The positive impacts of tick education on public awareness are well documented. Seifert et al. (2016) noted an enhancement in students' understanding of ticks, including tick biology, disease symptoms, and protective measures, as a result of training received in schools. Furthermore, community workshops have been effective in improving skills associated with tick identification, habitat management, and repellent application (Stafford et al., 2017). In general, increasing the knowledge of the public and professionals about ticks can greatly help to lower the incidence of TBDs (Riccò et al., 2020). New tools for engagement provide information, which is especially helpful in rural regions with limited internet access and formal health services (Beaujean et al., 2016).

11.5.4 Case Studies of Successful Community Health Resilience Programs

The public health response in New York State to the Asian longhorned tick was immediate and comprehensive. This invasive species, first discovered in Westchester county in 2018, raised concerns because of its asexual reproduction capability, as well as its potential to carry numerous human and animal pathogens, such as *Babesia, Rickettsia, Ehrlichia, Anaplasma,* and even the Heartland virus (Rochlin et al., 2023). Maine Tracking Network's approach to TBD surveillance, which is backed by the CDC, showcases the effectiveness of using environmental public health tracking systems. The network employs GIS technology to map out live data on Lyme disease, anaplasmosis, and babesiosis (Robinson, 2014). California serves as another great example. The state has famously been cited for the investment in risk communication and public outreach concerning the management of TBDs from the California Department of Public Health (CDPH, 2019; 2024a; 2024b).

11.6 Addressing Socio-economic Factors Influencing TBDs Resilience

Infection incidence, spread, and management are influenced by factors such as housing, occupational income, education level, and healthcare service accessibility (Bayles & Allan, 2014). In the case of TBDs, it is mostly the low-income population that is disproportionately affected due to limited access to health care services, adequate preventive measures, and educational materials (Parsons et al., 2024). Additionally, financial restrictions result in a delay in seeking care which can worsen outcomes of disease (Fix et al., 2000; Hirsch et al., 2020).

Race and ethnicity often intersect with socio-economic vulnerabilities. For example, some studies in the United States have documented the under-diagnosis of Lyme disease in Black and Hispanic populations attributable to a lack of language access, lower access to healthcare, and limited awareness of the disease (Nelson et al., 2016). People with lower levels of formal education may not have the skills needed to perform a basic tick-check, recognize the signs of an infection, or use a repellent (Kisomi et al., 2019). CSuch professions as agricultural workers, landscapers, forestry workers, and outdoor manual workers are at a higher risk because of

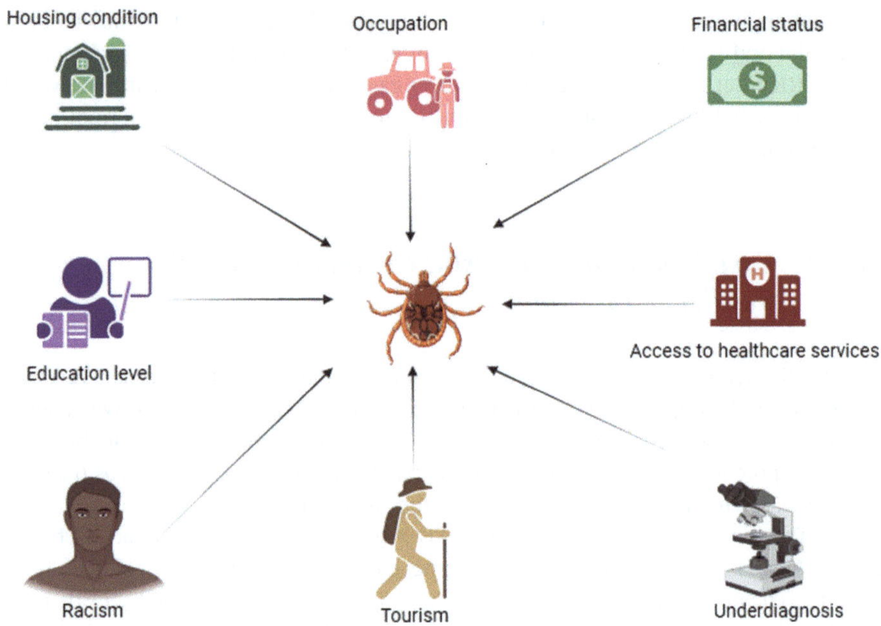

Fig. 11.3 Socio-economic factors influencing TBDs resilience

heavy contact with areas where ticks are located (Li et al., 2018). These populations are at risk because occupational safety cannot be ensured through proper PPE, tick check services during work hours, and medical care follow-up after tick bites (Nelson et al., 2016). Socio-economic factors influencing TBDs resilience have been given in Fig. 11.3.

11.7 Policy Recommendations for Strengthening Community Resilience Against TBDs

TBDs are a major public health concern in New York State as they result in expensive healthcare expenditures and affect the lives of thousands every year. Because of the increasing risk of ticks and their associated diseases, the New York State Department of Health (NYSDOH) has set up a public health response plan which is a 12-Point Plan aimed at strengthening coordinated public health responses. This initiative focuses on improving surveillance, education, research, partnerships, and overall enhancement of prevention-and-control measures for TBDs (Mead et al., 2015). Furthermore, there is a need to broaden educational outreach to other providers, which is one of the crucial goals. The training-and-information sharing improvement will better provider understanding of the clinical features of TBD, area of risk, diagnosis, and treatment as well as contemporary practices. At the community level, the need to elevate public awareness remains essential because a public well-educated on the subject tends to adopt behavior that decreases risk and exposure (Keesing et al., 2022).

Ticks are known to carry multiple pathogens contributing to a wide range of diseases, and their increasing presence across New York and the rest of the U.S. highlights the growing problem of TBDs. In these scenarios, in the state of New York, Lyme disease remains the most regularly reported disease with more than 7,500 cases every year, which adds to the need for improved and integrated comprehensive tick management strategies (Rochlin & Toledo, 2020). Lyme disease is most commonly transmitted through bites from infected black-legged ticks. This disease increases the chances of complication, untimely diagnosis, and treatment owing to the variability of symptoms and requirement of clinical suspicion. Moreover, the distribution of other TBDs as well as Lyme disease is not consistent throughout the state. Such variation means that healthcare practitioners must adapt their patient assessment and treatment strategies to a distinct and defined risk of illness within specific geographic areas (Rowan et al., 2023).

To achieve the goals put forward by the NYSDOH, there is a collaborative action plan toward enhancing laboratory testing and building effective liaisons with local public health departments. Integration of data from both human and animal surveillance systems provide health authorities with a more elaborate understanding of the disease dynamics (de la Fuente et al., 2023). It also assists several other research efforts directed toward determining the exact burden of Lyme disease in the Northeastern and Mid-Atlantic States as well as the attitudes of the people towards vaccination. Moreover, the initiative aids in the implementation of public education campaigns aimed at outdoor professionals and enthusiasts about Lyme disease. Furthermore, the measure provides for the active administration of tick control in state parks (Devchand et al., 2021).

Residents of New York State should keep track of local tick activity particularly in the warmer months when ticks are most active (Mead et al., 2015). For instance, yard work like mowing lawns, picking up leaves, clearing tall grass, and neatly stacking wood into piles can greatly cut down on the habitat areas for ticks. In addition to these measures, rodents are also discouraged which disrupts the tick life cycle (Biggs, 2016). Pet owners may wish to contact their vets regarding the proper steps for ensuring adequate control of ticks. Humans are put at risk because pets can introduce ticks to the home. Thus, the application of vet recommended repellents and routine tick examinations for pets must be done to prevent sickness in both animal and human (Iqubal, 2018).

Altogether, the integration of community-level actions with broader public health policy provides a robust pathway to reducing the burden of TBDs. The continued support for coordinated strategies among government agencies, healthcare professionals, and local communities will be vital for building long-term resilience against the growing threat of TBDs in an era of environmental change.

References

Adepoju, O. A., Afinowi, O. A., Tauheed, A. M., Danazumi, A. U., Dibba, L. B., Balogun, J. B., Flore, G., Saidu, U., Ibrahim, B., Balogun, O. O., & Balogun, E. O. (2023). Multisectoral perspectives on global warming and vector-borne diseases: A focus on Southern Europe. *Current Tropical Medicine Reports, 10*(2), 47–70.

Adhikari, B., Pell, C., Phommasone, K., Soundala, X., Kommarasy, P., Pongvongsa, T., Henriques, G., Day, N. P., Mayxay, M., & Cheah, P. Y. (2017). Elements of effective community engagement: Lessons from a targeted malaria elimination study in Lao PDR (Laos). *Global Health Action, 10*(1), Article e1366136.

Aenishaenslin, C., Bouchard, C., Koffi, J. K., & Ogden, N. H. (2017). Exposure and preventive behaviours toward ticks and Lyme disease in Canada: Results from a first national survey. *Ticks and Tick-Borne Diseases, 8*(1), 112–118.

Alale, T. Y., Sormunen, J. J., Nzeh, J., Agjei, R. O., Vesterinen, E. J., & Klemola, T. (2024). Public knowledge and awareness of tick-borne pathogens and diseases: A cross-sectional study in Ghana. *Current Research in Parasitology & Vector-Borne Diseases, 6*, Article 100228.

Bardosh, K. L., Ryan, S. J., Ebi, K., Welburn, S., & Singer, B. (2017). Addressing vulnerability, building resilience: Community-based adaptation to vector-borne diseases in the context of global change. *Infectious Diseases of Poverty, 6*, 1–21.

Bayles, B. R., & Allan, B. F. (2014). Social-ecological factors determine spatial variation in human incidence of tick-borne ehrlichiosis. *Epidemiology & Infection, 142*(9), 1911–1924.

Beard, R., Wentz, E., & Scotch, M. (2018). A systematic review of spatial decision support systems in public health informatics supporting the identification of high risk areas for zoonotic disease outbreaks. *International Journal of Health Geographics, 17*, 1–19.

Beaujean, D. J., Gassner, F., Wong, A., Steenbergen, J. E., Crutzen, R., & Ruwaard, D. (2016). Education on tick bite and Lyme borreliosis prevention, aimed at schoolchildren in the Netherlands: Comparing the effects of an online educational video game versus a leaflet or no intervention. *BMC Public Health, 16*(1), 1–10.

Behar, A., Yasur-Landau, D., & Leszkowicz-Mazuz, M. (2023). Vector-Borne Diseases in Ruminants. *Lester m. Shulman, 441*, 1–28.

Biggs, H. M., Behravesh, C. B., Bradley, K. K., Dahlgren, F. S., Drexler, N. A., Dumler, J. S., Folk, S. M., Kato, C. Y., Lash, R. R., Levin, M. L., Massung, R. F., Nadelman, R. B., Nicholson, W. L., Paddock, C. D., Pritt, B. S., & Traeger, M. S. (2016). Diagnosis and management of Tickborne Rickettsial diseases: Rocky mountain spotted fever and other spotted fever group Rickettsioses, Ehrlichioses, and Anaplasmosis–United States. *MMWR. Recommendations and Reports: Morbidity and Mortality Weekly Report. Recommendations and Reports, 65*(2), 1–44.

Bouchard, C., Dumas, A., Baron, G., Bowser, N., Leighton, P. A., Lindsay, L. R., Milord, F., Ogden, N. H., & Aenishaenslin, C. (2023). Integrated human behavior and tick risk maps to prioritize Lyme disease interventions using a "One Health" approach. *Ticks and Tick-Borne Diseases, 14*(2), Article 102083.

Bouchard, C., Dibernardo, A., Koffi, J., Wood, H., Leighton, P. A., & Lindsay, L. R. (2019). N Increased risk of tick-borne diseases with climate and environmental changes. *Canada Communicable Disease Report = Releve Des Maladies Transmissibles Au Canada, 45*(4), 83–89.

Carson, D. A., Kopsco, H., Gronemeyer, P., Mateus-Pinilla, N., Smith, G. S., Sandstrom, E. N., & Smith, R. L. (2022). Knowledge, attitudes, and practices of Illinois medical professionals related to ticks and tick-borne disease. *One Health, 15*, Article e100424.

Cassedy, A., Parle-McDermott, A., & O'Kennedy, R. (2021). Virus detection: A review of the current and emerging molecular and immunological methods. *Frontiers in Molecular Biosciences, 8*, Article 637559.

CDC. (2024). What you need to know about Asian longhorned ticks-a new tick in the United States.

California Department of Public Health (CDPH). (2019). California Vector Control Technician Certification and Training Program Guidelines. Vector-Borne Disease Section.

California Department of Public Health (CDPH). (2024a). Communications Toolkits.

California Department of Public Health (CDPH). (2024b). TBDs in California.

Chan, K., Marras, S. A. E., & Parveen, N. (2013). Sensitive multiplex PCR assay to differentiate Lyme spirochetes and emerging pathogens Anaplasma phagocytophilum and Babesia microti. *BMC Microbiology, 13*, 295.

Chua, A. P. B., Galay, R. L., Tanaka, T., & Yamazaki, W. (2020). Development of a loop-mediated isothermal amplification (LAMP) assay targeting the citrate synthase gene for detection of Ehrlichia canis in Dogs. *Veterinary Sciences, 7*(4).

Chung, I. H., Robinson, L. K., Stewart-Juba, J. J., Dasch, G. A., & Kato, C. Y. (2022). Analytically sensitive Rickettsia species detection for laboratory diagnosis. *The American Journal of Tropical Medicine and Hygiene, 106*(5), 1352–1357.

Coderre-Ball, A. M., Robertson, M., Matzinger, E., Durando, P., & Egan, R. (2024). Training initiatives that enhance knowledge, attitudes, and practices regarding the prevention, diagnosis and treatment of Lyme disease: A systematic review. *International Journal of Health Promotion and Education, 62*(3), 147–170.

Coyer, L., Sogan-Ekinci, A., Greutélaers, B., Kuhn, J., Saller, F. S., Hailer, J., Böhm, S., Brosch, R., Wagner-Wiening, C., & Böhmer, M. M. (2023). Knowledge, Attitudes and Behaviors Regarding Tick-Borne Encephalitis Vaccination and Prevention of Tick-Borne Diseases among Primary Care Physicians in Bavaria and Baden-Wuerttemberg, Germany, May–September 2022. *Microorganisms, 11*(4), 961.

Dantas-Torres, F., Chomel, B. B., & Otranto, D. (2012). Ticks and TBDs: A One Health perspective. *Trends in Parasitology, 28*(10), 437–446.

Das, S., Hammond-McKibben, D., Guralski, D., Lobo, S., & Fiedler, P. N. (2020). Development of a sensitive molecular diagnostic assay for detecting Borrelia burgdorferi DNA from the blood of Lyme disease patients by digital PCR. *PLoS ONE, 15*(11), Article e0235372.

de la Fuente, J., Estrada-Peña, A., Rafael, M., Almazán, C., Bermúdez, S., Abdelbaset, A. E., Kasaija, P. D., Kabi, F., Akande, F. A., Ajagbe, D. O., & Bamgbose, T. (2023). Perception of ticks and TBDs worldwide. *Pathogens., 12*(10), 1258.

Devchand, R., Koehler, L., Hook, S., Marx, G. E., Hooks, H., Schwartz, A., & Hinckley, A. (2021). Understanding consumer and clinician perceptions of a potential Lyme disease vaccine. *Health Education Research., 36*(5), 494–504.

Diuk-Wasser, M. A., VanAcker, M. C., & Fernandez, M. P. (2021). Impact of land use changes and habitat fragmentation on the eco-epidemiology of tick-borne diseases. *Journal of Medical Entomology, 58*(4), 1546–1564.

Diagnosis and management of tickborne rickettsial diseases: Rocky Mountain spotted fever and other spotted fever group rickettsioses, ehrlichioses, and anaplasmosis—United States. *MMWR. Recommendations and Reports, 65*.

Eastman, P., Awolola, T. S., Yoshimizu, M., Govellla, N., Chaki, P., & Zohdy, S. (2024). Community-based entomological surveillance and control of vector-borne diseases: A scoping review. *BioRxiv*, 2024–09.

Eikeland, R., Henningsson, A. J., Lebech, A.-M., Kerlefsen, Y., Mavin, S., Vrijlandt, A., Hovius, J. W., Lernout, T., Lim, C., Dobler, G., Fingerle, V., Gynthersen, R. M., Lindgren, P.-E., & Reiso, H. (2024). Tick-borne diseases in the North Sea region–A comprehensive overview and recommendations for diagnostics and treatment. *Ticks and Tick-Borne Diseases, 15*(2), Article 102306.

Eisen, L. (2022). Personal protection measures to prevent tick bites in the United States: Knowledge gaps, challenges, and opportunities. *Ticks and Tick-Borne Diseases, 13*(4), Article 101944.

Estrada-Peña, A., Ostfeld, R. S., Peterson, A. T., Poulin, R., & de la Fuente, J. (2014). Effects of environmental change on zoonotic disease risk: An ecological primer. *Trends in Parasitology, 30*(4), 205–214.

Fang, R., Li, S., Lyu, Y., Yang, X., Wang, T., & Li, S. (2024). Behavioral and cognitive factors influencing tick-borne disease risk in northeast China: Implications for prevention and control strategies. *One Health, 18*, Article 100736.

Fix, A. D., Pena, C. A., & Strickland, G. T. (2000). Racial differences in reported Lyme disease incidence. *American Journal of Epidemiology, 152*(8), 756–759.
Gay, J. (2015). No Time for Lyme: Public Health Approaches to Lyme Disease in Vermont.
Genchi, M., Mangia, C., Ferrari, N., & Loukeri, S. (2018). Evaluation of a rapid immunochromatographic test for the detection of low burden Dirofilaria immitis (heartworm) in dogs and cats. *Parasitology Research, 117*, 31–34.
Gerritzen, A., & Brandt, S. (2012). Serodiagnosis of Lyme borreliosis with bead-based immunoassays using multiplex technology. *Methods, 56*(4), 477–483.
Gilbert, L. (2021). The impacts of climate change on ticks and tick-borne disease risk. *Annual Review of Entomology, 66*(1), 373–388.
Golden, S. D., & Earp, J. A. L. (2012). Social ecological approaches to individuals and their contexts: Twenty years of health education & behavior health promotion interventions. *Health Education & Behavior, 39*(3), 364–372.
Gray, J. S., & Ogden, N. H. (2021). Ticks, human babesiosis and climate change. *Pathogens, 10*(11), 1430.
Gui, Z., Ren, Y., Guo, Q., et al. (2024). Development of a LAMP assay for the rapid visual detection of the emerging tick-borne Songling virus. *Parasites & Vectors, 17*, 447.
Hay, J., Routledge, I., & Takahashi, S. (2023). Serodynamics: a review of methods for epidemiological inference using serological data. *OSF Prepr, 10.*
Hirsch, A. G., Poulsen, M. N., Nordberg, C., Moon, K. A., Rebman, A. W., Aucott, J. N., Heaney, C. D., & Schwartz, B. S. (2020). Risk factors and outcomes of treatment delays in Lyme disease: A population-based retrospective cohort study. *Frontiers in Medicine (Lausanne), 7*, Article 560018.
Hussain, S., Hussain, A., Aziz, M. U., Song, B., Zeb, J., George, D., Li, J., & Sparagano, O. (2021). A review of Zoonotic Babesiosis as an emerging public health threat in Asia. *Pathogens, 11*(1), 23.
Inman, D. D., Van Bakergem, K. M., LaRosa, A. C., & Garr, D. R. (2011). Evidence-based health promotion programs for schools and communities. *American Journal of Preventive Medicine, 40*(2), 207–219.
Intirach, J., Lv, X., Sutthanont, N., Cai, B., Champakaew, D., Chen, T., Han, Q., & Lv, Z. (2024). Molecular and next-generation sequencing analysis of tick-borne pathogens of Rhipicephalus ticks (Acari: Ixodidae) in cattle and dogs. *Acta Tropica, 252*, Article 107138.
Iqubal, M. Z. (2018). *Serodiagnosis of Scrub Typhus in a Tertiary Care Hospital*. Doctoral dissertation, Rajiv Gandhi University of Health Sciences India.
Jain, K., Tagliafierro, T., Marques, A., Sanchez-Vicente, S., Gokden, A., Fallon, B., Mishra, N., Briese, T., Kapoor, V., Sameroff, S., Guo, C., Marcos, L. A., Hu, L., Lipkin, W. I., & Tokarz, R. (2021). Development of a capture sequencing assay for enhanced detection and genotyping of tick-borne pathogens. *Scientific Reports, 11*(1), 12384.
Keesing, F., Mowry, S., Bremer, W., Duerr, S., Evans, A. S., Jr., Fischhoff, I. R., Hinckley, A. F., Hook, S. A., Keating, F., Pendleton, J., & Pfister, A. (2022). Effects of tick-control interventions on tick abundance, human encounters with ticks, and incidence of tickborne diseases in residential neighborhoods, New York, USA. *Emerging Infectious Diseases., 28*(5), 957.
Killmaster, L. F., & Levin, M. L. (2016). Isolation and Short-Term Persistence of Ehrlichia ewingii in Cell Culture. *Vector Borne and Zoonotic Diseases (Larchmont, N.Y.), 16*(7), 445–448.
Kisomi, M. G., Wong, L. P., Tay, S. T., Bulgiba, A., & Nizam, Q. N. H. (2019). A qualitative study to explore farmworkers' knowledge, beliefs and preventive practices toward ticks and tick-borne diseases. *The Journal of Infection in Developing Countries, 13*(12), 1117–1126.
Kugeler, K. J., Griffith, K. S., Gould, L. H., Kochanek, K., Delorey, M. J., Biggerstaff, B. J., & Mead, P. S. (2011). A review of death certificates listing Lyme disease as a cause of death in the United States. *Clinical Infectious Diseases: An Official Publication of the Infectious Diseases Society of America, 52*(3), 364–367.

Levi, T., Kilpatrick, A. M., Mangel, M., & Wilmers, C. C. (2012). Deer, predators, and the emergence of Lyme disease. *Proceedings of the National Academy of Sciences of the United States of America, 109*(27), 10942–10947.

Lewis, J., Boudreau, C.R., Patterson, J.W., Bradet-Legris, J. & Lloyd, V.K. (2018), March. Citizen science and community engagement in tick surveillance—A Canadian case study. *Healthcare, 6*(1) 22.

Li, S., Juhász-Horváth, L., Trájer, A., Pintér, L., Rounsevell, M. D. A., & Harrison, P. A. (2018). Lifestyle, habitat and farmers' risk of exposure to tick bites in an endemic area of tick-borne diseases in Hungary. *Zoonoses and Public Health, 65*(1), e248–e253.

Mac, S., da Silva, S. R., & Sander, B. (2019). The economic burden of Lyme disease and the cost-effectiveness of Lyme disease interventions: A scoping review. *PLoS ONE, 14*(1), Article e0210280.

Marques, A. R. (2015). Laboratory diagnosis of Lyme disease—advances and challenges. *Infectious Disease Clinics of North America, 29*(2), 295–307.

Mattoon, S., Baumhart, C., Barsallo Cochez, A. C., MacQueen, D., Snedeker, J., Yancey, C. B., Gatch, M., & Mader, E. M. (2021). Primary care clinical provider knowledge and experiences in the diagnosis and treatment of tick-borne illness: A qualitative assessment from a Lyme disease endemic community. *BMC Infectious Diseases, 21*, 1–11.

Maxwell, S. P., McNeely, C. L., Brooks, C., & Thomas, K. (2022). Triangulating the new frontier of health geo-data: Assessing tick-borne disease risk as an occupational hazard among vulnerable populations. *International Journal of Environmental Research and Public Health, 19*(15), Article e9449.

Mead, P., Hinckley, A., Hook, S., & Beard, C. B. (2015). TickNET—a collaborative public health approach to tickborne disease surveillance and research. *Emerging Infectious Diseases., 21*(9), 1574.

Nchu, F., Nyangiwe, N., Muhanguzi, D., Nzalawahe, J., Nagagi, Y. P., Msalya, G., Joseph, N. A., Kimaro, E. G., Mollel, M., Temba, V., & Harouna, D. V. (2020). Development of a practical framework for sustainable surveillance and control of ticks and tick-borne diseases in Africa. *Veterinary World, 13*(9), Article e1910.

Nelson, C. A., Starr, J. A., Kugeler, K. J., & Mead, P. S. (2016). Lyme Disease in Hispanics, United States, 2000–2013. *Emerging Infectious Diseases, 22*(3), 522–525.

Norris-Roozmon, E.M. (2021). *Associations Among Self-Reported Tick-Borne Disease Symptoms, Treatments & Diagnoses* (Master's thesis, Queen's University (Canada). 1–14.

Nuttall, P. A. (2022). Climate change impacts on ticks and tick-borne infections. *Biologia, 77*(6), 1503–1512.

Pace, E. J., & O'Reilly, M. (2020). Tickborne diseases: Diagnosis and management. *American Family Physician, 101*(9), 530–540.

Parsons, E. S., Jowell, A., Veidis, E., Barry, M., & Israni, S. T. (2024). Climate change and inequality. *Pediatric research*, 1–8.

Pfäffle, M., Littwin, N., Muders, S. V., & Petney, T. N. (2013). The ecology of tick-borne diseases. *International Journal for Parasitology, 43*(12–13), 1059–1077.

Pinero de Plaza, M. A., Yadav, L., & Kitson, A. (2023). Co-designing, measuring, and optimizing innovations and solutions within complex adaptive health systems. *Frontiers in Health Services, 3*, 1154614.

Ramzan, M., Murtaza, G., Sattar, A., Munawar, N., Ullah, A., Ejaz, A., Ayaz, F., Anwar, S., Jameel, K., & Kamran, F. (2021). Techniques for managing ticks and tick-borne diseases under changing climate; A review. *Egyptian Academic Journal of Biological Sciences, B. Zoology, 13*(1), 117–128.

Randolph, S. E. (2013). Is expert opinion enough? A critical assessment of the evidence for potential impacts of climate change on tick-borne diseases. *Animal Health Research Reviews, 14*(2), 133–137.

Riccò, M., Bragazzi, N. L., Vezzosi, L., Balzarini, F., Colucci, M. E., & Veronesi, L. (2020). Knowledge, attitudes, and practices on tick-borne human diseases and tick-borne encephalitis vaccine among farmers from North-Eastern Italy (2017). *Journal of Agromedicine, 25*(1), 73–85.

Richardson, S. S., Mainville, C. A., Arguello-Marin, A., Whalley, D., Burton, W., Breitschwerdt, E. B., & Qurollo, B. A. (2023). A second-generation, point-of-care immunoassay provided improved detection of Anaplasma and Ehrlichia antibodies in PCR-positive dogs naturally infected with Anaplasma or Ehrlichia species. *Journal of Veterinary Diagnostic Investigation, 35*(4), 367–374.

Rivera, E. P., Arrivillaga, M. R., Juárez, J. G., De Urioste-Stone, S. M., Berganza, E., & Pennington, P. M. (2023). Adoption of community-based strategies for sustainable vector control and prevention. *BMC Public Health, 23*(1), 1834.

Robinson, S. (2014). Lyme disease in Maine: A comparison of NEDSS surveillance data and Maine Health Data Organization hospital discharge data. *Online Journal of Public Health Informatics, 5*(3), 231.

Rocha, S. C., Velásquez, C. V., Aquib, A., Al-Nazal, A., & Parveen, N. (2022). Transmission cycle of tick-borne infections and co-infections, animal models and diseases. *Pathogens, 11*(11), 1309.

Rochlin, I., & Toledo, A. (2020). Emerging tick-borne pathogens of public health importance: A mini-review. *Journal of Medical Microbiology, 69*(6), 781–791.

Rochlin, I., Benach, J. L., Furie, M. B., Thanassi, D. G., & Kim, H. K. (2023). Rapid invasion and expansion of the Asian longhorned tick (*Haemaphysalis longicornis*) into a new area on Long Island, New York, USA. *Ticks and Tick-Borne Diseases, 14*(2), Article 102088.

Roome, N. (2017). Developing environmental management strategies. *Corporate Environmental Responsibility, 323.*

Rowan, S., Mohseni, N., Chang, M., Burger, H., Peters, M., & Mir, S. (2023). From tick to test: A comprehensive review of tick-borne disease diagnostics and surveillance methods in the United States. *Life, 13*(10), 2048.

Sakamoto, J. M. (2018). Progress, challenges, and the role of public engagement to improve tick-borne disease literacy. *Current Opinion in Insect Science, 28*, 81–89.

Sanchez-Vicente, S., & Tokarz, R. (2023). Tick-Borne co-infections: Challenges in molecular and serologic diagnoses. *Pathogens, 12*(11), 1371.

Sanchez-Vicente, S., Jain, K., Tagliafierro, T., Gokden, A., Kapoor, V., Guo, C., Horn, E. J., Lipkin, W. I., & Tokarz, R. (2022). Capture sequencing enables sensitive detection of tick-borne agents in human blood. *Frontiers in Microbiology, 13*, Article 837621.

Schotthoefer, A. M., Meece, J. K., Ivacic, L. C., Bertz, P. D., Zhang, K., Weiler, T., Uphoff, T. S., & Fritsche, T. R. (2013). Comparison of a real-time PCR method with serology and blood smear analysis for diagnosis of human Anaplasmosis: Importance of infection time course for optimal test utilization. *Journal of Clinical Microbiology, 51*(7), 2147–2153.

Schwartz, A. M. (2017). Surveillance for lyme disease—United States, 2008–2015. *MMWR. Surveillance Summaries, 66*(22), 1–12.

Seifert, V. A., Wilson, S., Toivonen, S., Clarke, B., & Prunuske, A. (2016). Community Partnership Designed to Promote Lyme Disease Prevention and Engagement in Citizen Science. *Journal of Microbiology & Biology Education, 17*(1), 63–69.

Singh, S., Sharma, P., Pal, N., Sarma, D. K., Tiwari, R., & Kumar, M. (2024). Holistic one health surveillance framework: Synergizing environmental, animal, and human determinants for enhanced infectious disease management. *ACS Infectious Diseases, 10*(3), 808–826.

Slunge, D., & Boman, A. (2018). Learning to live with ticks? The role of exposure and risk perceptions in protective behaviour against tick-borne diseases. *PLoS ONE, 13*(6), Article e0198286.

Smit, P. W., Kurkela, S., Kuusi, M., & Vapalahti, O. (2015). Evaluation of two commercially available rapid diagnostic tests for Lyme borreliosis. *European Journal of Clinical Microbiology & Infectious Diseases, 34*, 109–113.

Somers, J. (2021). Tick Bites and Blood Relations: Regional Relationships and Interpretations of Lyme Disease Risk in New York's Hudson River Corridor. State University of New York at Albany.

Springer, A., Glass, A., Probst, J., & Strube, C. (2021). Tick-borne zoonoses and commonly used diagnostic methods in human and veterinary medicine. *Parasitology Research, 120*(12), 4075.

Stafford, K. C., III., Williams, S. C., & Molaei, G. (2017). Integrated pest management in controlling ticks and tick-associated diseases. *Journal of Integrated Pest Management, 8*(1), 28.

Taba, P., Schmutzhard, E., Forsberg, P., Lutsar, I., Ljøstad, U., Mygland, Å., Levchenko, I., Strle, F., & Steiner, I. (2017). EAN consensus review on prevention, diagnosis and management of tick-borne encephalitis. *European Journal of Neurology, 24*(10), 1214-e61.

Tagliafierro, T., Joshi, S., Sameroff, S., Marques, A., Dumler, J., Mishra, N., Sanchez-Vicente, S., Wormser, G., Marcos, L., Lipkin, W., & Tokarz, R. (2022). Detection of Antibodies to Anaplasma Phagocytophilum and Babesia Microti Using Linear Peptides. *Ticks and Tick-Borne Diseases, 13*(5), Article 101999.

Tambo, E., Fouad, A. M., & Khater, E. I. (2017). Strengthening community emergency preparedness and response in threats and epidemics disasters prevention and management in Saudi Arabia. *International Journal of Emergency Management, 13*(3), 288–303.

Thompson, K., Every, D., Rainbird, S., Cornell, V., Smith, B., & Trigg, J. (2014). No pet or their person left behind: Increasing the disaster resilience of vulnerable groups through animal attachment, activities and networks. *Animals, 4*(2), 214–240.

Tiffin, H. S., Rajotte, E. G., Sakamoto, J. M., & Machtinger, E. T. (2022). Tick control in a connected world: Challenges, solutions, and public policy from a United States border perspective. *Tropical Medicine and Infectious Disease, 7*(11), 388.

Valente, S. L., Wemple, D., Ramos, S., Cashman, S. B., & Savageau, J. A. (2015). Preventive behaviors and knowledge of tick-borne illnesses: Results of a survey from an endemic area. *Journal of Public Health Management and Practice, 21*(3), E16–E23.

Vasić, A., Bjekić, J., Veinović, G., Mihaljica, D., Sukara, R., Poluga, J., Filipović, S. R., & Tomanović, S. (2022). Knowledge, attitudes, and practices on tick-borne encephalitis virus and tick-borne diseases within professionally tick-exposed persons, health care workers, and general population in Serbia: A questionnaire-based study. *International Journal of Environmental Research and Public Health, 19*(2), 867.

Vernon-Wilson, E., Tetui, M., DeMarco, M., Grindrod, K., & Waite, N. M. (2024). Connect, collaborate and tailor: A model of community engagement through infographic design during the COVID-19 pandemic. *BMC Public Health, 24*(1), 2551.

Walker, A. R. (2011). Eradication and control of livestock ticks: Biological, economic and social perspectives. *Parasitology, 138*(8), 945–959.

Wong, K. H., Shapiro, E. D., & Soffer, G. K. (2022). A review of post-treatment Lyme disease syndrome and chronic Lyme disease for the practicing immunologist. *Clinical Reviews in Allergy & Immunology, 62*(1), 264–271.

Wright, W. F., Riedel, D. J., Talwani, R., & Gilliam, B. L. (2012). Diagnosis and management of Lyme disease. *American Family Physician, 85*(11), 1086–1093.

Chapter 12
Future Projections: Anticipating Climate-Driven Challenges in Tick Research

Muhammad Sohail Sajid, Hina tu Zahra, Urfa Bin Tahir, Saiful Islam, and Mahnoor Mahzhar Iqbal

Abstract The impacts of climate change (CC) on tick populations and diseases transmitted by them are likely to pose serious challenges to public health as well as to agriculture. Increases in global temperatures, variations in precipitation patterns, and alterations in humidity are altering the ecological mix of tick live habitats, allowing for expansion of populations of ticks outward into new areas. Moving forward, with the continued changes caused by climate, it is important that we take understanding of the tick–host–environment complex as a whole. Future research will be needed to determine how climate change will influence tick behavior, ability to live in habitats that they have not previously occupied, and the capacity of this organism to adapt in response. Additionally, integrating tick ecology studies with climate models will enable the prediction of future disease risks and control strategy for future years. Therefore, there is still a role for public awareness campaigns and community based interventions in tick-borne disease (TBD) management as they provide a means for individuals and communities to take preventative actions. To address these challenges finally, these need to be addressed through interdisciplinary collaborations among

M. S. Sajid (✉) · H. Zahra · U. B. Tahir
Department of Parasitology, Faculty of Veterinary Science, University of Agriculture, Faisalabad, Pakistan
e-mail: drsohailuaf@hotmail.com; drsohailuaf@uaf.edu.pk

U. B. Tahir
School of Natural and Environmental Sciences, Newcastle University, Newcastle Upon Tyne, UK

S. Islam
Department of Parasitology, Faculty of Veterinary, Animal and Biomedical Sciences, Sylhet Agricultural University, Sylhet, Bangladesh

M. M. Iqbal
Institute of Soil and Environmental Science, University of Agriculture, Faisalabad, Pakistan

M. S. Sajid
Climate Change Graduate Group, Institute of Soil and Environmental Science, University of Agriculture, Faisalabad, Pakistan

M. M. Iqbal
School of Environmental Studies, China University of Geosciences, Wuhan, China

H. M. Rizwan and M. S. Sajid (eds.), *Ticks in a Changing Climate*, Biometeorology 7,
https://doi.org/10.1007/978-3-032-12259-9_12

entomology, climate science, epidemiology and public health. Effective strategies for protecting human and livestock health and those intended to minimize the economic impacts of TBDs require a coordinated global effort that takes into account the impact of climate change.

Keywords Climate driven challenges · Climate models for tick research · Climatic determinants · Diseases transmission · Economic challenges · Data integration

12.1 Introduction

Climate change (CC) is a deep and complex process. It has impacted the world's perspective on ecosystems and its issues. Climate change works as a paradigm for describing changes in the climate of the whole world. It includes changes in individual regions, and local areas resulting from the addition of Greenhouse gases (GHGs) to the atmosphere (Hallegatte & Morlot, 2011). These gases include carbon dioxide (CO_2), Nitrogen Oxides (NO_X) and methane (CH_4). Human boring of fossil products has been significantly noted to contribute to the increase of GHGs. It began around industrial revolution period. The consequences of CC are global and embrace physical geographical features and social organizations or even economies (Sejian et al., 2015).

Global warming has been considered the major issue of whole world. The hurricanes, drought and wildfires at an alarming rate are all attributed to the current high temperatures (Stocker, 2011). CC also impacts various other aspects. It includes raising the sea level, changes in the acidity of oceans, changes in the seasons. It also affects the system of the environments with different biodiversity. However, global CC has a significant impact on ticks and related diseases (Barnett et al., 2004). Global warming influence development of ticks and increase the length of activity periods in year. It also impacts the geographical distribution which leads to increase in incidence of tick-borne diseases (TBDs). An insight into these dynamics is pivotal in the formulation of well-articulated public health frameworks as well as appropriate ecological interventions (Daniels & Endfield, 2009).

The climatic conditions have clearly altered the environment that is suitable for the existence of ticks. Resultantly it changes the way of disease transmission in humans and animals. Environmental factors including temperature, moisture, and vegetation cover affect the distribution and abundance of ticks (Dantas-Torres, 2015). Higher temperatures may also lengthen the tick host-seeking period. It also alters their geographical distribution by expanding the geographical range into new habitats. For instance, the life cycle, geographical distribution, and feeding habits of many of the tick species indicated that they have moved Northward due to changes in weather patterns by 35/55 km per year (Randolph & Rogers, 2007).

This condition is favored throughout host-animal movement and habitat partition due to deforestation and urbanization. When it comes to ticks this shift can have a bad influence and lead to an extension of the period of tick activity (Gilbert,

2021). Therefore, the period during which ticks can transmit a disease is also impacted due to changes in atmospheric conditions. Climatic stress may amplify host-seeking activities of ticks and hence can potentially raise the human-tick interaction frequency (Pfaffle et al., 2013) Development processes like deforestation and urban expansion break up enormous tracts of land into small parts. Such fragmentation might contribute to the formation of more favorable conditions for ticks' residence (Ogden & Lindsay, 2016). It facilitates ecological connection of natural and developed terrains. For instance, well managed gardens with canopy in the residential areas are perfect breeding grounds for the hosts and ticks that support the movement into urban cities (Estrada-Pena et al., 2013). Development of tick habitat to residential and recreation areas poses increased risk of human interaction with the ticks. Such an increased attention is understandable but also poses a threat to public health since diseases, transmitted by ticks including Lyme disease and tick-borne encephalitis expands into new areas (Cunze et al., 2022).

12.2 Climate Driven Challenges in Research on Ticks

Ticks and TBDs are progressively more difficult to study due to the different effects caused by CC. These challenges are related to many aspects of tick biology, ecology and disease transmission (Ostfeld & Brunner, 2015). These climatic alterations has raised the need to take a holistic approach in order to get more effective understanding of ticks in changing climate. It is important to manage the risks which are associated with TBDs (Bouchard et al., 2019). There are multiple difficulties associated with tick research due to CC issues including biological, ecological, and methodological constraints. Biology and ecology of ticks and TBDs remains a growing field of knowledge as CC advances new complexities to the study (Nuttall, 2022). Ticks are involved in the transmission of pathogen to both human and animals. They also exhibit high volatility relative to the environment in terms of lifecycle, geographic distribution, and population density. This phenomenon makes ticks the best example to study that how CC influences TBDs (Tiffin et al., 2022).

Ticks have four stages in their lifecycle; eggs, larva, nymphs, and adults are hatched and develop under the effects of temperature, humidity and photoperiod. CC can change these environmental conditions that would impact on ticks (Ostfeld & Brunner, 2015). It alters the survival rates, development rates and host-seeking behavior. For instance, higher temperatures increase the ticks' ability to reproduce. It allows the existence of several generations in a single year and, consequently, causes high tick populations in some places. Increased temperature is conducive to faster development and reproduction of ticks (Bouchard et al., 2019). It also expands their activity periods. However, these changes have got varying effects on the various types of the ticks and stages of their life cycle depending on the prevailing climatic condition. Sometimes tick survival can also be eliminated by extreme events such as heat waves or droughts, which can alter tick habitats or reduce tick populations (Nuttall, 2022).

CC may shift tick distribution to higher latitudes and altitudes. It results in causing TBDs to spread to previously unreported locations. This expansion is characterized by changes in temperature and precipitation, which create hospitable circumstances for ticks to survive and proliferate. Nonetheless, tick ranges in warmer places could be shortened as well, emphasizing the multidimensional nature of the tick dispersion shift (Tiffin et al., 2022). Ticks host-seeking behavior can also be changed as a result of climatic factors. It can influence the disease transmission. For example, higher temperatures may increase the number of days in a year or hours in a day that ticks have to search for appropriate hosts. It may increase the danger angle for TBDs to be transmitted. However, less host-seeking activity in some species can offset these effects respectively (Ogden et al., 2021).

Prediction of the consequences of CC on TBDs is a hard task because in this context, several biological subsystems interact with each other. Models must refer to different aspects of environment, ticks and their hosts, and interaction between pathogens and their hosts (Suss et al., 2008). There are two main approaches. The first one is ecological models that link environmental characteristics with distributions of ticks and the second one is biophysical models that explain the processes that support these distributions. However, these models mostly presuppose assumptions that have to be checked for meeting with actual data (Gilbert, 2021). CC affects tick distribution and seasonal activities too, human behavior might change as well. For example, while weather is warmer, people tend to spend more time outdoors, raising the chances of getting bitten by a tick. Also, such factors as alterations to natural habitats and farming practices, make human interact with tick environment and thus increase chances of getting infected (Beckage et al., 2018).

12.3 Climate Models for Tick Research

Ticks and climate models can be used to understand the changes in temperature, humidity, and precipitation since they contribute to studies of ticks without affecting their functioning. They enable an understanding of how changes in environment influence ticks and transmission of diseases (Kopsco et al., 2022). Fluctuations in temperature affect developmental changes within ticks and also the time that they are active. Relative humidity plays a role in tick survival, and host searching activity whereas, drought impacts on habitat favorability for the ticks. By integrating these factors, researchers can better understand the impacts of climate change on TBDs (Voyiatzaki et al., 2022).

12.3.1 Model for Interdisciplinary Research on Climate (MIROC)

The model under consideration is the Model for Interdisciplinary Research on Climate (MIROC). It is a massively used climate model in the ecological studies of species' distribution. The analysis includes the distribution of different sorts of ticks under future climatic conditions (Suzuki & Ishii, 2011). Assumptions of environmental alterations in the dynamics of tick population is an interesting research question in health and environment with the CC impacts. The model MIROC is an interactive coupled atmosphere–ocean general circulation model developed in Japan research consortium. The main purpose is to emulate the Earth's climate and to forecast the climate state in the given years under the given concentrations of GHGs. One of the number of variants of this model is MIROC5, which is most commonly used in the latest researches (Chikamoto et al., 2013). MIROC comprises atmospheric dynamics, ocean circulation, land surface processes, and sea ice dynamics as different components and overall provides a complete picture in this area.

Application of MIROC in Tick Distribution Studies

Given the knowledge about ticks, MIROC is the most helpful in thought of how climate changes would impact the habitat of ticks. Ticks are climatically favored animals, whose life, breed and feed are depending on temperature, humidity, and rain. Researchers integrate climatic data, which they obtain from MIROC, in order to predict how far the tick dispersal and the breadth of its environment will change in the future (Wing-Le & Abe-Ouchi, 2020). MIROC forecasts future climate scenarios of the Representative Concentration Pathways (RCPs). Greenhouse gases abatement paths (RCPs) describe levels of emissions and the potential outcomes for warming the planet (Mochizuki et al., 2012). RCP 4.5 are indicative of medium atmospheric emission package characterized by CO_2 contained in the atmosphere being brought to stabilization by 2100. High emissions package with increase of GHGs describe RCP 8.5. These scenarios are integrated into several studies that use MIROC in order to analyze how ticks may respond to different climatic changes in the future (Chan et al., 2011).

Methodology in Tick Distribution Modeling by MIROC

MIROC-based methods for tick distribution modeling can also be based on several approaches to estimate impacts of climate change on ticks. The first step involves downscaling of climate projection scenarios from the MIROC outputs in terms of emission pathways such as, RCP 4.5 or RCP 8.5 (Kopsco et al., 2022). Projecting these trends is then combined with Species Distribution Models (SDMs) like MaxEnt or GLM in order to forecast future tick distributions in changing climatic determinants such as temperature and precipitation. It has to be validated, using some metrics like ROC/AUC or the cross-validation techniques to confirm the results (Leger et al., 2013). Last, the geographical information system (GIS) applications for spatial data analysis and mapping are used in identifying the susceptive zones of

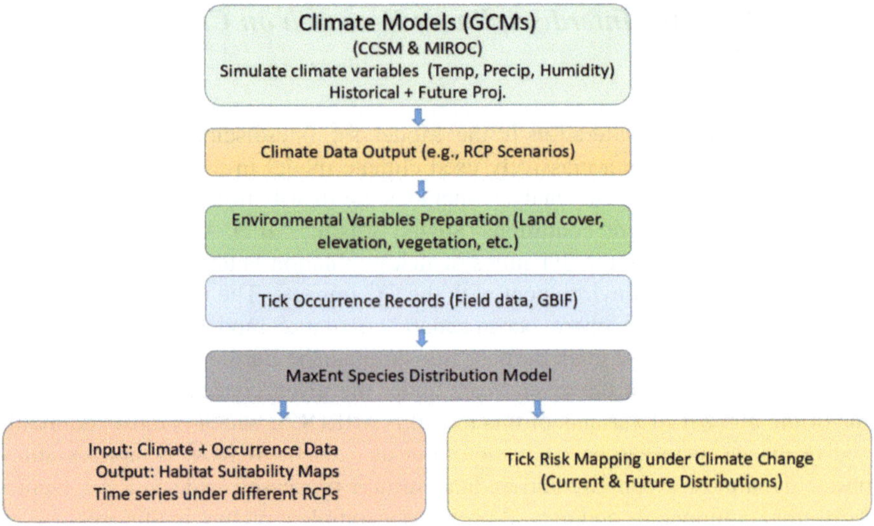

Fig. 12.1 Methodology in tick distribution modeling by MIROC and CCSM

TBDs (Fig. 12.1). Then the results are used to identify risks and guide the relevant prevention and control activities for tick and relative disease transmission. It helps to establish the appropriate prevention and management of tick population and TBDs (Williams et al., 2015).

Implications for Public Health

MIROC can be applied in public health in relation to tick information by giving the changes in the ticks distribution and diseases transmission. Incorporation of MIROC climate scenarios into species distribution models has resulted in the best estimates of impacts of changing climate (Watanabe et al., 2010). It also have provided a well estimated changes in the geographical distribution of tick species. All this information is critical in determining regions with high risk of TBDs. Due to its capability in estimating future tick distribution, MIROC is used when estimating the possibilities of risks related to TBDs (Nolzen et al., 2022). This makes it possible to prevent these risks by assigning more resources into preventive health care delivery to the public. It can be useful for determining regions requiring more attention to prevention and treatment of climate-sensitive diseases. This is especially useful for areas of the world that will witness a worsening of existing health problems due to changes in climate (Zannou et al., 2021).

Future Directions in Tick Research Using MIROC Model

Changes in climate are increasing gradually and the whole world is concerned about it. It has a significant impact on the study of tick in each aspect either its life cycle or its geographical distribution. It demands to be focused on several key areas (Alkishe et al., 2020). This would be accomplished through Longitudinal Monitoring that will follow the variation in population of ticks and the diseases spread. By adding Biotic Factors into the model, predictions will be improved with help of taking into consideration, the available hosts and their relations (Chikamoto et al., 2013). Regional Climate Models (RCMs) will give regional prognoses together with the general models as MIROC. Lastly, public engagement is important to acquaint the public with risks and to ensure that preventive measures are taken. By effectively putting the two approaches hand in hand, researchers will be in a better position to make their predictions more accurate and get the right overviews (Chan et al., 2011). It will help in developing the right strategies to fight TBDs. This integrated approach will build capacity and improve reaction capabilities to climate influences in tick distribution and phenology (Kopsco et al., 2022).

12.3.2 The Community Climate System Model (CCSM)

A sophisticated model appears in the form of the Community Climate System Model (CCSM). The research-based model goes by another name which is integrated coupled general circulation model (GCM). The model focuses on precise simulation of Earth's climate system (Blackmon et al., 2001). The CCSM develops through combined scientific research of leading climate experts who integrated complex global environmental relationships between air, ocean, ground surface and frozen ocean areas into a single unified climate simulation framework (Smith et al., 2010). The sophisticated framework that the CCSM provides for examining many climate phenomena. This is the main reason for scientists and researchers to use it as their highest advancement in climate change research. Using the model, scientists have gained much more insight into CC and its patterns because the model's competence to reproduce past and present and forecasted climates has been greatly expanded (Giles et al., 2014). This model depicts a wide range of factors as to how the GHGs emissions, oceanic processes and atmospheric systems interact with one another as correlated variables. CCSM scientists are able to get adequate tools to investigate the complex merger of nature based and human made elements as root causes to climate conditions (Wu et al., 2014). Extensively, the model is used in anthropogenic climate change research to help scientists know clearly the data through which they can conclude those human actions which are directly affecting worldwide temperature variations and rainfall distributions and these severe climate occurrence (Dome et al., 2016). CCSM plays a key role in analyzing not only inter-annual variability events and El Nino and La Nino phenomena, but also in being a good tool for the paleoclimate regime analysis to determine past climate state. To combat current climate changes worldwide, CCSM is a fundamental predictor tool to understand and create policies by which to deal with the challenges presented in climate changes (Porretta et al., 2013).

Relevance of CCSM to Tick Research

Habitat suitability predicted for changing climate environments is the main use of CCSM in tick related research. The approach represents a sophisticated modeling approach to the study of how tick populations and other climate elements relate and in combination, dictate future (Ogden et al., 2014). To help scientists to predict changes in tick habitats and to enable planning of prevention strategies and public health programs, a model was developed that simulates future climate conditions known as the CCSM. The presence of ticks is determined with the help of temperature. Additionally, with the warming global temperature, tick survival and reproduction occurs in new supportive areas leading to tick habitat being pushed further across the world (Chairman et al., 2003). Equally important to tick environment is the shift in precipitation pattern. Vegetation growth is directly affected by the amount of rainfall in that covering and environmental humidity is necessary for ticks. The changes of rainfall amount either make better conditions or insufficient for ticks to survive and reproduce (Blackmon et al., 2001). Researchers use CCSM forecasting systems to determine areas that would be at a higher risk from the growing tick populations and their continued spread of diseases such as Lyme disease. This valuable information is used by researchers and public health officials, as well as policymakers and ecologists which conduct their work to prevent the climate related health risks to humans and animals (Drake et al., 2005). With CCSM predictions, one uses current environmental conditions to make forward looking predictions of climate associated changes in tick populations.

CCSM is one essential use for it can reproduce host population variations. To these ends, climate models are informed with wildlife distribution records that incorporate critical host species (deer and rodents) to understand how changes in their populations will affect levels and activity rates of ticks as climate effects (Craig et al., 2005). The vital connection between hosts and ticks during life cycle stages and disease spreading requires thorough examination for better tick research. The Combatting Climate-Changed Spread of Disease model demonstrates its ability to forecast disease outbreak occurrences. Research assessments of tick-population forecasts alongside host-interactions enable scientists to anticipate when the diseases like Lyme disease and Rocky Mountain spotted fever will become more prevalent (Wu et al., 2014). Public health planning benefits from predictive power since official authorities can then deploy targeted interventions along with raising awareness in susceptible regions. Scientists gain insight into ecological factors of TBDs by applying CCSM outputs to develop better preventive measures (Ogden et al., 2014).

Methodology of CCSM for Tick Research

CCSM incorporates multiple components of climatic influences to simulate environmental parameters through complete simulation processes. The four core determination factors for tick survival and reproduction are atmospheric temperature, precipitation, humidity, and vegetation cover. When CCSM connects all its components it enables the creation of detailed projections at high resolution for different RCPs 4.5 and 8.5 scenarios often applied in ecological models (Kopsco et al., 2022).

The analysis of future habitable environments for ticks depends on CCSM as it enables researchers to predict climatic condition alterations through time simulation. CCSM is used to explain the predictions of the habitat shift of *Ixodes (I.) scapularis* (blacklegged tick) over two time periods: mid-century (2026–2045) and end-century (2086–2099). Climate data layers are combined with vegetation related data and climate change impacts are evaluated combined with precipitation alterations in order to potentially affect tick spread (Hahn et al., 2021). SDMs like Maximum Entropy Model (MaxEnt) and Random forest are used to merge the CCSM output to forecast tick distribution changes. Indeed, greenness metrics such as NDVI or BIOCLIM datasets are incorporated to increase the prediction accuracy (Shabani et al., 2016).

Case Studies Using Community Climate System Model

Researchers studied the western blacklegged tick (*I. pacificus*) by applying CCSM to produce future climate suitability models under RCP4.5 and RCP8.5. The study relied on records from a citizen science program to create binary habitat suitability maps which used ensemble modeling techniques for modeling three different time periods from 2026 to 2045 and 2086 to 2099. The research established that suitable tick habitats will grow in northern California but studies indicate they will decrease in southern California especially under high emission conditions. These findings demonstrate that temperature along with precipitation levels function as critical elements for understanding tick habitat suitability which supports essential public health planning about TBDs in California (Eisen & Lane, 2021).

Through CCSM, scientists developed a model to track *I. scapularis* population patterns which serves as the main carrier vector of Lyme disease across the United States. Scientists used CCSM outputs together with SDMs to understand how climate change will impact tick populations. The climate envelope model using 16 remote sensing dataset variables demonstrated a minor westward growth of suitable habitats with minor northern habitat expansion during the upcoming 2050 period. The case study demonstrates that using CCSM technology allows public health teams to develop better risk prediction models for Lyme disease risk areas in the face of climate change (Ogden et al., 2008).

Scientists in Europe utilized CCSM to model the distribution changes for *I. ricinus* ticks and their co-transmitted Lyme borreliosis pathogen throughout the region. Climate simulations conducted by the study demonstrated that Europe would experience major alterations in tick distribution zones that would benefit tick populations throughout northern European territories by the year 2050. The study's results showed climate warming will extend tick active seasons while boosting their survival rates thus increasing risks for people to contract TBDs. The analysis of the importance of CCSM to consider epidemiology of a TBDs related to the climate systems, and to prove the need for flexible public health policy, is demonstrated in this study (Klempa & Vasilenko, 2024).

These examples affirm the use of CCSM as a versatile tool in predicting how climate change is expected to affect tick numbers and TBDs. Researchers can use ecological information in conjunction with sophisticated climate models to create information that allows public health official to develop strategies to reduce the risk from newly emerging TBDs (Eisen et al., 2018; Porter et al., 2019).

12.3.3 Maximum Entropy Model (MaxEnt)

Maximum Entropy Model (MaxEnt) is a statistical approach that relies on machine learning principles, as a Maximum Entropy Principle (MEP). In the mechanical instrument, these distributions are chosen to satisfy all observed constraints according to information theory principles, using entropy maximizing distribution functions as the most suitable prediction (Phillips et al., 2006). Since MaxEnt does not enforce erroneous assumptions about the state model, it retains objectivity and manual influence. In an ecological analysis, the MaxEnt algorithm is the most convenient tool because it uses environmental variable data to model species occurrences across fields (Elith et al., 2011). In data minimal situations, the methodology works with present locations of species and not with the need of verifying empty locations. It is most appropriate to use the method in the research of hard to finds and difficult to understand species. In a way, the model works via an estimation process that seeks to predict probability distributions of target variable across predefined geographic territories or regions as well as minimizing possible ambiguity or uncertainty (Sobek-Swant et al., 2023). The model also allows environmental predictors such as temperature and precipitation to behave as limiting parameters and generate reasonable prediction results. Habitat suitability maps that depict the areas that allow for the survival and reproduction of the species given by the particular environmental factors which can be generated by MaxEnt (Zhou & Wang, 2018). As both an overfitting resistance and small sample data capability, MaxEnt helps to handle complex variable relationships. The use of this method is common among scientists when they undertake projects such as biodiversity protection, study of climate change effects, and measure for taking care of invasive species (Hernandez et al., 2006).

MaxEnt in Tick Research

Environmental variables can be used via the MaxEnt function as an essential research tool to study ticks and TBDs by predicting species distribution patterns. With presence of only data, MaxEnt performs well in building accurate ecological niches of tick species highlighting its potential suitable habitat areas and species distribution (Kleindienst and Gotz, 2022). As a tool to study the environmental factors, such as temperature, humidity and vegetation resulting in ticks activates their population, and their pathogen involvement, MaxEnt proves valuable in tick research. MaxEnt has been used by researchers to predict the range of *I. scapularis* which is the major Lyme disease transmitter in North America. They are assessing the Lyme disease risk by strongly relating tick occurrences to habitat and climatic information. For officials working in public health who must monitor and control specific areas and populations where ticks grow due to changing climate conditions (Zhou & Wang, 2023), the predictive functionality is necessary.

Researchers are able to understand how varied tick mortality and reproductive success occurs due to different bioclimatic elements. Scientists employ habitat analysis under both current and projected climate scenarios to predict potential shifts in tick population distributions. By integrating climate models with ecological data, researchers can estimate when and where suitable habitats for ticks are likely to expand, contract, or shift in response to changing environmental conditions (Liu et al., 2024). In summary, the research showed that MaxEnt can make a forecast of *I. ricinus* habitat suitability growth in northern European territories, because rising temperatures will increase the risks of TBDs. The integrating characteristic of MaxEnt permits researchers to assess intricate interaction between several environmental factors that directly impact tick population ecology (Miller & Hurst, 2021). The data obtained through these analyses are necessary to develop effective risk minimization strategies with respect to ticks and TBDs. Maximum Entropy Models have proven to be a great aid to the tick research and main source for producing useful data to support health planning of TBDs and vector control operations (Jiang & Zhang, 2016).

Methodology of MaxEnt for Tick Research

Maximum Entropy Model represents an effective analysis method which serves tick research by forecasting species distributions. The MaxEnt framework requires the collection of suitable data for its application. The research team obtains source data regarding tick occurrences by performing field surveys and accepting records from citizen scientists and museum collection databases. Researches gather occurrences of *I. scapularis* across specific geographical areas to create data collections (Phillips et al., 2006). The process includes two stages researchers must first discover environmental elements which affect tick distribution along with acquiring these ecological and climate variables. The collection of primary data involves World climate data as well as land cover maps along with other GIS layers (Elith et al., 2011). The prepared data needs analysis after the collection phase. The elimination of duplicate entries and execution of spatial uniqueness checks on presence data functions as a bias prevention measure. Realigning environmental layers at a resolution that matches the present data collection needs requires the process of environmental layer processing during data preparation (Kleindienst and Gotz, 2022).

Developing the MaxEnt model starts after preparing the datasets. MaxEnt software needs parameter configuration that includes deciding which features to use like linear, quadratic, hinge and establishing regularized conditions for avoidance of overfitting (Zhou & Wang, 2023). The MaxEnt algorithm executes to calculate the probable distribution patterns of target tick species habitats through analysis of present data and environmental elements. Careful consideration is needed for performance assessment of the model (Miller & Hurst, 2021). Validation methods that involve k-fold cross-validation and the test dataset evaluation can be utilized to assess model accuracy. The ability of the model to act with a predictive horizon is often evaluated via the Area under the Curve (AUC) measure which is obtained from a Receiver Operating Characteristic (ROC) analysis (Jiang & Zhang, 2016). Predictive habitats for ticks are predicted by statistical methods of negative binomial

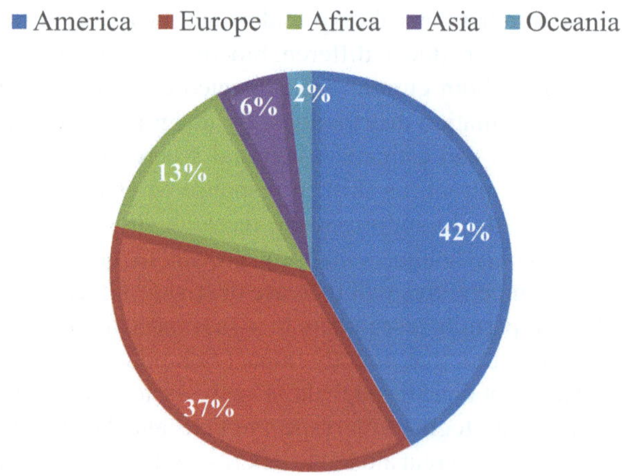

Fig. 12.2 Regional percentage of tick research in the world. The pie chart shows that the majority of tick research from 2011 to 2024 was conducted in America (42%) and Europe (37%), with significantly less focus in Africa, Asia, and Oceania. This highlights a regional imbalance in global tick research efforts

regression and Spearman's correlation on data that actual tick densities have taken. Careful thought must be given to performance assessment of the model. Validation methods using k-fold cross validation as well as test dataset evaluation can be used to assess the model accuracy. The AUC of the ROC is normally used to measure the model predictive performance (Ogden & Lindsay, 2016). The correlation between predicted suitable habitats and actual tick population distribution can be statistically tested by Spearman's correlation and negative binomial regression. Regional percentage of tick research in the world is given in Fig. 12.2.

Case Studies Using MAXENT

In Ontario, a research group studied MaxEnt to predict capability of estimating *I. scapularis* density. From Ottawa to Kingston, active tick surveys were performed at 60 locations in southern Ontario. At these locations, researchers developed predictions of tick habitat suitability using MaxEnt modeling and environmental variables of temperature, humidity and land cover. It was proved statistically that the predicted areas of suitable habitats correlated positively to measured tick densities with Spearman's correlation values between 0.517 and 0.707 in different research areas. According to the incidence rate ratio findings, the research showed that tick populations were more abundant within regions with suitable environments than within unsuitable environments. MaxEnt proved to be an economical and effective software tool which enables scientists to detect Lyme disease risk areas for enhancing public health tick control efforts (Kleindienst and Gotz, 2022).

Researchers used MaxEnt to create predictions regarding the geographic range of *Amblyomma (A.) americanum* (Lone Star ticks) across Kansas. The research used museum specimen and field collection records in combination with temperature variables and precipitation data as bioclimatic factors. According to the MaxEnt model the areas demonstrating maximum suitability for *A. americanum* distribution were revealed through identified associations between environmental factors and tick distribution patterns. Analysis results showed that climate change might expand the species range into new territories thus increasing human risk of encountering tick-caused diseases ehrlichiosis and tularemia. The tick distribution prediction based on MaxEnt demonstrates the crucial value of this model to forecast tick range modifications because of climate variations (Feria-Arroyo & Lutz, 2016).

12.4 Climatic Determinants and Tick Behavior

Environmental factors which create patterns of weather and control both ecosystems and animal population alongside human populations are known as climatic determinants. Each determinant in weather patterns consists of temperature alongside precipitation together with humidity and wind patterns as well as atmospheric pressure with solar radiation (Hahn et al., 2021). CC leads to warming global temperatures which alters ecosystem habitats and creates new heat hazards for living beings and produces more powerful weather phenomena. (Dell et al., 2008). Weather patterns as well as human comfort and disease transmission depend directly on the water content in the air (Lilleor and Van de Broeck, 2011). Heat stress occurs during high humid days but low humidity results in dry conditions that increase wildfire possibilities. The development of regional climates gets major influence from the wind patterns of trade winds and monsoons (Tai et al., 2010). Surface temperatures together with climate zones change according to the solar energy that reaches the planet. Weather conditions which include storms and droughts are controlled by atmospheric pressure systems between high and low intensities (Huang et al., 2017). These climatic determinants are needed in predicting weather patterns, mitigating climate change impacts, and formulating sustainable adaptation strategies (Porter et al., 2019). As ectoparasitic arthropods, ticks are very sensitive to environmental conditions. Survival, development and host seeking behavior of vectors is driven by temperature, humidity, precipitation and seasonal variations. As global climate change continues to shift these factors, it affects the distribution and activity patterns of ticks, and thereby disease transmission dynamics (Eisen et al., 2018).

12.4.1 Temperature and Ticks Development

When it comes to tick molting, reproducing, and questing, temperature is a very important factor. Ticks develop faster under warmer temperatures, because shorter time is needed before eggs hatch and then larvae molt to become nymphs and adults (Suss et al., 2008). As a result, tick populations increase and tick seasons are extended. On the other hand, extreme heat can be harmful, decreasing survival rate if moisture is not sufficient to minimize dehydration. With the input of global warming, some tick species have been able to expand their range northwards into formerly unsuitable, colder areas. This is a shift that is associated with more TBDs because the parasites fit into new hosts and ecological niches (Wu et al., 2010).

12.4.2 Humidity and Tick Survival

Survival of ticks is extremely dependent on humidity. For most of their life cycle, they spend it in the leaf litter or soil where they take in moisture from the environment. Desiccation and death will occur in species that are not adapted to arid conditions as a result of a drop in relative humidity (Rodgers et al., 2007). Rainfall is generally increased and humidity is higher in temperate and tropical climates and these conditions have increased the populations of ticks, making it easier for them to quest for hosts. Changes in humidity also influence tick behavior. Ticks are in a questing state when humidity levels are optimal, straining to cling to passing hosts on vegetation. Under dry conditions, they reduce their activity but prolong their survival by retreating to microhabitats (Herrmann & Gern, 2014).

12.4.3 Precipitation and Ick Survival

The suitability of tick habitat for survival is also dependent on rainfall, as a moist environment is needed for ticks to survive. The increase of precipitation leads to vegetation growth providing shelter and host availability (Nielebeck et al., 2023). Nevertheless, standing water from excessive rainfall can cause the tick population to drop temporarily after flooding because the insect is washed away from its environment. However, prolonged droughts also negatively affect tick survival, especially on species relying on humid microclimates. In defense, ticks may modify their questing activity by reducing it and seek locales that are more shaded and humid (Burtis et al., 2016).

12.4.4 Seasonal Variations and Host Availability

The activity of ticks vary with the season, although most species exhibit peak activity in spring and summer when there are plenty of hosts. Still, with warmer temperatures worldwide, ticks have started having autumn and even winter activity in milder climates (Dowell, 2001). The activity throughout this period puts all humans, livestock, and wildlife at a greater risk for transmission of TBDs. Climate change also alters migration patterns and population densities of the hosts, and hence indirectly affects tick distribution (Oorebeek & Kleindorfer, 2008). The availability of more mammalian hosts' behavior, including earlier emergence from hibernation, gives ticks additional chances to feed and reproduce. Although ticks are not particularly mobile, wind patterns can move them via hosts, birds and small mammals. Indirect effects of strong winds on tick populations can include shifting host movement, changing habitat structure, and thus tick distribution (Botto et al., 2005).

12.5 Major Diseases Transmitted by Ticks in a Changing Climate

Tick-borne diseases are emerging as a threat to public and animal health worldwide and climate change is affecting their spatial distribution and incidence as well as their severity (Gray et al., 2009). Increasing interaction between ticks and human and animal populations is attributed to rising global temperatures, shifted precipitation patterns, changing ecosystems, and increasing their geographical reach. With changing tick habitats, the prevalence of major TBDs is set to rise, as urgent research is warranted in terms of the further epidemiology, prevention and control of TBDs (Dantas-Torres, 2015).

The most common TBD in North America and Europe is caused by *Borrelia (B.) burgdorferi* and transmitted by *I. scapularis* and *I. ricinus.* Active season for these ticks is now extended on account of CC and they migrated into previously unaffected areas (Gray & Ogden, 2021). Ticks can be active all year round and are able to spread further, into Canada, Scandinavia, and even higher altitudes. Warmer winter reduces their mortality and a warming of the land (Marques et al., 2020). Therefore a higher density of tick can increase the exposure and infection of human. Lyme disease is on the rise and to reduce, early detection, climate adapted surveillance, and vaccination strategies are needed (Bouchard et al., 2019).

Anaplasmosis is caused by *Anaplasma phagocytophilum* and ehrlichiosis by *Ehrlichia (E.) chaffeensis* and *E. ewingii.* Both bacterial diseases usually transmitted by *I. scapularis* and *A. americanum*, respectively (Marques et al., 2020). Historically, these diseases had been typically restricted to temperate regions but are now being observed in new geographical regions. Expansion of distribution range of *A. americanum* from the southern U.S. has been associated with increased ehrlichiosis cases in the Midwest and Northeast. Increased overlaps between *I. scapularis* and *Amblyomma* species increase likelihood of suffering multiple TBDs at once (Bouchard et al., 2019).

Babesiosis is a protozoan infection transmitted by *I. scapularis* and caused by *Babesia microti.* A disease that infects red blood cells, it can be serious in immune-compromised individuals and reached popular attention as a growing public health problem (Homer et al., 2000). Climate induced shifts in tick seasonality increases the time frame during which *Babesia* specie is transmitted. The increase in the risk of babesiosis requires prioritizing on advancements in diagnostic tools and vector control strategies (Gray & Ogden, 2021).

Rocky Mountain Spotted Fever (RMSF), is cause *by Rickettsia (R.) rickettsii* that is transferred to man through *Dermacentor (D.) variabilis, D. andersoni*, and *Rhipicephalus (Rh.) sanguineus* (Snowden and Simonsen, 2025). If not treated promptly, RMSF has one of the highest fatality rates of the TBD. CC has been a limit to the reproductive success of *Rh. sanguineus*, the key vector of RMSF in urban areas, however, increased temperatures have improved its performance (Raghavan et al., 2016). CC associated tick adaptation has enabled this species to expand into human dominated areas and thus enhancing urban transmission. In addition, the rise in temperature may increase *R. rickettsii* virulence and result in more aggressive outbreaks. In the future, integrated tick control programs, climate resilient healthcare strategies as well as public awareness campaigns will be important in managing RMSF (Baumgarten et al., 2019).

Crimean Congo Hemorrhagic Fever (CCHF) is a dangerous viral disease transmitted by *Hyalomma* species of ticks. Traditionally the disease is endemic to parts of Africa, Middle East and Asia, but cc is allowing it to spread to new areas, most recently Southern Europe (Gale et al., 2010). *Hyalomma* ticks can now reproduce in Mediterranean countries and Central Asia as temperatures rise and warmer, moister conditions are conducive to tick virus persistence. *Hyalomma* ticks are also introduced into previously unaffected areas through the forces of globalization, trade and migratory bird movements (Maltezou and Papa, 2010). Preparations for future growing threat of CCHF need to include enhanced surveillance, early warning systems and research on use of vaccination targeted mostly at tick vectors (Duygu et al., 2018).

Tick-borne encephalitis (TBE) is an infection transmitted by *I. ricinus* and *I. persulcatus*. The central nervous system is damaged severely by the disease and may lead to death. CC is increasing TBE transmission by raising the tick survival rates and reproduction, and by warming the conditions that enable *I. ricinus* to expand its habitat to higher altitudes and latitudes, thus raising human exposure (Lindgren & Gustafson, 2001). Milder winters as well as early springs may raise the period of tick activity and subsequently the number of infections. Major efforts in widespread vaccination campaigns, tick control strategies and public health interventions will be needed to keep from seeing an increasing burden of TBE (Sumilo et al., 2007).

The Powassan virus (POWV) is demonstrated to be a rare but deadly TBD spread by the *I. Scapularis* and *I. cookei*. This can lead to long term complications or death and if it does occurs it can be severe such as encephalitis which is severe inflammation of the brain and covering of the brain (Hassett and Thangamani, 2021). Where Lyme disease needs tick attachment for a long time, POWV, a close relative of it, can be passed on in as little as 15 min after a tick bite. CC has expanded the range of *I. scapularis* and thus human exposure to POWV has increased while the effect of higher temperatures on viral evolution may result in more virulent strains (Hermance and Thangamani, 2017). Prevention of future outbreaks of POWV depends on investing in vaccine development and climate responsive disease surveillance.

Ticks are profoundly affected by CC and their epidemiology is being changed as such diseases spread into new geographic areas, extend their activity in the season, become more virulent (Raghavan et al., 2016). Predictive modeling, surveillance and control, public health initiatives, vaccine and treatment development are prioritized for the future of tick research. Efforts to develop climate based disease forecasting tools allows forecasting outbreak events, and enhancement of monitoring programs for tick populations and pathogen circulation will contribute to effective intervention (Dell et al., 2008).

12.6 Potential Evolutionary Responses and Resilience in Tick Populations

It is reported that CC is causing major alterations in ecosystems around the globe, and that tick populations are being extremely adaptable to those environmental changes. As ectoparasites, ticks are known to be dependent on external conditions to survive, develop and seek their host (Ogden & Lindsay, 2016). These changes have made such an impact on their life cycles that they are driving evolutionary responses to increase their resilience. Because these adaptive mechanisms also predict future tick population dynamics, spread of TBDs, and design of effective control strategies (Hahn et al., 2021). The thermal tolerance is one of the most significant evolutionary responses that we see in tick populations. With increasing global temperatures, ticks are spreading and approaching areas that used to be inhospitable (Chairman et al., 2003). For example, the main vectors of Lyme disease *I. scapularis* and *I. ricinus* have expanded into higher latitudes and elevations because of increased death rates in the winter (Craig et al., 2005). Warmer conditions allow ticks to stay alive for longer and therefore potentially make more ticks. Thus, natural selection will favor individuals with a greater heat resistance and populations will gradually evolve to be more thermally resilient (Porretta et al., 2013).

Evolutionary changes to questing activity in tick behavior also are linked to changes in climate. Ticks questing for hosts is dependent upon temperature, humidity, and carbon dioxide. Some tick species have adapted their questing behaviors to the warming temperatures by changing into cooler parts of the day, or behaviorally avoiding desiccation (Gent et al., 2011). In drier areas, ticks may evolve to tolerate lower humidity and therefore be better at surviving in a dry area (Smith et al., 2010). Another key mechanism used in tick adaptations to enhance resilience is reproductive adaptations. Increased numbers of reproductive cycles per year has resulted from prolonged tick activity seasons due to climate change (Wu et al., 2014). A few species evolve shorter and shorter times to complete their development because there is an advantage to completing the life cycle quickly and to use the brief time when conditions are favorable after the hatch (Drake et al., 2005). When they have higher reproductive output, there are population booms in suitable habitats, which greatly increases the risks of TBDs outbreaks. Genetic selection of higher fecundity individuals may also increase the persistence of populations in fluctuating climate (Giles et al., 2014).

Tick resilience also relies heavily on the ability of the host to adapt. Toward the end of their life cycle, ticks have to change to new hosts, as the host population change for traditional hosts due to climate change and fragmentation of habitats (Ogden et al., 2014). A few species have turned out to possess the capacity to eat an increasingly noticeable number of hosts and consequently chance upon of working in different ecosystems. The ability of ticks to alter host preference after encounters provides the basis for their continued success in altered stable climatic regions. Further, overlap of tick and wildlife, domestic animals and human populations increases and facilitates tick adaptation and increases the intensity of tick-borne pathogen transmission (Smith et al., 2010).

The emergence of traits which are impervious to present environment is being further influenced by changes in climate on genetic diversity within populations of ticks. Environmental changes are rapidly hurrying up, selectively pushing for the individual with genetic mutations that increase the resilience. These may be any of the mutations that lead to increased resistance to desiccation, altered metabolic rates, and improved immune evasion strategies against host defenses (Kopsco et al., 2022). The result of this natural selection process over time is that tick populations can develop an evolutionary resistance to the environmental stressors and control measures such as acaricides. Ticks are very resilient under CC, creating substantial challenges to tick management and to prevention of disease. This ability of ticks to adapt and change puts forth the need to continually develop and adjust control strategies to climate fluctuations (Williams et al., 2015).

The continuously shifting threats from tick populations require continuous improvement of integrated tick management approaches including habitat modification, targeted use of acaride, and host targeted interventions. Future work should focus on unveiling the genetic and physiological mechanisms underlying tick evolutionary resilience to predict future changes and discovery of new approaches to control the effects of TBDs (Alkishe et al., 2020). With CC speeding up, ticks are adapting quickly to survive in new and ever changing habits. These vectors of disease are able to adapt their questing behavior, modify the overwintering sequence, and increase reproductive success. Exploit new hosts with temperature extremes and some are arguably more successful in doing so than other Ixodid ticks (Chan et al., 2011). As global cc will influence future evolutionary trends in tick populations, forecasting future evolutionary trends in tick populations will be critical to the development and effective implementation of public health measures in response to these increasingly common and burdening TBDs (Nuttall, 2022).

12.7 Existing Strategies for Tick Control and Disease Monitoring

In response to the accelerating impacts of climate change, including the expansion of tick populations and the diseases they transmit, several strategies have been developed to mitigate tick infestations and reduce disease transmission. These are approaches that integrate chemical, biological, ecological and technological methods to reduce risk of tick to humans, livestock and wildlife (De La Rocque, et al., 2008). However, the efficacy of these strategies changes depending on conditions in the environment, availability of host and the adaptability of tick populations. Application of chemical acaricides is one of the most commonly used methods of tick control causing tick mortality at different stages of its life cycle. These tick population reducing agents are applied as chemicals directly onto livestock, companion animals or the environment.

Acaricides are effective in reducing infestations but their wide spread use has resulted in the development of resistance to most acaricides in many tick species (Johnson et al., 2022).

Consequently, producers are seeking alternative tick control methods due to these resistance and environmental concerns of pesticide residues. An example of biological control is using natural predators, parasites and pathogens that are predators of the tick to keep tick populations in check. *Metarhizium anisopliae* and *Beauveria bassiana* among the entomopathogenic fungi have the potential to infect and kill ticks in their natural environment (Ghosh et al., 2006). Nematodes and predatory insects for example, ants and beetles take a toll on the tick population in some ecosystems as well. Tick burdens can also be reduced without depending on chemical interventions by the introduction of tick resistant livestock breeds and wildlife management strategies (Stafford et al., 2017).

Modification of habitat is another important way to control ticks. Because ticks flourish in humid, vegetated habitats, having at least some tall grasses and leaf litter as well as well grazed pastures can limit tick habitats. The disruption of tick life cycles in agricultural landscapes have been achieved by using controlled burning and rotational grazing (Osbrink et al., 2022). Also, cultivating buffer zones between human dwellings and tick infested areas can help decrease the incidence of human tick bite. In host targeted interventions, the intervention effort is directed towards reducing tick infestations in key hosts, for example deer, rodents, and livestock (Johnson et al., 2022). Therefore, many areas of the country are engaged in deer population management as a means of controlling the spread of *I. scapularis*. Another promising disease prevention approach is vaccination efforts against tick-borne pathogens mainly for livestock diseases such as East Coast fever (Gale et al., 2010).

Tick population dynamics and the appearance of TBDs are monitored and surveyed through disease monitoring and surveillance systems. Passive surveillance is based on making reports of human and animal cases, whereas active surveillance includes field sampling of ticks and of the host species (Elith et al., 2011). Often infected ticks can only be detected by pathogen detection, usually using polymerase chain reaction (PCR) or next generation sequencing which help identify pathogens early enough to modify strategies to protect at risk populations. When predicting changes in tick distribution and future disease risks, remote sensing and climate modeling are also used more frequently (Liu et al., 2024). Public awareness and education programs are of great importance in tick control by presenting information on the preventive measures, for example personal protective clothing, repellents, and correct tick removal techniques (Stocker, 2011). In order to limit the effect of ticks on public health and agriculture, a mixture of integrated management strategies and technological progress will therefore be required as ticks increasingly spread into new areas (Barnett et al., 2004).

12.8 Tick Infestation and Economic Challenges

Shifts in tick distribution and dynamics of TBDs are driven by cc and present serious economic challenges in the area of public health and agriculture. Tick populations, that normally cause such diseases, are expanding into new areas and placing a higher burden on human and animal populations, resulting in higher healthcare costs, greater losses in livestock raising, and reduced agricultural production (Caminade et al., 2019). In particular, the serious intensification of these challenges makes it critical to understand their long run economic and social consequences in order to effectively develop mitigation strategies. Increases in tick range also mean increased healthcare burden. The rise in the TBDs like Lyme disease, Rocky Mountain spotted fever, CCHF demand more investments on medical care, diagnostics and public health interventions (Ghafar et al., 2021). Therefore, in locations where TBDs appear to be emerging, the healthcare infrastructure may not have the resources to deal with increasing numbers of cases. Medical treatment, hospitalization, long term complications and illness resulting in loss of productivity place enormous financial burden on individuals as well as healthcare providers. For instance, chronic Lyme disease cases, which require drawn out treatment, decrease work ability and give ascent to expanded disability cases, include in the economic weights on national and localized medical care frameworks (Ogden & Lindsay, 2016).

Tick infestations are of great importance in agricultural sector due to their significant impact on livestock health and productivity. A wide range of diseases such as bovine anaplasmosis, babesiosis, and theileriosis as well as other relevant diseases are transmitted by ticks and causes milk production decline, weight loss, reproductive failure, and in severe cases death. These diseases reduce agronomic value of the herd through loss of valuable breeding stock and incur veterinary expenses. Finally, outbreaks of TBDs in livestock have ramifications for trade restrictions that later, negatively affect the economic stability of livestock dependent communities and economies of nations that rely on meat and dairy for export (Rocklov and Dubrow, 2020).

Apart from the direct health effects, agricultural productivity is influenced as there are climate driven changes in tick activity. Ticks can be very heavy parasites of cattle, as heavily infected animals become anemic, lethargic and have a poor feed conversion efficiency. TBDs lead to loss of livestock which hampers food security in areas where cattle, goats or sheep are the livelihood of cattlemen (Gray et al., 2009). Unfortunately for the farmers, whose profit margins are already slim, the money lost battling tick infestations also reduces their profit (Caminade et al., 2019). Another economic problem is that there is a growing reliance on chemical tick control measures. Acaricides have been widely used as a primary method for controlling tick infestations, but increasing its use has developed resistance in tick population requiring higher doses and more frequent applications. The increase in chemical dependence has a high cost for farmers, is an environmental threat that threatens environmental sustainability and has effects on non-target organisms such as soil microbes and pollinators that are critical to agricultural health (Sumilo et al., 2007).

As cc is shifting tick distribution and disease transmission dynamics, unchecked tick infestations will have an increasingly more profound economic impact. Sustainable strategies to decrease healthcare costs, protect livestock, and minimize the negative impact on agricultural productivity in the context of emerging TBD threats need to be developed in consultation of all stakeholders, agricultural, researchers, and finally policymakers (Gale et al., 2010).

12.9 Importance of Public Awareness Campaigns and Community-Based Interventions

Tick-borne disease outbreaks and the rising tick populations driven by climate change have highlighted the urgent need for robust public health interventions. In addition to these measures, public awareness campaigns and community-based initiatives are becoming increasingly vital in effectively addressing the risks posed by ticks (Sakamoto, 2018). These strategies enable the populations and communities to be proactive in reducing tick exposure, preventing disease, and create awareness about the best ways of promoting health. Public awareness campaigns have been effective not only in informing the general public, but also can play a role in making the public feel like TBD prevention is shared responsibility. Once aware of the risks associated with tick exposure, people can take precautions to prevent the spread of diseases from ticks (Potes et al., 2023). Many of these campaigns can be done through the media, educational materials or outreach to schools, community centers and workplaces. These campaigns can provide information regarding tick identification, symptoms of TBDs and how to correctly remove a tick so that individuals may take precautions. For instance, people can be taught to use tweezers to correctly remove ticks, to wear protective clothing when entering tick infested areas and/ or apply tick repellents to stay away from infection (Mowbray et al., 2012).

Alongside awareness efforts, especially those focused on prevention, these campaigns can significantly reduce the burden on healthcare systems. By encouraging early diagnosis and treatment, they help prevent tick-borne diseases from progressing to more severe stages and reduce associated complications. Public awareness campaigns are one of the primary means of reaching the various corners of the population. Additionally, public awareness campaigns can encourage coordinated response from the public health agencies, local authorities, and community organizations to respond to the increasing threat of TBDs. Dissemination of information is helped and more people are engaged in decisions about preventive measures. Tick control and disease prevention strategies at the local level are supported by community based interventions (Sakamoto, 2018). Local communities should be involved in tick management, so that they feel that the initiatives are theirs. Local surveillance programs for residents trained to identify ticks and report their presence, use of environmentally sustainable practices (i.e. habitat modification, integrated pest management) comprise of community based interventions. Engagement of community members in monitoring the spread of ticks and identification of cases of TBDs is one method for local authorities to react quickly to outbreaks and target resource allocation where it is needed most (Johnson et al., 2022).

In addition, community based interventions allow for use of local knowledge and cultural practices, especially in realms were conventional tick control techniques may not be applicable. For example, there are plant based repellents that some communities have traditional knowledge, or control strategies of livestock and wildlife that are complementary to the modern scientific methodology (Lindgren and Gustafson, 2001). Specifically, public health campaigns that utilize these local practices also incorporating evidence-based interventions may allow for a more holistic approach to TBD prevention. A second benefit of community based interventions is that they can decrease the stigmatization of people who have been afflicted by TBDs. In many cases, people in rural and remote areas might be afraid of social isolation and not having the adequate knowledge about the existence of health resources (Duygu et al., 2018). These barriers can be reduced by community interventions promoting health seeking behavior, changing the social norm to discuss TBDs and promote early intervention. Therefore, community leaders, health workers, and local organizations should be the trusted source of information for the people so individuals would feel free to look for medical advice and help.

12.10 Critical Areas for Future Research

Changing climates should also be studied through research into tick physiology and genetic adaptations in order to determine how ticks will react to these changes. Evidently, physiological or genetic adaptions may develop in ticks to cope with new environmental pressures, which consequently allow them to live in new and different habitats (Gray et al., 2009). We will also look at the potential of tick genomics and study of gene environment interactions to give us insight into evolutionary responses to such changes, for instance, in increased resistance to environmental or acaricide stressors. It will also help to understand how ticks adapt their life cycle, reproduction rate, and host preferences in expected climate change scenarios to define more targeted and sustainable control strategies (Obaid et al., 2022). Another important area of research is on the manner in which ticks interact with their host species in changing climates. Ticks need to obtain blood meals from a variety of hosts including livestock, wildlife and humans. Given that climate change will alter ecosystems and species' distributions, and hence potentially the availability of suitable tick hosts. Further research is needed to determine whether these species will continue to provide significant host habitat to the ticks (Hussain et al., 2024). Host migration patterns, changes in host densities, and the consequences of these shifts to tick population are essential to understanding how climate change impact on the transmission dynamics of TBDs. Finally, determining the roles of wildlife in tick ecology and disease transmission will help to better understand disease reservoirs and their relations to human populations (Deshpande et al., 2023).

Another critical area for future research is the development of novel diagnostic tools for the early detection of TBDs. Therefore, early and rapid detection of pathogens in both tick vectors and host populations will become more important as ticks extend into new regions. More sensitive and specific diagnostic advancements such as CRISPR based technologies could lead to more rapid and accurate identification of tick-borne pathogens (Garcia et al., 2022). Better diagnostic measures would aid in better surveillance of the diseases as well as promote more effective treatments and interventions to stop the spread of the same. Research in these areas will be interdisciplinary and therefore fundamental. A whole host of tick research falls in the realm of entomology, ecology, epidemiology, climate science, veterinary science and public health (Sejian et al., 2015). Combining these disciplines will assist researchers to exploit information on environmental change, vector biology and disease dynamics so as to develop integrated strategies for tick and disease control and prevention. An example is the use of ecologist and climate scientists in making predictions about how shifts in temperature and precipitation patterns will affect tick populations and host and interactions. Epidemiologists and medical researchers can combine forces in order to track the spread of TBDs and assess its public health impacts due to climate change (Stocker, 2011). Finally, collaborations between nations are required because these diseases are global. Given that ticks are expanding over borders, sharing data, and resources and expertise will be imperative in addressing the tick challenges. Tick population expansion can be mitigated through global coordination of surveillance, research and development of tick control strategies and vaccines (Barnett et al., 2004).

References

Alkishe, A., Cobos, M. E., Peterson, A. T., & Samy, A. M. (2020). Recognizing sources of uncertainty in disease vector ecological niche models: An example with the tick *Rhipicephalus sanguineus sensu lato*. *Perspectives in Ecology and Conservation, 18*, 91–102.

Barnett, T. P., Malone, R., Pennell, W., Stammer, D., Semtner, B., & Washington, W. (2004). The effects of climate change on water resources in the west: Introduction and overview. *Climatic Change, 62*, 1–11.

Baumgarten, E. B., Enger, M., Hickey, B. V., Kennedy, R., Matthews, T., & Qualls, S. (2019). Effects of climate change on Rocky Mountain spotted fever and public health implications in Western Montana. Undergraduate Theses, Professional Papers, and Capstone Artifacts (Vol. 243). https://scholarworks.umt.edu/utpp/243

Beckage, B., Gross, L. J., Lacasse, K., Carr, E., Metcalf, S. S., Winter, J. M., & Hoffman, F. M. (2018). Linking models of human behavior and climate alters projected climate change. *Nature Climate Change, 8*, 79–84.

Blackmon, M., Boville, B., Bryan, F., Dickinson, R., Gent, P., Kiehl, J., & Washington, W. (2001). The community climate system model. *Bulletin of the American Meteorological Society, 82*, 2357–2376.

Botto-Mahan, C., Cattan, P. E., Canals, M., & Acuña, M. (2005). Seasonal variation in the home range and host availability of the blood-sucking insect *Mepraia spinolai* in wild environment. *Acta Tropica, 95*(2), 160–163.

Bouchard, C., Dibernardo, A., Koffi, J., Wood, H., Leighton, P. A., & Lindsay, L. R. (2019). Climate change and infectious diseases: The challenges: Increased risk of tick-borne diseases with climate and environmental changes. *Canada Communicable Disease Report, 45*, 83–89.

Burtis, J. C., Sullivan, P., Levi, T., Oggenfuss, K., Fahey, T. J., & Ostfeld, R. S. (2016). The impact of temperature and precipitation on blacklegged tick activity and Lyme disease incidence in endemic and emerging regions. *Parasites & Vectors, 9*, 1–10.

Caminade, C., McIntyre, K. M., & Jones, A. E. (2019). Impact of recent and future climate change on vector-borne diseases. *Annals of the New York Academy of Sciences, 1436*, 157–173.

Chairman, J. T. K., Bretherton, N. C. S., Chang, P., & Bitz, C. (2003). *Community Climate System Model science plan*. National Center for Atmospheric Research.

Chan, W. L., Abe-Ouchi, A., & Ohgaito, R. (2011). Simulating the mid-Pliocene climate with the MIROC general circulation model: Experimental design and initial results. *Geoscientific Model Development, 4*, 1035–1049.

Chikamoto, Y., Kimoto, M., Ishii, M., Mochizuki, T., Sakamoto, T. T., Tatebe, H., & Imada, Y. (2013). An overview of decadal climate predictability in a multi-model ensemble by climate model MIROC. *Climate Dynamics, 40*, 1201–1222.

Craig, A. P., Jacob, R., Kauffman, B., Bettge, T., Larson, J., Ong, E., & He, Y. (2005). CPL6: The new extensible, high performance parallel coupler for the community climate system model. *The International Journal of High Performance Computing Applications, 19*, 309–327.

Cunze, S., Glock, G., Kochmann, J., & Klimpel, S. (2022). Ticks on the move. Climate change-induced range shifts of three tick species in Europe: Current and future habitat suitability for *Ixodes ricinus* in comparison with *Dermacentor reticulatus* and *Dermacentor marginatus*. *Parasitology Research, 121*, 2241–2252.

Daniels, S., & Endfield, G. H. (2009). Narratives of climate change: Introduction. *Journal of Historical Geography, 35*, 215–222.

Dantas-Torres, F. (2015). Climate change, biodiversity, ticks and tick-borne diseases: The butterfly effect. *International Journal for Parasitology: Parasites and Wildlife, 4*, 452–461.

De La Rocque, S., Rioux, J. A., & Slingenbergh, J. (2008). Climate change: Effects on animal disease systems and implications for surveillance and control. *Review Scientifique Technique, 27*(2), 339–354.

Dell, M., Jones, B. F., & Olken, B. A. (2008). Climate change and economic productivity: Evidence from temperature stress in the U.S. *Journal of Environmental Economics, 101*(6), 116–128.

Deshpande, G., Beetch, J. E., Heller, J. G., Naqvi, O. H., & Kuhn, K. G. (2023). Assessing the influence of climate change and environmental factors on the top tick-borne diseases in the United States: A systematic review. *Microorganisms, 12*(1), 50.

Dome, C., Sandor, A. D., & Mihalca, A. D. (2016). Climate change and species distribution: Possible scenarios for thermophilic ticks in Romania. *Geospatial Health, 11*(2).

Dowell, S. F. (2001). Seasonal variation in host susceptibility and cycles of certain infectious diseases. *Emerging Infectious Diseases, 7*(3), 369–374.

Drake, J. B., Jones, P. W., & Carr, G. R., Jr. (2005). Overview of the software design of the community climate system model. *The International Journal of High Performance Computing Applications, 19*(2), 177–186.

Duygu, F., Sari, T., Celik, H. (2018). Effects of platelet function on the haemorrhagic manifestations and mortality in Crimean-Congo haemorrhagic fever. *Infections in Medicine, 26*(4), 341–346.

Eisen, L., & Lane, R. S. (2021). Modeling future climate suitability for the western blacklegged tick, *Ixodes pacificus*, in California with an emphasis on land access and ownership. *Ticks and Tick-Borne Diseases, 12*, Article 102746.

Eisen, R. J., Eisen, L., & Lane, R. S. (2018). Modeling climate suitability of the western blacklegged tick in California. *Journal of Medical Entomology, 55*(5), 1133–1142.

Elith, J., Phillips, S. J., Hastie, T., Dudík, M., & Chee, Y. E. (2011). A statistical explanation of MaxEnt for ecologists. *Diversity and Distributions, 17*(1), 43–57.

Estrada-Peña, A., Gray, J. S., Kahl, O., Lane, R. S., & Nijhof, A. M. (2013). Research on the ecology of ticks and tick-borne pathogens: Methodological principles and caveats. *Frontiers in Cellular and Infection Microbiology, 3*, 29.

Feria-Arroyo, T. P., & Lutz, R. (2016). Maximum entropy-based ecological niche model and bio-climatic determinants of *Amblyomma americanum* in Kansas. *Journal of Vector Ecology, 41*(1), 99–111.

Gale, P., Estrada-Peña, A., Martinez, M., Ulrich, R. G., Wilson, A., Capelli, G., Phipps, P., De La Torre, A., Muñoz M.J., Dottori M., Mioulet V., & Fooks A. R. (2010). The feasibility of developing a risk assessment for the impact of climate change on the emergence of Crimean-Congo haemorrhagic fever in livestock in Europe: A review. *Journal of Applied Microbiology, 108*, 1859–1870. https://doi.org/10.1111/j.1365-2672.2009.04638.x

Garcia, K., Weakley, M., Do, T., & Mir, S. (2022). Current and future molecular diagnostics of tick-borne diseases in cattle. *Veterinary Sciences, 9*(5), 241.

Gent, P. R., Danabasoglu, G., Donner, L. J., Holland, M. M., Hunke, E. C., Jayne, S. R., & Zhang, M. (2011). The community climate system model version 4. *Journal of Climate, 24*(19), 4973–4991.

Ghafar, A., Gasser, R. B., & Jabbar, A. (2021). Impact of climate change on tick-borne diseases of livestock in Pakistan–looking ahead. In: Nuttal, P. (ed.), Climate, ticks and disease. CABI.

Ghosh, S., Azhahianambi, P., & de la Fuente, J. (2006). Control of ticks of ruminants, with special emphasis on livestock farming systems in India: Present and future possibilities for integrated control—a review. *Experimental & Applied Acarology, 40*(1), 49–66. https://doi.org/10.1007/s10493-006-9022-5

Gilbert, L. (2021). The impacts of climate change on ticks and tick-borne disease risk. *Annual Review of Entomology, 66*, 373–388.

Giles, J. R., Peterson, A. T., Busch, J. D., Olafson, P. U., Scoles, G. A., Davey, R. B., & Wagner, D. M. (2014). Invasive potential of cattle fever ticks in the southern United States. *Parasites & Vectors, 7*, 1–11.

Gray, J. S., Dautel, H., Estrada-Peña, A., Kahl, O., & Lindgren, E. (2009). Effects of climate change on ticks and tick-borne diseases in Europe. *Interdisciplinary Perspectives on Infectious Diseases, 2009*, Article 593232.

Gray, J. S., & Ogden, N. H. (2021). Ticks, human babesiosis, and climate change. *Pathogens, 10*(11), 1430.

Hahn, M. B., Feirer, S., Monaghan, A. J., Lane, R. S., Eisen, R. J., Padgett, K. A., & Kelly, M. (2021). Modeling future climate suitability for the western blacklegged tick, *Ixodes pacificus*, in California with an emphasis on land access and ownership. *Ticks and Tick-Borne Diseases, 12*, Article 101789.

Hallegatte, S., & Morlot, J. C. (2011). Understanding climate change impacts, vulnerability and adaptation at city scale: An introduction. *Climatic Change, 104*, 1–12.

Hassett, E. M., & Thangamani, S. (2021). Ecology of powassan virus in the United States. *Microorganisms, 9*(11), 2317. https://doi.org/10.3390/microorganisms9112317

Hermance, M. E., & Thangamani, S. (2017). Powassan virus: An emerging arbovirus of public health concern in North America. *Vector-Borne and Zoonotic Diseases, 17*(7), 453–462. https://doi.org/10.1089/vbz.2017.2110

Hernandez, P. A., Graham, C. H., Master, L. L., & Albert, D. L. (2006). The effect of sample size and species characteristics on performance of different species distribution modeling methods. *Ecological Applications, 16*(1), 1750–1760.

Herrmann, C., & Gern, L. (2014). Survival of *Ixodes ricinus* (Acari: Ixodidae) under challenging conditions of temperature and humidity is influenced by *Borrelia burgdorferi* sensu lato infection. *Journal of Medical Entomology, 47*(6), 1196–1204.

Homer, M. J., Aguilar-Delfin, I., Telford, S. R., III., Krause, P. J., & Persing, D. H. (2000). Babesiosis. *Clinical Microbiology Reviews, 13*(3), 451–469.

Huang, Y., Cheng, H., & Yang, J. (2017). Seasonal variations in urban air pollution in relation to meteorological conditions: Evidence from case studies in the U.S. *Journal of the Air & Waste Management Association, 67*(5), 1235–1243.

Hussain, S., Hussain, A., Ho, J., Li, J., George, D., Rehman, A., Zeb, J., & Sparagano, O. (2024). An epidemiological survey regarding ticks and tick-borne diseases among livestock owners in Punjab, Pakistan: A One Health context. *Pathogens, 10*(3), 361.

Jiang, J., & Zhang, H. (2016). Maximum entropy-based ecological niche model and bio-climatic determinants of Lone Star Tick (*Amblyomma americanum*) niche in Kansas, USA. *Ecological Informatics, 36*, 107–115.

Johnson, N., Phipps, L. P., Hansford, K. M., Folly, A. J., Fooks, A. R., Medlock, J. M., Mansfield, K. L. (2022). One Health approach to tick and tick-borne disease surveillance in the United Kingdom. *International Journal of Environmental Research and Public Health, 19*(10), 5833. https://doi.org/10.3390/ijerph19105833

Kleindienst, S., & Götz, M. (2022). The utility of a maximum entropy species distribution model for *Ixodes scapularis* in predicting the public health risk of Lyme disease in Ontario. *Canada. Ticks and Tick-Borne Diseases, 13*, Article 102746.

Klempa, I., & Vasilenko, A. (2024). Species distribution modeling of *Ixodes ricinus* (Linnaeus) under current and future climate scenarios in Europe. *Sustainability, 12*, 184.

Kopsco, H. L., Smith, R. L., & Halsey, S. J. (2022). A scoping review of species distribution modeling.

Leger, E., Vourc'h, G., Vial, L., Chevillon, C., & McCoy, K. D. (2013). Changing distributions of ticks: Causes and consequences. *Experimental and Applied Acarology, 59*, 219–244.

Lilleør, H. B., & Van de Broeck, K. (2011). Economic development and environmental migration: Views from anthropology. *Journal of Development Studies, 14*(2), 34–47.

Lindgren, E., & Gustafson, R. (2001). Tick-borne encephalitis in Sweden and climate change. *Lancet, 358*(9275), 16–18. https://doi.org/10.1016/S0140-6736(00)05250-8

Liu, C., White, M., & Newell, G. (2024). Predicting *Ixodes ricinus* habitat suitability growth in northern Europe using MaxEnt modeling techniques.

Maltezou, H. C., & Papa, A. (2010). Crimean-Congo hemorrhagic fever: Risk for emergence of new endemic foci in Europe? *Travel Medicine and Infectious Disease, 8*(3), 139–143. https://doi.org/10.1016/j.tmaid.2010.04.008

Marques, R., Krüger, R. F., Peterson, A. T., de Melo, L. F., Vicenzi, N., & Jiménez-García, D. (2020). Climate change implications for the distribution of the babesiosis and anaplasmosis tick vector, *Rhipicephalus (Boophilus) microplus. Veterinary Research, 51*, 1–10.

Miller, M. J., & Hurst, C. J. (2021). Modeling the distribution of medically important tick species using maximum entropy: Implications for public health planning and vector control strategies. *Insects, 12*, 190.

Mochizuki, T., Chikamoto, Y., Kimoto, M., Ishii, M., Tatebe, H., Komuro, Y., & Mori, M. (2012). Decadal prediction using a recent series of MIROC global climate models. *Journal of the Meteorological Society of Japan. Ser. II, 90*, 373–383.

Mowbray, F., Amlôt, R., Rubin, G. J. (2012). Ticking all the boxes? A systematic review of education and communication interventions to prevent tick-borne disease. *Vector-Borne and Zoonotic Diseases, 12*(9), 817–825. https://doi.org/10.1089/vbz.2011.0774

Nielebeck, C., Kim, S. H., Pepe, A., Himes, L., Miller, Z., Zummo, S., & Monzón, J. D. (2023). Climatic stress decreases tick survival but increases rate of host-seeking behavior. *Ecosphere, 14*(1), Article e4369.

Nolzen, H., Brugger, K., Reichold, A., Brock, J., Lange, M., & Thulke, H. H. (2022). Model-based extrapolation of ecological systems under future climate scenarios: The example of *Ixodes ricinus* ticks. *PLoS ONE, 17*, Article e0267196.

Nuttall, P. A. (2022). Climate change impacts on ticks and tick-borne infections. *Biologia, 77*, 1503–1512.

Obaid, M. K., Islam, N., Alouffi, A., Khan, A. Z., da Silva Vaz, I. Jr., Tanaka, T., & Ali, A. (2022). Acaricides resistance in ticks: Selection, diagnosis, mechanisms, and mitigation. *Frontiers in Cellular and Infection Microbiology, 12*, 941831.

Ogden, N. H., Beard, C. B., Ginsberg, H. S., & Tsao, J. I. (2021). Possible effects of climate change on ixodid ticks and the pathogens they transmit: Predictions and observations. *Journal of Medical Entomology, 58*, 1536–1545.

Ogden, N. H., & Lindsay, L. R. (2016). Effects of climate and climate change on vectors and vector-borne diseases: Ticks are different. *Trends in Parasitology, 32*, 646–656.

Ogden, N. H., Radojevic, M., Wu, X., Duvvuri, V. R., Leighton, P. A., & Wu, J. (2014). Estimated effects of projected climate change on the basic reproductive number of the Lyme disease vector *Ixodes scapularis. Environmental Health Perspectives, 122*, 631–638.

Ogden, N. H., St-Onge, L., Barker, I. K., Brazeau, S., Bigras-Poulin, M., Charron, D. F., Francis, C. M., Heagy, A., Lindsay, L. R., Maarouf, A., et al. (2008). Risk maps for range expansion of the Lyme disease vector, *Ixodes scapularis*, in Canada now and with climate change. *International Journal of Health Geographics, 7*, 24.

Oorebeek, M., & Kleindorfer, S. (2008). Climate or host availability: What determines the seasonal abundance of ticks? *Parasitology Research, 103*, 871–875.

Osbrink, W. L. A., Thomas, D. B., Lohmeyer, K. H., & Temeyer, K. B. (2022). Climate change and alternative hosts complicate the eradication of cattle fever ticks (Acari: Ixodidae) in the Southern United States. A review. *Annals of the Entomological Society of America, 115*(1), 39–55. https://doi.org/10.1093/aesa/saab034

Ostfeld, R. S., & Brunner, J. L. (2015). Climate change and *Ixodes* tick-borne diseases of humans. *Philosophical Transactions of the Royal Society b: Biological Sciences, 370*, 20140051.

Pfäffle, M., Littwin, N., Muders, S. V., & Petney, T. N. (2013). The ecology of tick-borne diseases. *International Journal for Parasitology, 43*, 1059–1077.

Phillips, S. J., Anderson, R. P., & Schapire, R. E. (2006). Maximum entropy modeling of species geographic distributions. *Ecological Modelling, 190*, 231–259.

Porretta, D., Mastrantonio, V., Mona, S., Epis, S., Montagna, M., Sassera, D., & Urbanelli, S. (2013). The integration of multiple independent data reveals an unusual response to Pleistocene climatic changes in the hard tick *Ixodes ricinus. Molecular Ecology, 22*, 1666–1682.

Porter, W. T., Monaghan, A. J., Eisen, L., Moore, C. M., Delorey, M. J., Biggerstaff, B. J., & McAllister, J. C. (2019). Predicting climate-driven distribution shifts in *Hyalomma marginatum. Scientific Reports, 9*(1), 1249.

Potes, L., Bouchard, C., Rocheleau, J. P., Richard, L., Leighton, P., Pelletier, J., et al. (2023). Evaluation of a community-based one health intervention to reduce the risk of Lyme disease in a high-incidence municipality. *CABI One Health*, ohcs202300017

Raghavan, R. K., Goodin, D. G., Neises, D., Anderson, G. A., & Ganta, R. R. (2016). Hierarchical bayesian spatio-temporal analysis of climatic and socio-economic determinants of Rocky Mountain spotted fever. *PLoS One, 11*(3), e0150180. https://doi.org/10.1371/journal.pone.0150180

Randolph, S. E., & Rogers, D. J. (2007). Ecology of tick-borne disease and the role of climate. In *Crimean-Congo hemorrhagic fever: A global perspective* (pp. 167–186). Springer Netherlands.

Rocklov, J., & Dubrow, R. (2020). Climate change: an enduring challenge for vector-borne disease prevention and control. *Nature Immunology, 21*(5), 479–483. https://doi.org/10.1038/s41590-020-0648-y. Erratum in: *Nature Immunology, 21*(6), 695. https://doi.org/10.1038/s41590-020-0692-7

Rodgers, S. E., Zolnik, C. P., & Mather, T. N. (2007). Duration of exposure to suboptimal atmospheric moisture affects nymphal blacklegged tick survival. *Journal of Medical Entomology, 44*(2), 372–375.

Sakamoto, J. M. (2018). Progress, challenges, and the role of public engagement to improve tick-borne disease literacy. *Current Opinion in Insect Science, 28*, 81–89. https://doi.org/10.1016/j.cois.2018.05.011

Sejian, V., Bhatta, R., Soren, N. M., Malik, P. K., Ravindra, J. P., Prasad, C. S., & Lal, R. (2015). Introduction to concepts of climate change impact on livestock and its adaptation and mitigation. In *Climate Change Impact on Livestock: Adaptation and Mitigation*. Springer India.

Shabani, F., Kumar, L., & Ahmadi, M. (2016). A comparison of absolute performance of different correlative and mechanistic species distribution models in an independent area. *Ecology and Evolution, 6*, 5973–5986.

Smith, R., Jones, P., Briegleb, B., Bryan, F., Danabasoglu, G., Dennis, J., & Yeager, S. (2010). *The parallel ocean program (POP) reference manual: Ocean component of the community climate system model (CCSM) and community earth system model (CESM)* (LAUR-01853).

Snowden, J., & Simonsen, K. A. (2025). Rocky Mountain spotted fever (Rickettsia rickettsii) [Updated 2023 Jul 17]. In: StatPearls [Internet]. Treasure Island (FL): StatPearls Publishing. Available from: https://www.ncbi.nlm.nih.gov/books/NBK430881/

Sobek-Swant, K., Boulanger, J., & Kearney, M. (2023). An application of maximum entropy model to predict the distribution of invasive species in a changing climate. *Journal of Biogeography, 50*, 765–779.

Stafford, K. C. III., Williams, S. C., & Molaei, G. (2017). Integrated pest management in controlling ticks and tick-associated diseases. *Journal of Integrated Pest Management, 8*, 28.

Stocker, T. (2011). *Introduction to climate modelling*. Springer Science & Business Media.

Sumilo, D., Asokliene, L., Bormane, A., Vasilenko, V., Golovljova, I., & Randolph, S. E. (2007). Climate change cannot explain the upsurge of tick-borne encephalitis in the Baltics. *PLoS ONE 2*, e500. https://doi.org/10.1371/journal.pone.0000500

Suss, J., Klaus, C., Gerstengarbe, F. W., & Werner, P. C. (2008). What makes ticks tick? Climate change, ticks, and tick-borne diseases. *Journal of Travel Medicine, 15*, 39–45.

Suzuki, T., & Ishii, M. (2011). Regional distribution of sea level changes resulting from enhanced greenhouse warming in the Model for Interdisciplinary Research on Climate version 3.2. *Geophysical Research Letters, 38*, L02601.

Tai, A. P. K., Mickley, L. J., & Jacob, D. J. (2010). Correlations between fine particulate matter (PM2.5) variability and temperature changes. *Environmental Research Letters, 5*(1), 014013.

Tiffin, H. S., Rajotte, E. G., Sakamoto, J. M., & Machtinger, E. T. (2022). Tick control in a connected world: Challenges, solutions, and public policy from a United States border perspective. *Tropical Medicine and Infectious Disease, 7*, 388.

Voyiatzaki, C., Papailia, S. I., Venetikou, M. S., Pouris, J., Tsoumani, M. E., & Papageorgiou, E. G. (2022). Climate changes exacerbate the spread of *Ixodes ricinus* and the occurrence of Lyme borreliosis and tick-borne encephalitis in Europe—how climate models are used as a risk assessment approach for tick-borne diseases. *International Journal of Environmental Research and Public Health, 19*, 6516.

Watanabe, M., Suzuki, T., O'ishi, R., Komuro, Y., Watanabe, S., Emori, S., & Kimoto, M. (2010). Improved climate simulation by MIROC5: Mean states, variability, and climate sensitivity. *Journal of Climate, 23*, 6312–6335.

Williams, H. W., Cross, D. E., Crump, H. L., Drost, C. J., & Thomas, C. J. (2015). Climate suitability for European ticks: Assessing species distribution models against null models and projection under AR5 climate. *Parasites & Vectors, 8*, 1–15.

Wing-Le, C., & Abe-Ouchi, A. (2020). Pliocene Model Intercomparison Project (PlioMIP2) simulations using the Model for Interdisciplinary Research on Climate (MIROC4m). *Climate of the past, 16*, 1523–1545.

Wu, T., Song, L., Li, W., Wang, Z., Zhang, H., Xin, X., & Zhou, M. (2014). An overview of BCC climate system model development and application for climate change studies. *Journal of Meteorological Research, 28*, 34–56.

Wu, X., Duvvuri, V. R., & Wu, J. (2010). Modeling dynamical temperature influence on tick *Ixodes scapularis* population. *International Journal of Environmental Health Research, 20*(6), 407–418.

Zannou, O. M., Ouedraogo, A. S., Biguezoton, A. S., Abatih, E., Coral-Almeida, M., Farougou, S., & Saegerman, C. (2021). Models for studying the distribution of ticks and tick-borne diseases in animals: A systematic review and a meta-analysis with a focus on Africa. *Pathogens, 10*, 893.
Zhou, Y., & Wang, Y. (2018). Using Maximum Entropy modeling to predict the potential distribution of invasive species: A case study of *Aedes albopictus* in China. *Ecological Modelling, 371*, 66–74.
Zhou, Y., & Wang, Y. (2023). An application of maximum entropy model to predict the distribution of invasive species: A case study of *Aedes albopictus* in China. *Ecological Modelling, 471*, Article 110130.